MANAGEMENT OF MYCORRHIZAS
IN AGRICULTURE, HORTICULTURE AND FORESTRY

Developments in Plant and Soil Sciences

VOLUME 56

Management of Mycorrhizas in Agriculture, Horticulture and Forestry

Proceedings of an International Symposium on Management of Mycorrhizas in Agriculture, Horticulture and Forestry
28 September – 2 October 1992, Perth, WA, Australia

Edited by
A. D. ROBSON, L. K. ABBOTT and N. MALAJCZUK

Partly reprinted from *Plant and Soil*, Volume 159, No. 1 (1994)

Kluwer Academic Publishers
DORDRECHT / BOSTON / LONDON

Library of Congress Cataloging-in-Publication Data

Management of mycorrhizas in agriculture, horticulture, and forestry / edited by A.D. Robson, L.K. Abbott and N. Malajczuk
p. cm. -- (Developments in plant and soil sciences ; v. 56)
Papers from the International Symposium on the Management of Mycorrhizas in Agriculture, Horticulture, and Forestry, held at the University of Western Australia, 28 September - 2 October 1992.
ISBN 0-7923-2700-4 (hb : acid free paper)
1. Mycorrhizas--Congresses. I. Robson, A.D. II. Abbott, L.K. III. Malajczuk, N. IV. International Symposium on the Management of Mycorrhizas in Agriculture, Horticulture, and Forestry (1992 : University of Western Australia) V. Series.
QK604.2.M92M35 1994
631.4'6--dc20 94-685

ISBN 0-7923-2700-4

Published by Kluwer Academic Publishers,
P.O. Box 17, 3300 AA Dordrecht, The Netherlands.

Kluwer Academic Publishers incorporates
the publishing programmes of
D. Reidel, Martinus Nijhoff, Dr W. Junk and MTP Press.

Sold and distributed in the U.S.A. and Canada
by Kluwer Academic Publishers,
101 Philip Drive, Norwell, MA 02061, U.S.A.

In all other countries, sold and distributed
by Kluwer Academic Publishers Group,
P.O. Box 322, 3300 AH Dordrecht, The Netherlands.

Printed on acid-free paper

Printed in the Netherlands

Contents

* Chapters indicated with an asterisk are reprinted from *Plant and Soil*, Volume 159, No. 1 (1994).

Preface

The International Symposium on the Management of Mycorrhizas in Agriculture, Horticulture and Forestry was held at The University of Western Australia in September, 1992. The purpose of the symposium was to evaluate the current and potential roles of arbuscular, ecto and ericoid mycorrhizas in agriculture, horticulture and forestry including the role of mycorrhizas in natural ecosystems and the revegetation of degraded lands.

Leading international scientists were invited to review the biology of the fungi, the formation and functioning of mycorrhizas and the development of practices for maximising benefits.

It was not intended that the reviews would be exhaustive accounts of the whole area but that each review would provide an overview of existing knowledge and would identify priorities for further research.

Each paper was refereed by two peers and we thank Dr D J Read, Dr M Brundrett, Dr K Dixon, Dr D P Stribley, Dr J P Thompson, Dr K Helyar, Dr W B Kendrick, Dr J M Trappe, Dr G A Chilvers, Dr R Koske, Dr M F Allen, Dr P A McGee, Dr A E Ashford, Dr V Gianinazzi-Pearson, Dr B Daniels-Hetrick, Dr N J Barrow, Dr S M Schwab, Dr D T Clarkson, Dr G P Sparling, Dr B D Kay, Dr D J Bagyaraj, Dr P Reddell, Dr J Garbaye, Dr M H Miller, Dr M A Adams, Dr D H Marx, Dr R M Miller, Dr D A Jasper, Dr I R Hall, Dr P A O'Brien, Mr P F Schweiger, Dr R Howeler, Dr A J Termorshuizen, Dr P D Millner, Dr I Tommerup, Dr S E Smith, Dr R Molina, Dr M N Hunter, Dr R M Danielson and Dr F Buscot for their willingness to referee the manuscripts.

We thank all the members of the Organizing Committee of the Symposium for their valuable contributions and the organizations that provided financial support for the Symposium.

Finally, we are very grateful to Pauline and Chris Gazey who formatted and corrected the entire manuscript, to Maureen Davison and Karen Lukin for excellent secretarial assistance, and to Valerie Landgrebe and Judy Hurst for providing excellent support in organizing and running the conference.

October 1993

A D Robson
L K Abbott
N Malajczuk

International Symposium on Management of Mycorrhizas in Agriculture, Horticulture and Forestry

The meeting was supported financially by

Australian International Development Assistance Bureau
Australian Tourist Bureau
Qantas Airways Ltd
Australian Airlines Ltd
Australian Mining Industries Council
Land and Water Resources Research and Development Corporation
Grain Research and Development Corporation
Wool Research and Development Corporation
Rural Industries Research and Development Corporation
Horticultural Research and Development Corporation
Alcoa of Australia Ltd
Bunnings TreeFarms Pty Ltd
CSBP and Farmers Ltd
RGC Mineral Sands Ltd
INCITEC Ltd

International Symposium on Management of Mycorrhizas in Agriculture, Horticulture and Forestry

Organizing Committee

L K Abbott, Convenor
Soil Science and Plant Nutrition, School of Agriculture, The University of Western Australia, Nedlands WA 6009

I C Tommerup, Secretary
CSIRO, Division of Forestry and Forest Products, Private Mail Bag, Post Office, Wembley WA 6014

A E Ashford
School of Biological Science, University of NSW, Kensington NSW 2033

M D A Bolland
Department of Agriculture Western Australia, Baron-Hay Court, South Perth WA 6151

B Dell
School of Environmental and Biological Sciences, Murdoch University, Murdoch WA 6150

C Gazey
Soil Science and Plant Nutrition, School of Agriculture, The University of Western Australia, Nedlands WA 6009

D A Jasper
Soil Science and Plant Nutrition, School of Agriculture, The University of Western Australia, Nedlands WA 6009

N Malajczuk
CSIRO, Division of Forestry, Private Mail Bag, Post Office, Wembley WA 6014

A D Robson
Soil Science and Plant Nutrition, School of Agriculture, The University of Western Australia, Nedlands WA 6009

J P Thompson
Queensland Wheat Research Institute, PO Box 2282, Toowoomba Queensland 4350

B Thomson
CSIRO, Division of Forestry, Private Mail Bag, Post Office, Wembley WA 6014

The mycorrhizal association: just one of many nutrient acquiring specializations in natural ecosystems

J.S. PATE
Botany Department, University of Western Australia, Nedlands, WA 6009, Australia

Key words: natural ecosystems, nutrient acquisition, roots

Abstract

This paper examines the general biology of mycorrhizal associations alongside the wide range of alternative trophic adaptations which higher plants may employ when competing for limited resources of specific nutrients within an ecosystem. All examples described come from highly nutrient-impoverished heathlands or open woodlands of the kwongan of southwest Australia. An account is given of the general patterns of rooting morphology and their association with various mycorrhizal and non-mycorrhizal nutrient-acquiring strategies, including various forms of parasitism, epiparasitism, autotrophy with or without mycorrhizal association. Taxonomic affinities of each grouping are examined alongside growth and life form characteristics.

A case study of patterns of utilization of a specific nutrient, nitrogen in a Banksia woodland ecosystem is presented to illustrate how a multifaceted approach can be used for studying species responses and interactions. The study categorizes species according to nitrate-utilizing ability and suggests how ^{15}N natural abundance of soil and plant components and organic solutes of nitrogen is xylem might be utilized to separate species into different trophic categories. Response of the ecosystem to fire is examined in respect of the nutritional interrelationships of component species as the ecosystem changes from being nitrate dominant immediately after fire to increasingly ammonium-producing thereafter.

The paper concludes by examining generally trophic relationships within whole ecosystems and outlines some of the challenges for future research in this connection.

Introduction

This chapter examines the mycorrhizal phenomenon alongside the range of other mechanisms employed by plants when competing for limited resources of nutrients within an ecosystem. To provide local flavour to the conference attention is focussed on the heathlands and open woodlands of the sandplain (kwongan) of southwest Australia. The physical characteristics of such ecosystems are outlined and the rooting morphologies encountered considered in relation to growth and life form, response to fire and seasonal patterns of growth and nutrient uptake. Commonly occurring mycorrhizal and non-mycorrhizal specializations enhancing nutrient uptake are listed and related to root morphology. A case study follows on patterns of utilization of a specific nutrient—nitrogen—in a southwest Australian woodland ecosystem to illustrate a multifaceted approach to study of species responses and interactions. The final section discusses information regarding flow pathways for nutrients amongst different trophic specializations and outlines how one might proceed when attempting to quantify the mutualistic or competitive interactions which may apply when nutrients are cycled between living and non-living components of an ecosystem.

A.D. Robson, L.K. Abbott & N. Malajczuk (eds.), Management of mycorrhizas in agriculture, horticulture and forestry, 1–10.

Physical characteristics of the ecosystem

The sandplain soils of SW Australia comprise the products of complex long term weathering and erosion of a land mass into a current structure of uniformly low relief covered by a mantle of the products of weathering up to 50 m deep. Whether relatively ancient and of lateritic origin or more recently formed and of coastal location, kwongan soils are heavily leached and extremely depauperate in macronutrients, especially N and available P (Bettenay, 1984). Organic matter contents are very low and organically-derived nutrients accordingly concentrated largely within the A horizon. The intractable surface litter which accumulates between fires is markedly depauperate in essential nutrients due to extraordinarily efficient presenescence retrieval of elements such as N, P and K (see Pate and Dell, 1984). However, hot canopy fires consume green foliage fully loaded with nutrients, leading to deposition of ash rich in all elements except N. Most of the latter escapes as ammonia or oxides of nitrogen. Fire kills heat sensitive 'obligate seeder' species, which add further nutrient capital, as their biomass decomposes below and above ground. These releases of nutrients, together with improved penetration of light and reduced competition for water, create a temporary niche exploited by a suite of fast growing fire ephemeral species (see Pate et al., 1985). Seedlings of longer-lived seeder and fire-resistant 'sprouter' species also establish successfully in the immediate post fire environment (Bell et al., 1984). Seeder species of legumes dominate early after fire in many habitats (see Monk et al., 1981; Hansen and Pate, 1987) since the ecosystem at this stage is specifically deficient in nitrogen. At other times, however, productivity of southwest Australian ecosystems is considered to be driven by availability of phosphorus (see Lamont, 1984; Pate and Dell, 1984).

The Mediterranean-type environment experienced by kwongan vegetation is typified by severe water stress from mid summer (November/December) to early autumn (March). Wetter but colder conditions prevail in late autumn through winter, while increasing warmth and uncertain rainfall are characteristic of spring and early summer (see Beard, 1984). Soil moisture contents typically show drying out from soil surface to water table from September to April, a progressive recharge of the profile from April say to late June, followed by only one or two months (July/August) when the whole profile is at or close to field capacity (see Dodd et al., 1984).

Rooting morphologies in relation to growth and life form and seasonal growth and water relations

Several descriptions have been made of the rooting morphologies of southwest Australian species, including accounts of 43 woody species (Dodd et al., 1984), a selection of 429 species embracing all life and growth forms (Pate et al., 1984), a suite of 200 species of geophytes (Pate and Dixon, 1982), a range of south coastal heathland flora and 80 or so members of the Restionaceae (Pate et al., 1991). The data bank includes information on rooting depth and lateral spread, branching patterns of roots, presence or absence of specialized modifications such as sand-binding roots, cord roots, (Pate et al., 1989) root suckers, lignotubers, fleshy storage organs, cluster roots, (Lamont, 1982; 1984) and stilt roots (Pate, 1989). Matching information is also available in most cases on longevity, growth form and presence or absence of trophic specializations such as symbiotic nitrogen fixation, hemi- or holo-parasitism and carnivory (see Pate et al., 1984, Lamont, 1984). Unfortunately information on mycorrhizal status is available for only a relatively small portion of the study species, although data provided by Lamont (1982, 1984), Brundrett and Abbott (1992) and Stewart et al. (1993) suggest that it may now be possible to generalize at genus and family level for many taxa. The resulting data can then be compared with that already available for other ecosystems worldwide (see Warcup, 1980; Newman and Reddell, 1987; Trappe, 1987; Tommerup, 1988; Brundrett, 1991; Mitchell, 1992).

Four principal types of rooting morphology may be recognized amongst kwongan flora (Fig. 1). The commonest pattern (Type R_1, Fig. 1),

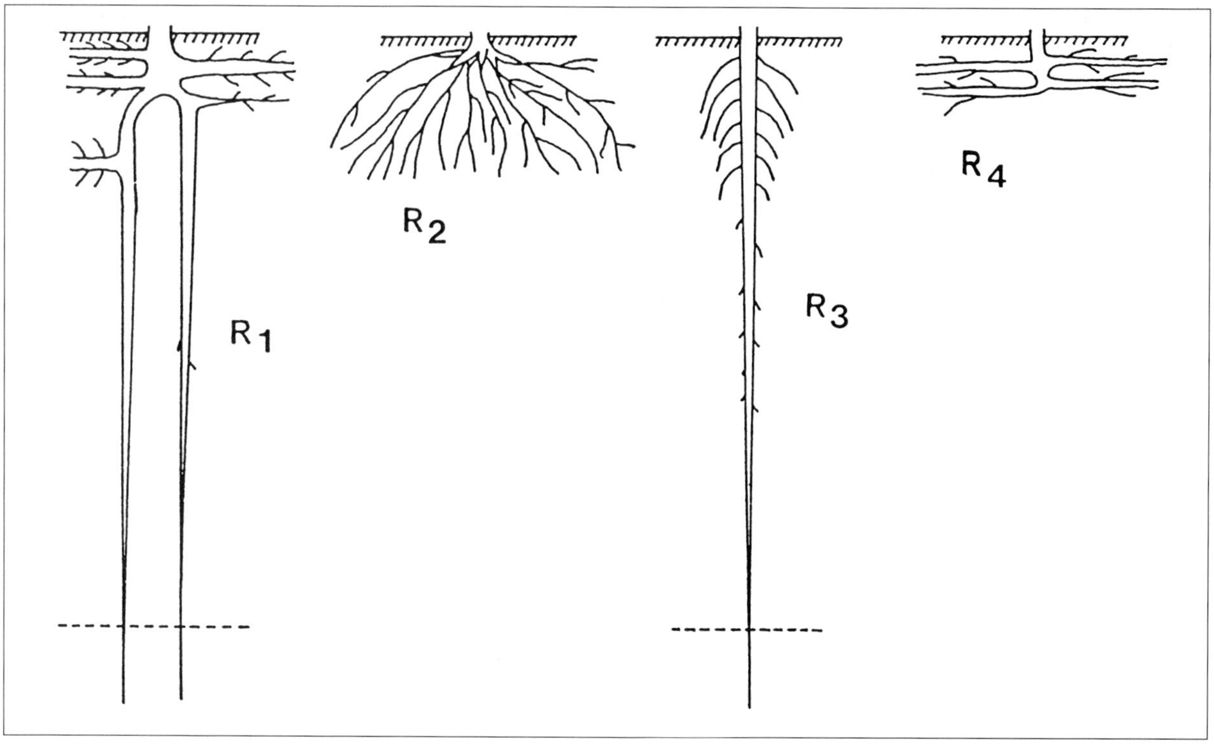

Fig. 1. Principal types of rooting morphologies exhibited by plants of sandplain (kwongan) ecosystems. The position of the water table is indicated by a broken line for Types R_1 and R_3.

typical of large trees and resprouter shrubs, shows a clear demarcation into superficial lateral feeding roots and deep 'sinker' roots. The latter are usually devoid of laterals and penetrate to the water table, thereby permitting the species to grow and reproduce during summer (Bell and Stephens, 1984). Water stress is moderate throughout the year for these species (Dodd et al., 1984). A recent study of seedling development of obligate seeder and resprouter representatives of woody species of Proteaceae (Bowen and Pate, 1991) has shown that seedlings penetrate to the water table (2.3 m below soil surface at the study site) by means of a single main root during their first season of growth and that it is not until the second season that specialized nutrient-acquiring cluster roots form on the near surface lateral roots. Significant increases in nutrient levels then occur above those initially supplied by the well-stocked cotyledons (Pate et al., 1986; Stock et al., 1990). By their third year fire sensitive obligate seeder species are distinguished by higher proportions of root biomass in the top 30 cm of the root profile resprouters and significantly higher shoot : root biomass ratios than in (Bowen, 1991).

The second most common rooting pattern (Type R_2, Fig. 1) is shallow and of fibrous character. Most geophytes, (see Pate and Dixon, 1982) conform to this type of root system as do rhizomatous hemicrytophytes and a range of winter-active small shrubby species (e.g. Epacridaceae and certain Myrtaceae). Such plants do not normally penetrate to the water table, so become dormant and rely on stored water (geophytes) or become severely stressed and fail to increase in dry matter (non geophytes) during summer. In both cases nutrient uptake is strictly confined to times when surface layers of the soil are adequately moist. The third rooting category (Type R_3, Fig. 1) is characterized by a noticeably thickened tap root which carries seasonal investments of upper feeding roots. It is typical of are long lived sprouters (Pate et al.,

(A) HERBACEOUS HABIT

Trophic type	Root modification	Examples
AUTOTROPHY	— Non Mycorrhizal	— Fire ephemerals and certain geophytes (mostly, R_2)
	VA Mycorrhizal	— Many Dicot. and Monocot. species (R_2 or R_3)
	Orchid-type Mycorrhizal	— Orchidaceous geophytes (R_2)
NITROGEN-FIXING	— VA Mycorrhizal	— Fabaceae (R_2 or R_3)
CARNIVORY	— Non Mycorrhizal	— Droseraceae, Byblidaceae, Utriculariaceae (R_2)
HOLOPARASITISM	— Non mycorrhizal	— On roots – *Orobanche* (R_2)
		Subepidermal – *Pilostyles* (no root)
		On stems – *Cuscuta, Cassytha* (no root)
HETEROTROPHIC EPIPARASITISM	— Shared *Rhizoctonia* mycorrhiza (*Rhizanthella* (no roots) on host *Melaleuca* (R_1))	
MYCOAUTOTROPHY	— Ectomycorrhizhal	— *Lobelia* spp. (R_1)

(B) WOODY HABIT

Trophic type	Root modification	Examples
AUTOTROPHY	— Ecto + VA Mycorrhizal	— Range of tree and shrub genera (R_1–R_3)
	Cluster roots (non mycorrhizal)	— Proteaceae (R_1–R_3)
	VA mycorrhizal	— Large number of shrubby taxa (R_1–R_3)
	Ericoid mycorrhizal	— Epacridaceae (R_1 or R_3 (hair roots))
NITROGEN-FIXING	— VA Mycorrhizal	— *Macrozamia*, certain Fabaceae (R_1 or R_3)
	Ecto + VA mycorrhizal	— Fabaceae, Mimosaceae, Casuarinaceae (R_1–R_3)
	Cluster Roots (non mycorrhizal)	— *Daviesia, Viminaria, Acacia* (few species) R_1 or R_3)
ROOT HEMIPARASITISM	— Non mycorrhizal	— *Olax, Exocarpos, Leptomeria, Anthobolus* (R_4) *Nuytsia, Santalum* (R_1)
AERIAL HEMIPARASITISM	— Non mycorrhizal	— *Amyema* spp., *Lysiana* spp., (no root)

Fig. 2. Classification of trophic types encountered amongst herbaceous (A) and woody (B) species of sandplain (kwongan) ecosystems. Mycorrhizal associations or other forms of root modification, rooting morphology (R_1–R_4, see Fig. 1) and examples of taxa are given against each trophic type.

1984). Seasonal water relations of members of this group are similar to those indicated above for root type R_1. The fourth type of rooting morphology (R_4, Fig. 1) is very shallow (all roots sited within the top 30 cm of soil) but laterally very extensive. It is typical of shrubby or herbaceous root hemiparasites (e.g. *Olax, Leptomeria, Exocarpos, Anthobolus*) all of which exhibit prolific haustorial development on a range of host species (e.g. see studies on *Olax phyllanthi* by Pate et al., 1990a, b). For example a third year plant of *O. phyllanthi* (see Pate et al., 1990b) 1 m in height may possess up to 60 m of roots carrying and an average of 8 haustoria per metre of root length. Haustoria comprise 0.7 –3.5% of plant dry weight and the laterally extensive root system explores a catchment area up to 50 times that of the shoot canopy and may comprise almost half of the total dry matter of the plant. Hosts of both shallow (Type R_3) and deep (Types R_1 and R_2) rooting morphology are regularly parasitized by *Olax*. Haustoria are initiated mostly in autumn. Those on deep rooted hosts persist for over a year, while those on shallow-rooted hosts senesce the following summer once the host develops water potentials more negative than those of *Olax* (Pate et al., 1990c). Seedlings fail to survive their first summer of heat and drought stress unless able to parasitize deep rooted hosts (J S Pate, unpublished results). Comparison of the balance of nutrients in seedling dry matter at the end of the first 1–2 month period of fully autotrophic (pre-parasitic) growth with that after a following 4–6 months of parasitism suggests that the main benefit from attachment to hosts derives from haustorial acquisition of N, P and K (Pate et al., 1990b).

Trophic typings, growth and life form and rooting morphology

Judging from published information on the floras of a number of kwongan ecosystems (e.g. see Lamont, 1982, 1984; Pate et al., 1984; Brundrett and Abbott, 1992; Stewart et al., 1993), interesting correlations hold between taxonomic status and specific forms of trophic behaviour, mycorrhizal status and rooting morphology. As shown in Fig. 2 some of these apply generally at family level others relate to unusual trophic specializations encountered in one or relatively few species.

Many of the trophic affinities listed in Fig. 2 find parallels in other ecosystems, but two mycotrophic relationships deserve special mention, namely the epiparasitism totally of underground orchid *Rhizanthella gardneri* (see Dixon et al., 1991) the usual mycotrophic behaviour of certain fleshy rooted species of *Lobelia* (e.g. *L. gibbosa*) (Warcup, 1988; G Neilson, K W Dixon and J S Pate, unpublished results). *Rhizanthella gardneri* is a stem-tuberous rootless orchid deriving all its carbon and mineral nutrients from the host shrub *Melaleuca uncinata* via a shared mycorrhizal partner (*Rhizoctonia* sp.) (cf. the situation reported by Bjorkman (1960) for the N. Hemisphere species *Monotropa hypopitys*). *R. gardneri* bears underground capitula-like inflorescences apparently pollinated by termites. Seeds are exceptionally large for an orchid and the dispersal agent is unknown (see Dixon et al., 1991). In *Lobelia gibbosa* the first half of the life cycle is mycorrhizal and fully subterranean. Seed germinates deeply below ground in autumn and following infection with a basidiomycete mycorrhiza a coralloid root mass develops of up to 20 g fresh weight and 10 cm diameter (J S Pate, unpublished results). The following spring shoot buds develop to form an above-ground leafy vegetative phase of several months duration. The plant then flowers, sets seeds and dies. The root mass, becomes progressively exhausted of its highly concentrated stock of nutrients during shoot development flowering and fruiting.

A number of the features illustrated in Fig. 2 call for comment. Firstly, non-mycorrhizal herbaceous species are widespread taxonomically, especially in fire ephemeral and geophytic species of habitats well supplied with nutrients from fire or seasonal mineralization of litter. Secondly, non-mycorrhizal species are probably confined amongst woody species to root hemiparasites or species bearing cluster (proteoid) roots (Lamont, 1984). Thirdly, VA mycorrhizae alone or in company with ectotrophic mycorrhizae are the most widespread form of mycorrhizal relationship in woody taxa of kwongan ecosystems whereas VA mycorrhiza alone is the commonest association encountered in herbaceous non-orchidaceous species (see Lamont, 1984; Brundrett and Abbott, 1992). Fourthly, all N_2-fixing taxa carry some prominent form of root specialization (see Fig. 2) of presumed significance in absorption of phosphorus and thereby making them competitive over other species lacking N_2 fixing capacity. Fifthly, the apparently universal association of Australian species of Epacridaceae with ericoid-type mycorrhizae (see Reed, 1987), suggests, nutritional similarities with the Ericaceae. However, N. Hemisphere ericoids are normally associated with peats or forest litter of extremely high organic matter whereas epacrids of the kwongan, and ericoids of the fynbos of S. Africa, occur mostly on seasonally arid sands of

Table 1. Comparison of $\delta^{15}N$ and shoot nitrate concentrations for 34 common species from mixed Banksia woodland in Kings Park, Perth, Western Australia. The species are compared in terms of mode of nutrient acquisition. Data for each species in the analysis of variance (AOV) are mean values for ^{15}N natural abundance or shoot nitrate across 8 sites of varying age since last burn (data from Pate, et al. 1993)

Nutrient acquisition group	$\delta^{15}N$	AOV	Shoot nitrate (μm/g dw)	AOV
Non-mycorrhizal species (n = 11)	1.81	A[a]	3.38	nsd[b]
Mycorrhizal species (n = 13)	0.02	B	1.39	nsd
N_2 fixing species (n = 10)	-0.36	B	1.02	nsd

[a] significantly different at P < 0.001.
[b] not significantly different.

extremely low organic content. Finally, carnivorous and parasitic species which acquire a significant proportion of their nutrient resources from another macroorganism are mostly found to be non-mycorrhizal (e.g. root parasites (Pate et al., 1990a, b, c) and carnivorous species (see Dixon et al., 1980; Schulze et al., 1991; Karlsson and Pate, 1992).

Case study of patterns of acquisition and utilization of nitrogen within a southwest Australian woodland ecosystem

This primary data relating to this study are found in two recent publications (Stewart et al., 1993 and Pate et al., 1993) investigating how habitat sources of nitrogen are exploited by 34 commonly occurring species within fire-prone *Banksia* woodland in Kings Park, Perth, Western Australia. Examining each species at 8 sites spanning 0.5–23 years since last burn measurements are recorded of in vivo nitrate reductase activities (NRA) of shoots and roots, capacities of root and shoot to store nitrate and the relative proportions of nitrate as opposed to organic sources of N which are transported from root to shoot in the transpiration stream. Parallel with these measurements levels of nitrate and ammonium in the soil solution are measured and the ^{15}N natural abundance of plant total N, soil nitrate and total soil N assessed. Data are then examined in relation to life and growth form of species and presence or absence of mycorrhizal or other forms of specialized root systems.

A number of interesting features emerge from the study. Nitrate and ammonium occur in appreciable amounts at all sites during the winter growing season but nitrate predominates and mean shoot NRA values are especially high at recently burnt sites. In sites not burnt for many years NRA values are generally low but supplying nitrate to shoots through the transpiration stream leads to induction of shoot NRA to levels typical of recently burnt sites.

As a general rule shallow rooted short-lived herbaceous species, show high NRA values and high NO_3-storing capacity, whereas perennial woody species with cluster or mycorrhizal roots are generally weak in nitrate reduction and therefore probably feeding on ammonium or organic forms of N. Certain species, notably the carnivore *Drosera stolonifera,* two species of *Daviesia* and the epacrids *Conostephium pendulum* and *Astroloma macrocalyx* exhibit barely detectable NRA, even after shoot feeding of nitrate via the transpiration stream.

Most species show higher shoot NRA than root NRA and, as might be expected, species especially active in reduction in their shoots, but not their roots, carry unusually high proportions of their xylem N as nitrate.

The principal organic solutes of nitrogen transported in xylem prove to be species specific, including cases in which asparagine, glutamine, g-methylene glutamine, citrulline or arginine comprise the dominant solute. Nitrogen-fixing species all utilize asparagine as major xylem translocate, glutamine transport is typical of all nitrate utilizing herbaceous species but no consistent correlations are encountered amongst woody non-N_2 fixing species in respect of compounds transported and type of mycorrhizal association.

Turning to the data for ^{15}N natural abundance a grouping of species according to presence or

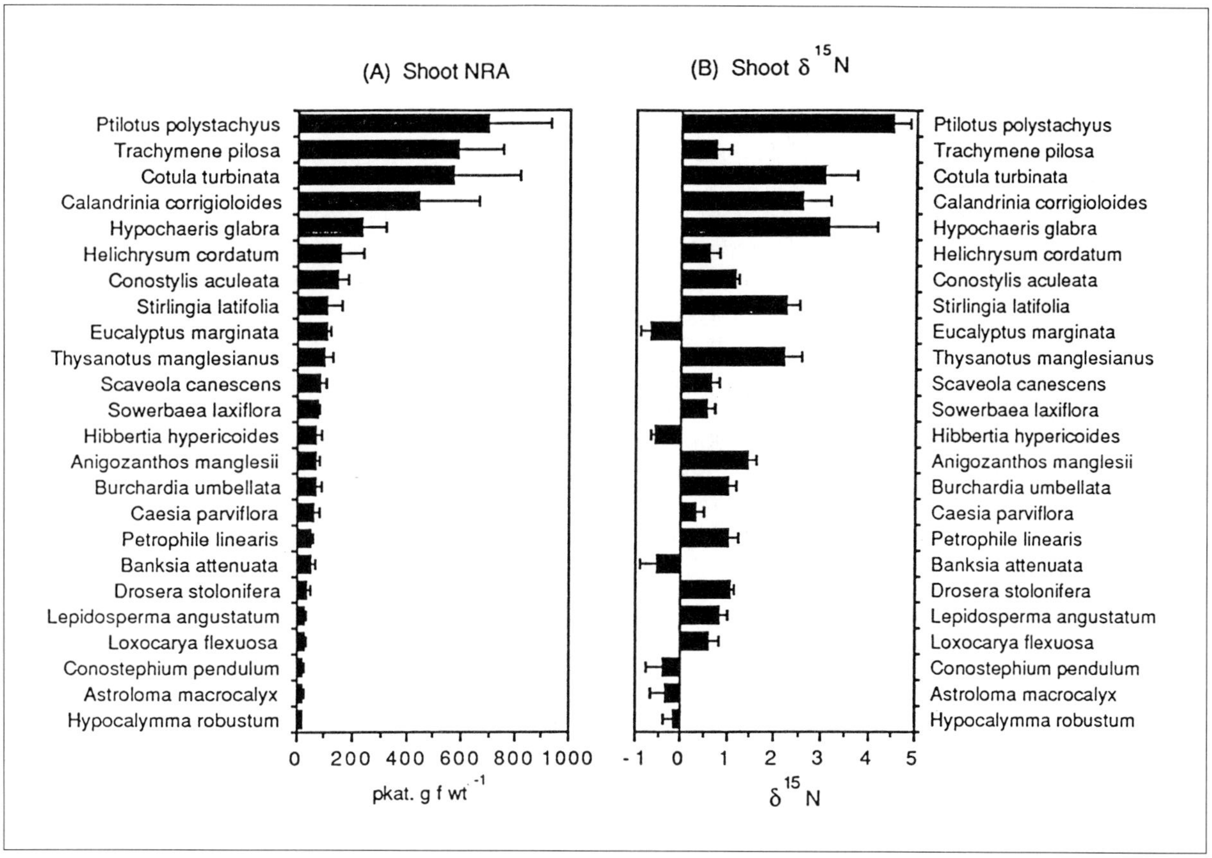

Fig. 3. Relationship between shoot nitrate reductase study (A) and ^{15}N natural abundance of shoot total dry matter (B) for 24 common non N_2-fixing species of mixed *Banksia* woodland, Kings Park, Perth, Western Australia. Data in each case are given as means (± standard error) of measurements at eight sites of varying age since last burn in the woodlands. Data from Pate et al. (1993).

absence of mycorrhizae and absence or presence of N_2 fixing capacity shows non-mycorrhizal non-N_2 fixers to have $\delta^{15}N$ values significantly higher, (by almost +2‰) than the other groups (see Table 1). Further examination within the group of 24 non N_2-fixing species shows species of high shoot NRA capacity to have the most elevated $\delta^{15}N$ values (+3 to +5). However the correlation between NRA and $\delta^{15}N$ holds poorly in species of lower $\delta^{15}N$ value and medium or low NRA (see Fig. 3).

Trophic relationships within whole ecosystems—some challenges for future research

It is deeply entrenched in the literature on mycorrhizal associations that the principal benefit to the host plant in non-ericoid type associations derives principally from efficaceous access via the fungal mycelium to less available forms of phosphorus (Harley and Smith, 1983; Finlay and Read, 1986; Bolan et al., 1987), while that in ericoid-type associations relates to the ability of the fungus to assimilate intractable forms of nitrogen such as peptides, proteinaceous residues etc (see Read, 1983; Read et al., 1985; Abuzinadah and Read, 1986; Leake et al., 1989). These concepts are well supported experimentally but the possibility remains that mycorrhizal transfer to the host of nutrients other than the ones signified above might also be important. This may well apply for example to the NO_3^- or NH_4^- assimilating capacities of

VAM or ECM (see Marschner and Dell, Chapter 9; Smith et al., Chapter 10). In similar fashion, while good evidence exists on the effectiveness of cluster roots as localized sites of intensive dissolution and uptake of insoluble forms of phosphorus (see Lamont, 1984), this root specialization may also be active in uptake of nitrogen. This has been indicated recently by Pate and Jeschke (unpublished results) in the finding of much higher concentrations of organic N as well as phosphate in xylem sap collected from cluster roots of species of Proteaceae than in xylem sap collected from other parts of the root system. In any event it generally holds that the extent to which a specialized mode of nutrition is developed and the degree to which it contributes to total plant nutrition are inversely related to nutrient availability (see Jasper et al., 1979; Lamont, 1984; Abbott and Robson, 1991). Accordingly, at very low nutrient levels unusually large amounts of plant resource are channelled towards nutrient acquisition. For example, when sand cultures of the southwest Australian legume *Viminaria juncea* are supplied with critically low levels of phosphate (2.5 mg P. L^{-1}) and no added mineral N, up to 45% of total dry weight of plants consists of cluster roots and a further 2–5% is comprised of nitrogen-fixing root nodules (Walker and Pate, 1986). With further costs in maintenance and operation (see Pate and Layzell, 1990) of such root specializations it comes as no surprise to find extremely low rates of growth under such circumstances and lower cost : benefit ratios than where combined N and P are readily available in the soil solution. While it has proved difficult to define comparable costs, for construction and maintenance of mycorrhizal associations, these may also comprise an appreciable expense to the host plant component in comparison say with conventional uptake via root hairs (see Warcup, 1980; Trappe, 1987; Harris and Paul, 1987; Raich and Nadehoffer, 1990; Jakobsen and Rosendahl, 1990).

Turning to trophic interrelationships within natural ecosystems information is clearly lacking on a number of crucial aspects, even where much is already known concerning the nutrient-acquiring propensities of the common species (see St John and Coleman, 1983; Brundrett, 1991). Firstly, at broadest level, one needs to assess the relative extents to which different mycorrhizal or non-mycorrhizal mechanisms contribute to flow of potentially-limiting nutrients within the ecosystem. Secondly, one should determine what proportions of the nutrient uptake budgets of individual species are consummated via specialized roots or symbiotic activity as opposed to non specialized forms of root uptake and how these proportions change seasonally in relation to phenologies of plant and microorganism and fluctuations in temperature and availability of nutrients and water. Thirdly, the extent and nature of interactions of species should be prescribed, particularly whether outright competition for the same nutrient source is involved or whether complementary patterns of utilization occur through absorption from mutually exclusive rooting zones or from different forms of a specific nutrient element. The possibility also exists of synergism or partial parasitism through common access to nutrient resources via the same mycorrhizal mycelium (see Francis et al., 1986; Newman, 1988; Haystead et al., 1988; Ashford et al., 1989) or additive effects of several independent types of specialization on nutrient release from relatively unavailable sources (see Bolan et al., 1987). Finally, a large number of pertinent questions must be asked at species level regarding mechanisms of uptake and cycling of specific nutrients. Prominent among these are the types of nutrient resource (e.g. organic or inorganic insoluble forms of P and N as opposed to soluble phosphate NH_4^+ or NO_3^-) utilized by the plant and its microbial partner (see Finlay and Read, 1986), the forms in which an element is transported from sites of uptake to sites of storage or utilization in the plant (see Ashford et al., 1989; Brundrett, 1991; Mitchell, 1992) and the nature of the metabolic support provided by a host to its symbiotic associate(s) (see Lambers, 1987).

The possibility exists that specific organic solutes are produced as a result of the symbiosis (e.g. as shown to be the case for citrulline and glutamine in the N_2-fixing coralloid roots of cycads (see Pate et al., 1988) arginine from carnivory in *Drosera* spp. (Dixon et al., 1980) and glutamine following N uptake by cluster roots of *Banksia* spp. (J S Pate and W D Jeschke, unpublished results). Appearance of such indicator molecules, say in xylem exudates, would then herald the commencement of a particular form of specialized nutritional activity. One would thus be able to assess, for example, when seedlings establishing after fire first begin to utilize mycorrhizae as a major supplier of nutrients when sprouter species in the same situation re-establish their symbioses and become progressively less dependent on ash bed nutrients and how specialized forms of nutrition wax and wane between and within seasons in relation to factors such as temperature and availability of water and nutrients.

References

Abbott L K and Robson A D 1991 Factors influencing the occurrence of vesicular-arbuscular mycorrhizas. J. Agric. Ecos. Environ. 35, 121–150.

Abuzinadah R A and Read D J 1986 The role of proteins in the nitrogen nutrition of ectomycorrhizal plants. II. Utilization of protein by mycorrhizal plants of *Pinus contorta*, New Phytol. 103, 495–506.

Ashford A E, Allaway W G Peterson C A and Carney J W G 1989 Nutrient transfer and the fungus-root interface. Aust. J. Plant Physiol. 16, 85–97.

Beard J S 1984 Biogeography of the kwongan. *In* Kwongan—Plant Life of the Sandplain. Eds. J S Pate and J S Beard. pp 1–26. University of Western Australia Press, Nedlands.

Bell D T and Stephens L J 1984 Seasonality and phenology of kwongan species. *In* Kwongan—Plant Life of the Sandplain. Eds. J S Pate and J S Beard. pp 205–256. University of Western Australia Press, Nedlands.

Bell D, Hopkins A J M and Pate J S 1984 Fire in the Western Australian kwongan. *In* Kwongan—Plant Life of the Sandplain. Eds. J S Pate and J S Beard. pp 178–204. University of Western Australia Press, Nedlands.

Bettenay E 1984 Origin and nature of the sandplains. *In* Kwongan—Plant Life of the Sandplain. Eds. J S Pate and J S Beard. pp 51–68. University of Western Australia Press, Nedlands.

Bjorkman E 1960 *Monotropa hypopitys* L—an epiparasite on tree roots. Physiol. Plant. 13, 308–327.

Bolan N S, Robson A D and Barrow N J 1987 Effects of vesicular arbuscular mycorrhiza on the availability of iron phosphates to plants. Plant Soil. 99, 401–410.

Bowen B J 1991 Fire response within the family Proteaceae: A comparison of plants displaying the seeder and resprouter mode of recovery. Ph.D thesis. University of Western Australia, Perth.

Bowen B J and Pate J S 1991 Adaptations of S W Australian members of the Proteaceae; allocation of resources during early growth. *In* Proc. International Protea Association Sixth Biennial Conference. pp 289–301. Promaco Conventions Pty Ltd, Perth, Western Australia.

Brundrett M.C 1991 Mycorrhizas in natural ecosystems. Adv. Ecol. Res. 21, 171–313.

Brundrett M C and Abbott L K 1992 Roots of jarrah forest plants I Mycorrhizal associations of shrubs and herbaceous plants. Aust. J. Bot. 39, 445–457.

Dixon K W, Pate J S and Bailey W J 1980 Nitrogen nutrition of the tuberous sundew *Drosera erythrorhiza* Lindl. with special reference to catch of arthropod fauna by its glandular leaves. Aust. J. Bot. 28, 283–297.

Dixon K W, Pate J S and Kuo J 1991 The Western Australian subterranean orchid *Rhizanthella gardneri* Rogers. *In:* Orchid Biology, Reviews and Perspectives. Ed. J Arditti. pp 37–63. Timber Press, Oregon, Canada.

Dodd J, Heddle E M, Pate J S and Dixon KW 1984 Rooting patterns of sandplain plants. *In* Kwongan—Plant Life of the Sandplain. Eds. J S Pate and J S Beard. pp 146–177. University of Western Australia Press, Nedlands.

Finlay R D and Read D J 1986 The uptake and distribution of phosphorus by ectomycorrhizal mycelium. *In* Physiology and Genetical Aspects of Mycorrhizae. Eds. V Gianinazzi-Pearson and S Gianinazzi. pp 351–355. INRA, Paris, France.

Francis R, Finlay R D and Read D J 1986 Vesicular-arbuscular mycorrhizal in natural vegetation systems IV. Transfer of nutrients in inter- and intra-specific combinations of host plants. New Phytol. 102, 103–111.

Hansen A P and Pate J S 1987 Comparative growth and symbiotic performance of seedlings of *Acacia pulchella* and *A. alata* in defined pot culture or as natural understorey components of a eucalypt forest ecosystem in S W Australia. J. Exp. Bot. 38, 13–25.

Harley J L and Smith S.E 1983 Mycorrhizal symbiosis. Academic Press, New York. 368 p.

Harris D and Paul E A 1987 Carbon requirements of vesicular arbuscular mycorrhizae. *In* Ecophysiology of VA Mycorrhizal Plants. Ed. G Sasfir. pp 93–105. CRC Press, Boca Raton, Florida, USA.

Haystead A, Malajczuk N and Grove T S 1988 Underground transfer of nitrogen between positive plants infected with vesicular-arbuscular mycorrhizal fungi. New Phytol. 108, 417–423.

Jakobsen I and Rosendahl L 1990 Carbon flow into soil and external hyphae from roots of mycorrhizal cucumber plants. New Phytol. 115, 77–83.

Jasper D A, Robson A D and Abbott L K 1979 Phosphorus and the formation of vesicular-arbuscular mycorrhizal fungi. Aust. J. Bot. 35, 641–652.

Karlsson P S and Pate J S 1992 Contrasting effects of supplementary feeding of insects or mineral nutrients on the growth and nitrogen and phosphorus economy of pygmy species of *Drosera*. Oecologia 92, 8–13.

Lamont B B 1982 Mechanisms for enhancing nutrient uptake in plants, with particular reference to Mediterranean South Africa and Western Australia. Bot. Rev. 48, 597–689.

Lamont B B 1984 Specialized modes of nutrition. *In* Kwongan—Plant Life of the Sandplain. Eds. J S Pate and J S Beard. pp 236–245. University of Western Australia Press, Nedlands.

Lambers H 1987 Growth, respiration exudation and symbiotic

associations: the fate of carbon translocated to the roots. *In* Root Development and Function. Ed. P J Gregory, J V Lake and D A Rose. pp 125–145. Cambridge University Press, Cambridge, UK.

Leake J R, Shaw G and Read D J 1989 The role of ericoid mycorrhizal in the ecology of ericaceous plants. Agric. Ecos. Environ. 29, 237–250.

Mitchell D T 1992 Mycorrhizal associations. *In* The Exploitation of Microorganisms in Applied Biology. Ed. D G Jones. Chapman and Hall (*in press*).

Monk D, Pate J S and Loneragan W A 1981 Biology of *Acacia pulchella* R.Br. with special reference to symbiotic nitrogen fixation. Aust. J. Bot. 29, 579–592.

Newman E I 1988 Mycorrhizal links between plants: their functioning and ecological significance. Adv. Ecol. Res. 18, 243–270.

Newman E I and Reddell P 1987 The distribution of mycorrhizas among families of vascular plants. New Phytol. 106, 745–751.

Pate J S 1989 Australian micro stilt plants. Trends Ecol. Evol. 4, 45–49.

Pate J S and Dell B 1984 Economy of mineral nutrients in sandplain species. *In* Kwongan—Plant Life of the Sandplain. Eds. J S Pate and J S Beard. pp 227–252. University of Western Australia Press, Nedlands.

Pate J S and Dixon K W 1982 Tuberous, Cormous and Bulbous Plants. University of Western Australia Press, Nedlands. 268 p.

Pate J S and Layzell D B 1990 Energetics and biological costs of nitrogen assimilation. *In* The Biochemistry of Plants, Vol. 16. Eds. P K Stumpf and E E Conn. pp 1–42. Academic Press, Inc, California.

Pate J S, Dixon K W and Orshan G 1984 Growth and life form characteristics of kwongan species. *In* Kwongan—Plant Life of the Sandplain. Eds. J S Pate and J S Beard. pp 84–100. University of Western Australia Press, Nedlands.

Pate J S, Casson N E, Rullo J and Kuo J 1985 Biology of fire ephemerals of the sandplains of the kwongan of southwestern Australia. Aust. J. Plant Physiol. 12, 641–655.

Pate J S, Rasins E, Rullo J and Kuo J 1986 Seed nutrient reserves of Proteaceae with special reference to protein bodies and their inclusions. Ann. Bot. 57, 747–770.

Pate J S, Lindblad P and Atkins C A 1988 Pathways of assimilation and transfer of fixed nitrogen in coralloid roots of the cycad—*Nostoc* symbiosis. Planta 176, 461–471.

Pate J S, Kuo J, Dixon K W and Crisp M D 1989 Anomalous secondary thickening in roots of *Daviesia* (Fabaceae) and its taxonomic significance. Bot. J. Linn. Soc. 99, 175–193.

Pate J S, Kuo J and Davidson N J 1990a Morphology and anatomy of the haustorium of the root hemiparasite *Olax phyllanthi* (Olacaceae), with special reference to the haustorial interface. Ann. Bot. 65, 425–436.

Pate J S, Pate S R, Kuo J and Davidson N J 1990b Growth, resource allocation and haustorial biology of the root hemiparasite *Olax phyllanthi* (Olacaceae). Ann. Bot. 65, 437–449.

Pate J S, Davidson N J, Kuo J and Milburn J A 1990c Water relations of the root hemiparasite *Olax phyllanthi* (Labill) R.Br. (Olacaceae) and its multiple hosts. Oecologia 84, 186–193.

Pate J S, Meney K A and Dixon K W 1991 Contrasting growth and morphological characteristics of fire-sensitive (obligate seeder) and fire-resistant (resprouter) species of Restionaceae (S. Hemisphere restiads) from southwestern Western Australia. Aust. J. Bot. 39, 505–525.

Pate J S, Stewart G R and Unkovich M J 1993 ^{15}N natural abundance of plant and soil components of a *Banksia* woodland ecosystem in relation to nitrate utilization, life form, mycorrhizal status and N_2-fixing abilities of component species. Plant Cell Env. 16, 365–373.

Raich J W and Nadelhoffer K J 1990 Below ground carbon allocation in forest ecosystems: global trends. Ecology 70, 1346–1354.

Read D J 1983 The biology of mycorrhiza in the Ericales. Can. J. Bot. 61, 985–1004.

Read D J, Francis R and Finlay R 1985 Mycorrhizal mycelia and nutrient cycling in plant communities. *In* Ecological Interactions in Soil: Plants, Microbes and Animals. Ed. A H Fitter, D Atherson, D J Read and M B Usher. pp 193–217. Blackwell, Oxford, UK.

Reed M L 1987 Ericoid mycorrhiza of Epacridaceae in Australia. *In* Mycorrhizae in the Next Decade: Practical Applications and Research Priorities, Proc. 7th NACOM. Eds. D M Sylvia, L L Hung and J F Graham. p 335. Institute of Food and Agriculture Science, University of Florida, Gainsville, Florida.

Schulze E-D, Gebauer G, Schulze W and Pate J S 1991 The utilization of nitrogen from insect capture by different growth forms of *Drosera* from southwest Australia. Oecologia 87, 240–246.

St John T V and Coleman D C 1983 The role of mycorrhizae in plant ecology. Can. J. Bot. 61, 1005–1014.

Stock W D, Pate J S and Delfs J C 1990 Influence of seed size and quality on seedling development under low nutrient conditions in five Australian and South African members of the Proteaceae. J. Ecol. 78, 1005–1020.

Stewart G R, Pate J S and Unkovich M J 1993 Characteristics of inorganic nitrogen assimilation of plants in fire-prone Mediterranean type vegetation. Plant. Cell. Env. 16, 351–363.

Tommerup I C 1988 The vesicular-arbuscular mycorrhizas. Adv. Plant. Pathol. 6, 81–91.

Trappe J M 1987 Phylogenetic and ecological aspects of mycotrophy in the Angiosperms from an evolutionary standpoint. *In* Ecophysiology of VA Mycorrhizal Plants. Ed. G R Safir. pp 5–25. CRC Press, Boca Raton, Florida, USA.

Walker B A and Pate J S 1986 Morphological variation between seedling progenies of *Viminaria juncea* (Schrad. and Wendl.) Hoffmans. (Fabaceae) and its physiological significance. Aust. J. Plant Physiol. 13, 305–319.

Warcup J H 1980 Ectomycorrhizal associations of Australian indigenous plants. New Phytol. 85, 531–535.

Warcup J H 1988 Mycorrhizal associations and seedling development in Australian Lobeliadeae (Campanulaceae). Aust. J. Bot. 36, 461–472.

Reprinted from *Plant and Soil*, Volume 159, No. 1 (1994), pp. 1–10.

The contributions of mycorrhizal fungi to the determination of plant community structure

R. FRANCIS and D.J. READ
Department of Animal and Plant Sciences, University of Sheffield, P.O.Box 601, Sheffield, S10 2UQ, UK

Key words: ectomycorrhizal fungi, non-hosts, plant interactions, vesicular-arbuscular fungi

Abstract

While it is now widely accepted, even by ecologists, that most plants in the majority of ecosystems are infected by mycorrhizal fungi, few experiments have been designed to investigate the function of the mutualism at the community level. Those involved with mycorrhizal research have been largely pre-occupied with questions of the mineral, particularly phosphorus, nutrition of individual plants, while plant community ecologists have too often found it convenient, even when acknowledging the presence of infection, to ignore its possible function in the ecosystem.

This presentation examines a selected number of seminal papers written by plant community ecologists and highlights some of 'the most striking mysteries' which they reveal. It describes experiments designed to determine whether knowledge of the presence and activity of the mycorrhizal mycelium can help us to unravel the 'mysteries' which they define.

It is revealed that by having direct adverse effects upon seedlings of many 'r' selected species, while at the same time being beneficial, if not essential, to those that are 'K' selected, the activities of the mycelium of VA fungi have a direct bearing upon community composition. The extent to which 'turf compatibility' is actually a reflection of the compatibility of plant species with the VA mycorrhizal mycelium is discussed and the possible role of the mycelium in consigning some species to the ruderal habit is considered.

It is concluded that those attempting scientifically to understand, or managerially to manipulate, plant communities, without recognizing the role of the mycorrhizal mycelium, do so at their peril, and it is recommended that scientists involved in research on mycorrhiza extend their vision beyond the limited horizons which are currently so often defined by considerations of the phosphorus nutrition of individual host plants.

Introduction

Whereas mycorrhizal associations evolved in complex and relatively stable natural environments which characteristically support mixed assemblages of plant species, man seeks to manage the symbiosis in simplified systems often involving monocultures of agricultural, horticultural or forest crops. Research on mycorrhizal function has likewise largely been carried out under standardized laboratory or glasshouse conditions specifically designed to exclude the complexities associated with natural soil as well as interspecific interactions between 'hosts' or fungal symbionts. It is partly for this reason that while we are able to manipulate the mycorrhizas of single fungal and 'host' species under sterile conditions, or in those lacking indigenous inoculum, attempts to introduce or manage fungal symbionts in natural conditions have often failed. We know little about the factors determining competitive and other

A.D. Robson, L.K. Abbott & N. Malajczuk (eds.), Management of mycorrhizas in agriculture, horticulture and forestry, 11–25.

interactions between mycorrhizal fungi below ground and even less about the impact of the symbiosis upon interactions between 'host' or 'non-host' species as revealed above-ground in plant community structure.

This paper describes progress towards an understanding of some of these complex issues and stresses the need for research under more realistic conditions.

The VA mycorrhizal symbiosis in natural ecosystems

The majority of species of temperate, sub-tropical and tropical plant communities are infected by VA mycorrhizal fungi Trappe (1987) and it is evident that infection of individual plants germinating in species rich natural communities occurs within a few days of radicle emergence (Read and Birch, 1988). There is every indication that in such communities a vigorous semi-permanent population of fungal symbionts with low 'host' specificity is involved in an infection process which effectively integrates compatible species into extensive mycelial networks. That this infection occurs with the same rapidity whether seedlings of a given species are germinating in intra- or inter-specific situations adds support to the view that in many natural communities a large number of plant species are interconnected by a relatively small number of infecting fungal species. The advantage to the plant of such low 'host' specificity is that it increases the chance that infection by a compatible fungus will occur rapidly and so facilitate its integration into the potentially enormous nutrient catchment provided by the mycorrhizal mycelium. As shown below, this process can be important for establishment and survival of compatible species but such observations leave unanswered two fundamental questions which go right to the heart of any consideration of the role of mycorrhizal infection in plant communities. First, what is the impact of fungi upon plant species which are not susceptible to infection? Second, what is the impact of infection upon interactions between compatible plants, in the post-establishment phase. These two questions are addressed below.

The impact of VA fungi upon establishment and survival of 'non-host' species

We know that whereas the majority of species in natural, semi-natural and agricultural plant communities are susceptible to infection by mycorrhizal fungi, members of some plant families, notably the Cruciferae, Caryophyllaceae, Chenopodiaceae and Polygonaceae, so-called 'non-host' plants, are not normally mycorrhizal. It is of ecological interest that the majority of species in this category are ruderals of disturbed habitats, which are excluded from most closed plant communities. It is also of economic importance that amongst these plants are some of the most pernicious and widespread weeds of agricultural systems. There is therefore a need to understand the factors which determine their success in open and their failure in closed habitats. Many ruderal weeds are annual or biennial species the life cycles of which can be completed in the brief period of time between a disturbance event and recolonization by perennial species, most of which are infected by vesicular-arbuscular (VA) mycorrhizal fungi. The success of ruderal weeds in disturbed habitats has been attributed to their ability to germinate rapidly and exploit bare ground (Grime, 1979). They can therefore capitalize on the pulses of nutrient release produced by the disturbance event. Their failure, conversely, to survive in stable environments dominated by turf-forming species, their turf incompatibility, has been attributed simply to a lack of competitive ability (Fenner, 1978). None of these conventional explanations recognizes that disturbance events reduce the vigour of mycorrhizal propagules (Read and Birch, 1988) or that a characteristic of the stable turf is the presence of an active VA mycelial network. While Grubb (1986) postulated that mycorrhizal fungi might 'tip the balance' against turf incompatible species in the field no mechanisms for any such action have been suggested.

There have been studies of the impact of VA mycelium on 'non-host' plants (Ocampo et al., 1980, 1981; Glenn et al., 1985; Allen et al., 1989) but the main thrust of these has been towards an analysis of the nature of any infection processes. Allen et al. (1989) reported adverse effects of the fungus on the 'non-host' plant *Salsola kali* which were seen in the form of localized lesions on the roots around aborted penetration points. In general, however, these experiments do not simulate the natural process in which seed germination and early growth of the 'host' or 'non-host' plant takes place in the presence or absence of a pre-established VA mycelium, and are therefore not able to throw light on the substantive question concerning the role of the fungus in determining patterns of recruitment into plant communities.

Table 1. Survivorship (%) of forbs after six months in mycorrhizal (M) and non-mycorrhizal (NM) microcosms

Species	M	NM
Centaurium erythraea	64	2
Galium verum	58	11
Hieraceum pilosella	49	6
Leontodon hispidus	42	13
Plantago lanceolata	71	10
Sanguisorba minor	53	6
Scabiosa columbaria	84	16
Arabis hirsuta	8	42
Rumex acetosa	11	60

Grime et al. (1987) observed that when plant communities consisting initially of turf compatible and turf incompatible ruderal species were grown from seed in microcosms with (M) or without (NM) mycorrhizal inoculum the ruderal species grew less vigorously and had poorer survivorship in the presence of VA inoculum (Table 1). Overall diversity was increased in the M microcosms because the majority of species, which were compatible and became mycorrhizal benefited strongly from the infection. These results suggest that mycorrhizal inoculum plays a role in determining species composition in the community by influencing plant fitness at the establishment phase but provide no indication of the basis of the differential responses of 'host' and 'non-host' plants.

The microcosm experiment has been followed up with studies of the early responses of several ruderal weed species to the presence of an established VA mycelium of the kind likely to

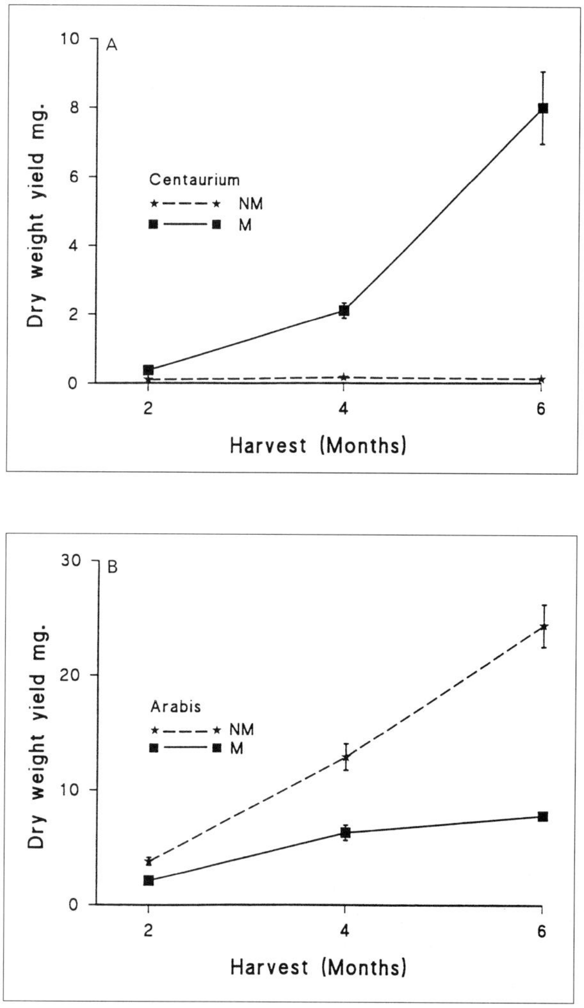

Fig. 1. Dry weight yield of the 'host' plant *Centaurium erythraea* (a) and the 'non-host' plant *Arabis hirsuta* (b) grown in a grass–legume community with (M), and without (NM), the presence of VA mycorrhizal fungi.

occur in most soils of low to moderate fertility under stable conditions. They were carried out using shallow trays of nutrient poor sand in which was established a grass–legume (*Festuca ovina–Medicago lupulina*) mixture grown in either the M or NM condition. After these plants had become established, pre-imbibed seeds of the 'host' species *Centaurium erythraea* and the 'non-host' plants *Arabis hirsuta* (Cruciferae) and *Arenaria serpyllifolia* (Caryophyllaceae) were sown onto the open sand surface. Early growth, establishment, as well as patterns of root

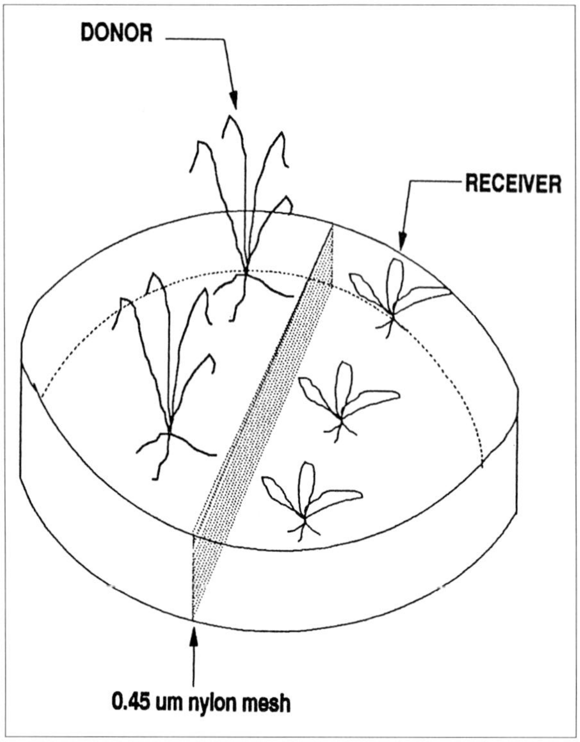

Fig. 2. Diagram of split-chamber system used to separate roots of 'donor' plants (*Plantago lanceolata*) from 'host' or 'non-host' 'receivers' and to obtain discrimination between mycelial and root competition effects.

therefore designed in which two-compartment chambers were used to facilitate separation of roots of mycorrhizal and 'non-host' plants while enabling the VA mycelium to develop in both compartments. Separation was achieved using a nylon barrier with a mesh size sufficient to enable mycelial penetration, while blocking penetration of roots (Fig. 2). *Plantago lanceolata* was grown as the 'host' plant in the mycorrhizal (donor) or non-mycorrhizal condition in one compartment. After sufficient time had elapsed (c 4 weeks) to enable growth of colonization and survivorship were monitored. As observed by Grime et al. (1988), 'host' plants e.g. *Centaurium erythraea* (Fig. 1a) benefited from the presence of inoculum, while 'non-host' plants e.g. *Arabis hirsuta*, (Fig. 1b) grew slowly and were weak in appearance. These effects could not be reversed by addition of phosphate (P) or complete nutrient solution. Careful examination of the roots of these plants at various stages of development revealed the presence of VA mycelium in contact with the root surface and root hairs but there was no evidence of penetration by hyphae or of lesions of the kind reported by Allen et al. (1989).

It was considered possible that root-competition with the more vigorous 'host plants' in the mycorrhizal system rather than fungal effects was disadvantaging the 'non-host' species. Supplementary experiments were

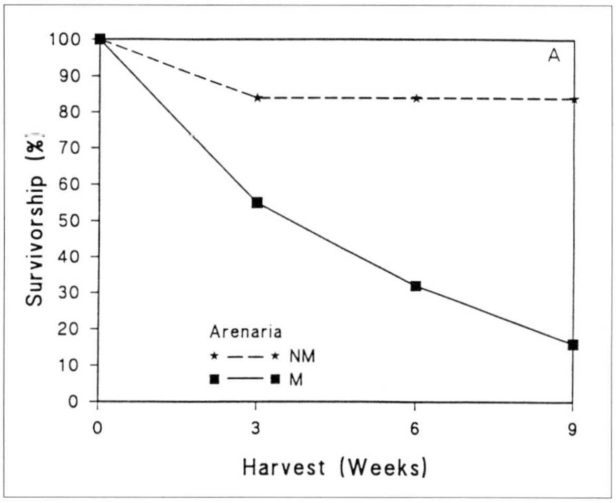

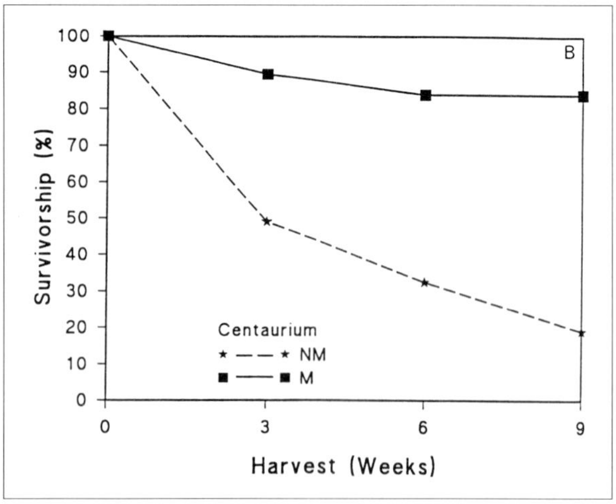

Fig. 3. Survivorship of the 'non-host' *Arenaria* (a) and the 'host' plant *C. erythraea* (b) sown at the time of radicle emergence onto the receiver side of split-chambers, with (M), and without (NM), the presence of VA mycelium.

mycelium through the barrier, three pre-imbibed seeds of the 'host' plant *C. erythraea* or of the 'non-host' *A. hirsuta* or *A. serpyllifolia* were planted into the opposite (receiver) compartment of a series of divided chambers (Fig. 2).

The responses of these species were comparable with those seen in the open community. In the presence of mycelium of VA fungi, growth of *C. erythraea* was stimulated while that of the two weed species was severely reduced the impacts being visibly evident within a few days of germination. After 9 weeks more than 80% of *Arenaria* seedlings had died in the M systems (Fig. 3a). *Centaurium* in contrast showed high mortality in the NM treatment (Fig. 3b).

Closer comparative examination of M and NM plants of *Arenaria* (Table 2) demonstrated that root extension, root branching and, most notably of all, root hair development were strongly inhibited in the presence of VA mycelium. There was no evidence that P levels were depleted by the VA mycelium. For this reason a search was commenced for inhibitory compounds. When soil water from M and NM systems was extracted in vacuo and used directly in seedling bioassays some of the inhibitory effects seen in the M soil were reproduced in the M soil extracts (Table 2). This observation has been confirmed in subsequent assays.

If, as seems likely, it is subsequently shown that such laboratory based observations of direct antagonistic interactions between the VA mycelium and 'non-host' species can be repeated in the field, the role of these symbionts in determining plant-community structure will have to be re-appraised. While the predominance of non-mycorrhizal weeds in the early stages of secondary succession and the failure of 'obligate' mycotrophs have both been attributed to the absence of infective VA propagules (Reeves et al., 1979; Miller, 1987; Allen and Allen, 1980; Janos, 1980) these features have generally been interpreted in a nutritional context, a progressive increase of selection in favour of plants with VA mycorrhizal infection being envisaged as availability of mineral ions decreases with time. The possibility now emerges that it is the prevention of regeneration of the largely short-lived ruderal weed species by the VA fungi which leads to their elimination and consequently to the maintenance of a stable sward or 'turf' or VA compatible species. The selective advantages to the fungus of maintaining a high proportion of compatible species to act as carbon sources, at the expense of incompatible species, is evident. The observed above ground stability is only changed by perturbations such as disturbance or phosphate enrichment which are known (Abbott et al., 1984) directly to reduce the growth of the mycelium, and would therefore be likely to reduce its antagonistic effects.

The short life span of many ruderals, while facilitating rapid exploitation of gaps created in turf by disturbance events, also carries the disadvantage that it necessitates regeneration by seed and hence exposure of the seedlings of new generations to re-encroaching VA mycelium. In small gaps this encroachment is likely to be sufficiently rapid to exclude 'non-host' plants and so maintain the integrity of the turf-compatible community. In large gaps, in contrast, there may be time for ruderals to complete several generations and so establish a seed-bank in the soil of sufficient magnitude to enable recolonization when a future disturbance

Table 2. Mean root lengths, root hair lengths and root hair widths of *Arabis hirsuta* and *Arenaria serpyllifolia* after 96 h growth of the newly emerged radicle in aqueous extracts of sand supporting (M), or free of (NM), living mycelium of VA fungi

		Root length (mm)		Root hair length (mm)		Root hair width (mm)	
A. hirsuta	M	13.66	NS	584	$p < 0.05$	11.64	$p < 0.05$
	NM	15.40		1182		8.53	
A. serpyllifolia	M	3.80	$p < 0.05$	693	$p < 0.05$	11.61	$p < 0.05$
	NM	6.20		1044		8.98	
		n = 12		n = 300		n = 300	

event adversely effects the VA mycelium. There is evidence for the production of such seed banks in *A. hirsuta*, *A. serpyllifolia* (Grime et al., 1988). It is noteworthy in this context that the small numbers of non-host species that are able to persist in closed turf, for example members of the Cyperaceae, are perennials which have a well-developed ability to spread by vegetative means.

The impact of infection upon interactions between mature compatible species

When pairs of species are grown together with or without the addition of VA inoculum the outcome of the interaction between them is known to be very different. Fitter (1977) observed that when *Lolium perenne* was grown with *Holcus lanatus* in the presence of VA fungi its yield was reduced by 46% relative to that seen when the two grasses were grown together in the non-mycorrhizal condition. Similarly Hetrick et al. (1989), examining two grass species, *Andropogon gerardii* and *Koeleria pyramidata,* which co-exist in tall-grass prairie, found that whereas *K. pyramidata* was little influenced by the presence of *A. gerardii* in sterile soil, the addition of VA inoculum to the pair led to yield reduction in *Koeleria* of up to 91%. These authors suggest that temporal separation of vegetative activities might enable the early season C3 species *Koeleria* to avoid the main impact of *A gerardii* which is a late season C4 grass. In the case of grass–legume mixtures it has been shown that addition of inoculum produces a significant positive effect upon the legume at the expense of the grass (Hall, 1978; Buwalda, 1980). Since the plants used in all of these experiments are normally infected in nature it is implicit in the observations arising from them that the balance between plant species in the field is at least in part a reflection of the impact of infection. To obtain clear-cut evidence of any such impacts on plants growing under natural conditions is, however, difficult because we are as yet unable selectively to remove mycorrhizal fungi from natural vegetation systems.

The microcosm approach in which communities are reconstructed and grown in the presence or absence of inoculum for periods of several months, also provides useful indications of some of the likely impacts of infection on mature plants in the field. In their microcosms Grime et al. (1987) observed that the grass *Festuca ovina* which was used as the turf-forming species showed a reduction of biomass in the presence of inoculum despite the fact that it was heavily infected, as indeed, it normally is in the field (Read et al., 1976; Harley and Harley, 1987). The majority of VA compatible forbs, in contrast, responded positively to infection and produced significantly higher yields in the M than the NM condition after one year. The responses may be partly explained by the fact that C3 grasses in general have better developed root systems than C3 forbs (Hetrick et al., 1988). The fibrous root systems of *Festuca* are likely to capture nutrients with sufficient effectiveness through most of the life of the host to nullify any advantageous effects of VA infection. In cases, such as this in which the costs of the mycorrhiza exceed the benefits, the host might be expected, assuming it was able, to reject the infection. However, to assess 'fitness' of wild plants especially those of nutrient stressed natural communities on the basis of gains or losses of vegetative biomass may be convenient but inappropriate.

Costs in terms of reduced vegetative growth may be outweighed by benefits accruing to plants at other stages of their lives. The importance of compatibility with VA fungi for recruitment of seedlings into the community has been described above. Some studies suggest that infection may also play a key role in seed production itself. Both the numbers of seeds produced (Koide et al., 1988) and the nutrient reserves of the individual seeds themselves (Lewis and Koide, 1990; Koide and Lu, 1991) can be enhanced if the maternal plants are mycorrhizal. Seeds of comparable weight produced by infected parent plants of wild oat *Avena fatua* L contained significantly more organic P than those produced by uninfected plants (Koide and Lu, 1991). The same may be true of other grasses such as *Festuca*, and would be expected to improve the 'fitness' of the next generation by increasing its competitive ability (Parrish and Bazzaz, 1985) or by permitting survival during the period between germination and recruitment into the mycorrhizal mycelial network.

It is clearly inappropriate to attempt to categorize mycorrhizal 'dependence' of a given plant on the basis of vegetative responses during a short window of time at some ill-defined stage of its life cycle. Even the use of terms such as 'facultative' and 'obligate' must be viewed with caution especially if it is the intention to extrapolate from observations made in pots to real community situations.

The ectomycorrhizal symbiosis in natural ecosystems

There has been much emphasis in the past upon the comparability of function of VA and ectomycorrhiza, and the ability of both types of mutualism to provide improvement of access to and storage of phosphate ions is well established (Harley and Smith, 1983). This emphasis, however, overlooks the fact that the two mutualisms assume predominant importance in plant communities that are spatially, structurally and nutritionally distinct (Read, 1991). Thus, while the characteristic nutritional limitations of temperate and sub-tropical herbaceous of communities dominated by VA mycorrhizal plants may well be primarily phosphorus, the productivities of most of those forest communities of the world that are dominated by ectomycorrhizal plants are limited by nitrogen availability. It is predictable under these circumstances that whereas mutualisms providing enhancement of access to phosphate

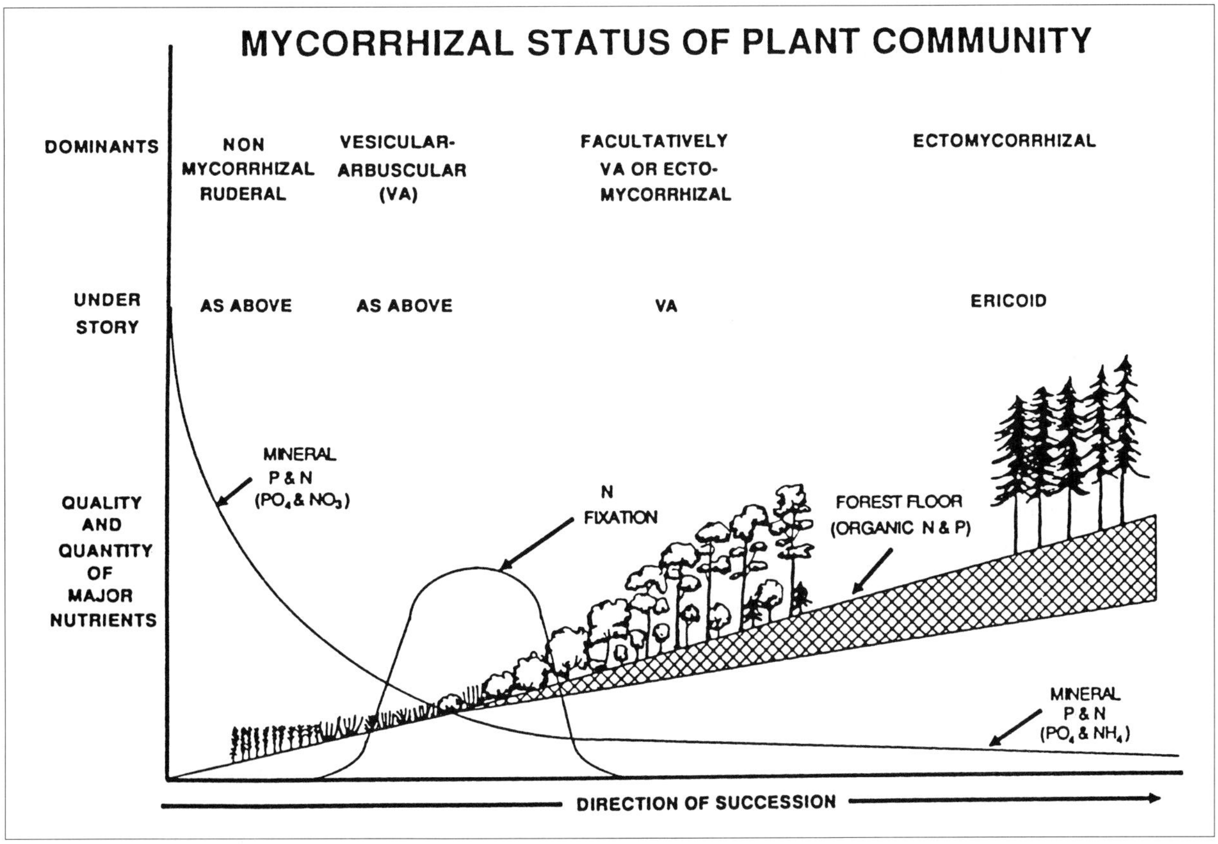

Fig. 4. Showing the proposed relationships between changing of physico-chemical qualities of the rooting environment in the course of succession and selection of differing mycorrhizal types. This succession refers to a boreal forest climax but similar processes are likely to be involved in many temperate forests and in the most nutrient impoverished tropical forests.

would have a selective advantage in the former case, those providing access to nitrogen would be favoured in the latter. Increasingly the evidence points in this direction and we need to examine the extent to which plant community structure is indeed determined by the need of host plants to form compatible associations with mutualists capable of supplying them with the major growth limiting nutrients.

The conventional view of both systems was largely based upon studies of the physiology of the mycorrhizal root in pot or solution culture in the laboratory. While having the advantages that experimental conditions can be controlled these approaches have the disadvantage that the results obtained will often be of limited relevance to the real world in which the different types of mycorrhiza evolved and diverged. Amongst the most serious limitations of experiments based upon pot or solution cultures is that the mycelial phase of the symbiosis is restricted in its development, or worse still, eliminated completely.

The more we examine the distribution and function of intact mycorrhizal systems in natural and semi-natural substrates the more we realise that whereas the mycelium of VA fungi can legitimately be seen spatially and functionally to be an extension of the root system which supports it, the ectomycorrhizal mycelium fulfils a far more fundamental role. It not only widely subsumes the function of the root but in the cases of many fungal species, it also has the biochemical attributes necessary to provide the host with access to nutrients contained in organic resources that are otherwise unavailable to it (Abuzinadah and Read, 1986a, b, 1989a, b). It is this feature which can be expected to provide ectomycorrhizal plants with increasingly large selective advantages in natural ecosystems as with time, progressively greater proportions of the key nutrient elements N and P are locked up in surface organic horizons and may determine the structure of plant communities as succession proceeds. (Fig. 4).

Studies of the distribution of ectomycorrhizal roots indicate that in contrast to their VA counterparts they are predominantly concentrated in or immediately under litter horizons (Harvey et al., 1976, Persson, 1980, Scheirose, 1990). Indeed, in the case of *Eucalyptus* it has been suggested (Chilvers and Pryor, 1965) that the formation of such horizons is an essential prerequisite for ectomycorrhiza formation, and there is evidence (Reddell and Malajczuk, 1984) that individual plants of *E. marginata* will form VA mycorrhiza in mineral soil and ectomycorrhiza in litter.

Not only do the two types of mycorrhiza characteristically occur in substrates that we qualitatively distinct but also the quantities of roots produced in the two associations is different. Root densities of ectomycorrhizal trees are generally lower than those produced by fibrous rooted herbaceous species that are typically 'hosts' to VA fungi (see e.g. Newman, 1969). Only recently have studies using unsterile forest soil or peat in microcosms revealed the extent to which the mycelium of ectomycorrhizal roots can fulfil what are normally considered to be the functions of roots and so provide explanation for the apparent anomaly.

A number of typical ectomycorrhizal basidiomycetes have now been shown (Read and Finlay, 1986a, b; Coutts and Nicoll, 1990a, b) to produce mycelial systems that are so extensive as to suggest that the mycorrhizal roots themselves are of little significance as absorptive structures (Read, 1992). This view is encouraged by an awareness achieved from studies in the field and in microcosms, (Read, 1992) that the majority of the ectomycorrhizal roots from which these mycelia develop are formed in air filled pores where they have little contact with soil.

Measurements of the ratio of mycelial length to total length of infected mycorrhizal root provide a convenient and physiologically meaningful indication of the importance of the mycelium. In microcosms ratios of 10^4 and even 10^5:l are achieved after 10^4:1 and even 10^5:1 are achieved after 3 months growth (Fig. 5). Such values are one to two orders of magnitude greater than those normally quoted for VA mycorrhizal systems (Sanders and Tinker, 1973; Tisdall and Oades, 1979; Abbott and Robson, 1985; Sylvia, 1990).

While the growth of the mycelium takes the form of a fan which is capable of growing through soil at 2–4 mm day it is not entirely unconstrained or unselective in its foraging.

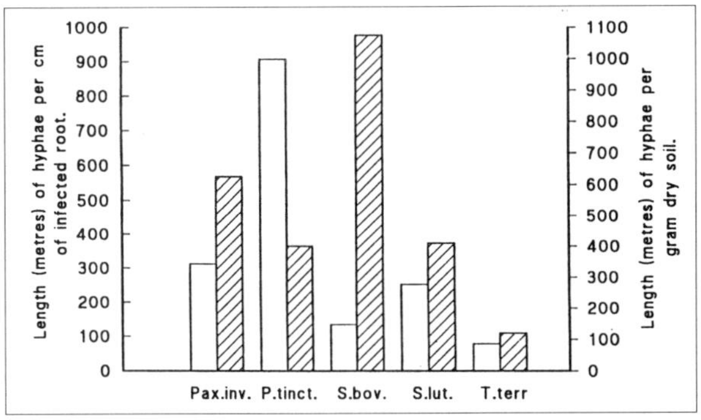

Fig. 5. Mean hyphal lengths expressed per unit of infected root (open bars) and of dry soil (hatched bars) of some representative ectomycorrhizal fungi growing from Pine in microcosms of unsterile organic soil (Pax.inv. = *Paxillus involutus*, P.tinct. = *Pisolithus tinctorius*, S.bov. = *Suillus bovinus*, S.lut. = *Suillus luteus*, T.terr = *Thelephora terrestris*).

Incompatibility between the mycelia of species or even strains of a single species, demonstrated in laboratory studies using pure cultures (Dahlberg and Stenlid, 1990) can also be seen in mycelia growing from mycorrhizal plants. While the hyphae of compatible mycelia appear to fuse on contact so producing a single mycelium interconnecting individual host plants (Plate 1a), when two different species of comparable growth rate e.g. *Suillus bovinus* and *P. tinctorius* approach each other (Plate 1b, c) their incompatibility is shown by inhibition of extension growth at their mycelial fronts and the formation of a bare zone comparable with that seen in pure cultures. The resulting stasis can be broken by growth of a root across the zone of antagonism (Plate 1c). The uninfected tip of a primary root normally grows slightly more rapidly (5–6 mm day) than does the advancing hyphal front and, being unaffected by antagonism between the heterotrophs it can penetrate the domains of other fungi. Here, laterals emerging from it may be infected by the new fungus or by the fungus originally infecting that plant (Plate 1c) which uses the root surface as a 'bridge' upon which to cross into the new domain. In cases where a particularly vigorous fungus encounters a slow growing mycelium of another species antagonism may not be evident, the more vigorous species passing uninfected through the slower mycelium 'capturing' any uninfected roots as it does so.

While the fan like growth of the mycelial front can be effective but generalized foraging strategy there is also evidence of selective foraging. Intensive patches of mycelial development remains behind the advancing mycelial front where substrates of a particular resource quality are encountered. These 'patches' can be induced by introducing appropriate organic substrates into otherwise homogeneous peat or humus. Carleton and Read (1991), in attempts to simulate the forest floor environments of the conifer–feather moss ecosystems that cover so much of the boreal forest zone, showed that mycorrhizal mycelium proliferated selectively around senescing parts of the shoots of *Pleurozium schreberi* (Plate 2a). Similar selective proliferation can be induced by introducing material from the FH layer of coniferous woodland into microcosms of relatively homogeneous humified material (Plate 2b). These are the types of substrate in association with which mycorrhizal roots and mycelia proliferate in the field and bearing in mind the ability of the fungi forming these patches to mobilize organic N there is every reason to believe that the proliferation in involved with N mobilization. Evidence for a direct role of these fungi in N mobilization in 'patches' is still required but enhancement of needle N content in association with 'patch' formation has been demonstrated (Read, 1991). These patterns of selective foraging find close parallels in the roots of herbaceous plants, a feature which emphasizes the comparability of the roles of the 'conventional root' and the ectomycorrhizal mycelium. There are reports of the proliferation of VA mycelium around organic particles, but in the absence of any evidence that these fungi have the ability directly to attach such substrates it must be assumed that they are scavenging for minerals released by the decomposer population.

At the soil–litter interface of Douglas-fir forests discrete hyphal mats, reminiscent of large patches, are produced by mycorrhizal fungi such as *Hysterangium setchellii* (Cromack et al., 1979) which are of narrow host-range (Griffiths et al., 1991). In the mats levels of proteolytic,

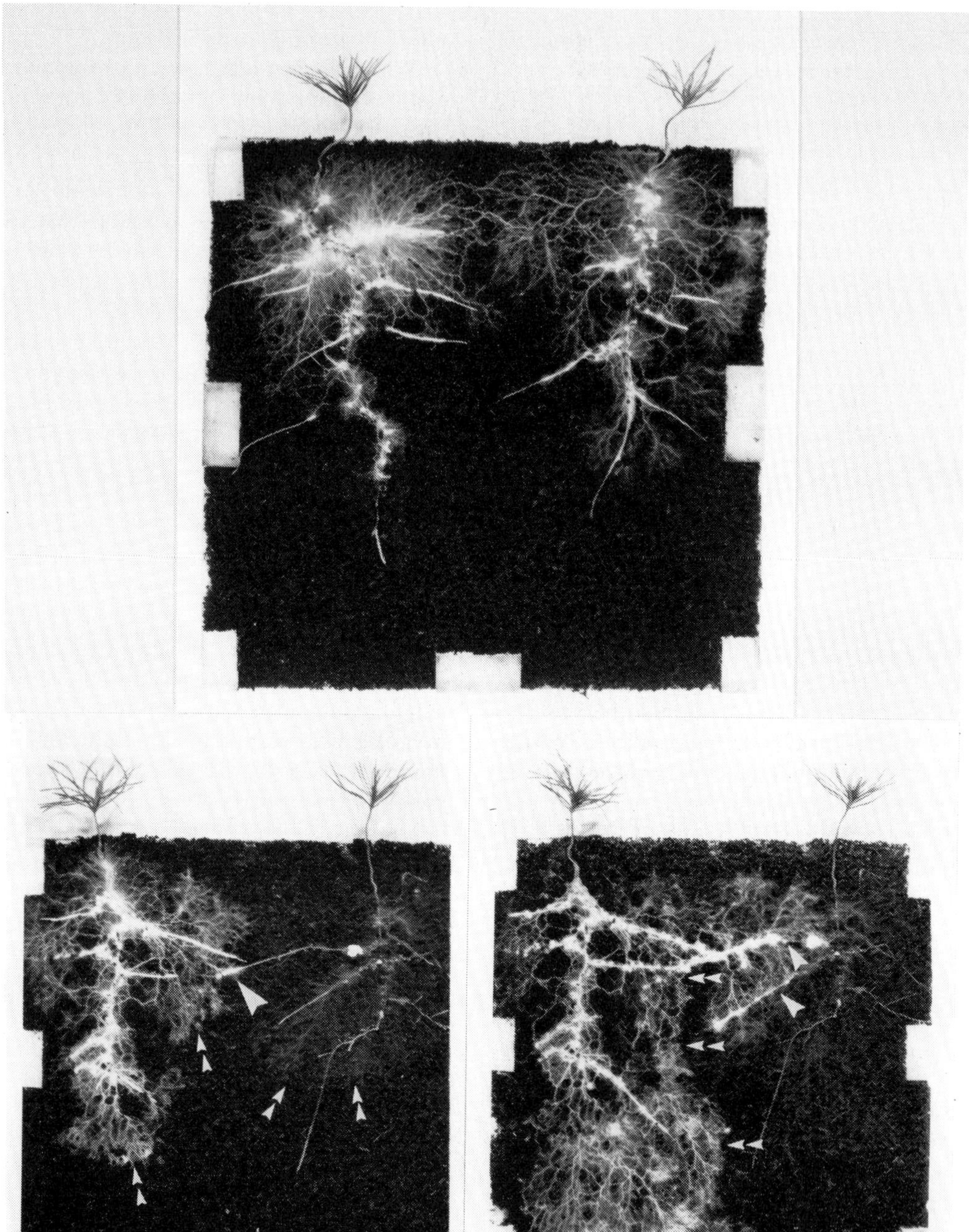

Plate 1a. Transparent observation chamber supporting seedlings of *Pinus sylvestris* each of which were introduced into the chamber infected with this an identical race of the fungus *Suillus bovinus*. The individual mycelia have merged to form a common absorptive structure.
Plate 1b. Transparent observation chamber supporting seedlings of *Pinus sylvestris* each of which was introduced into the chamber infected by a different ectomycorrhizal fungus, either *Suillus bovinus* (left) or *Pisolithus tinctorius* (right). Early stages of growth of the respective mycelia are seen. The hyphal fronts are arrowed (double arrows) and early stages of capture of an extending primary root of the *P. tinctorius* plant by *S. bovinus* are evident (single arrow).
Plate 1c. Later stage of chamber sown in 2b. A zone of antagonism (double arrows) has developed along the points of convergence of the two mycelia. Invasion of territory previously occupied by *P. tinctorius* has occurred where *S. bovinus* has grown along the roots which it has 'captured' from the *P. tinctorius* 'host' (single arrows).

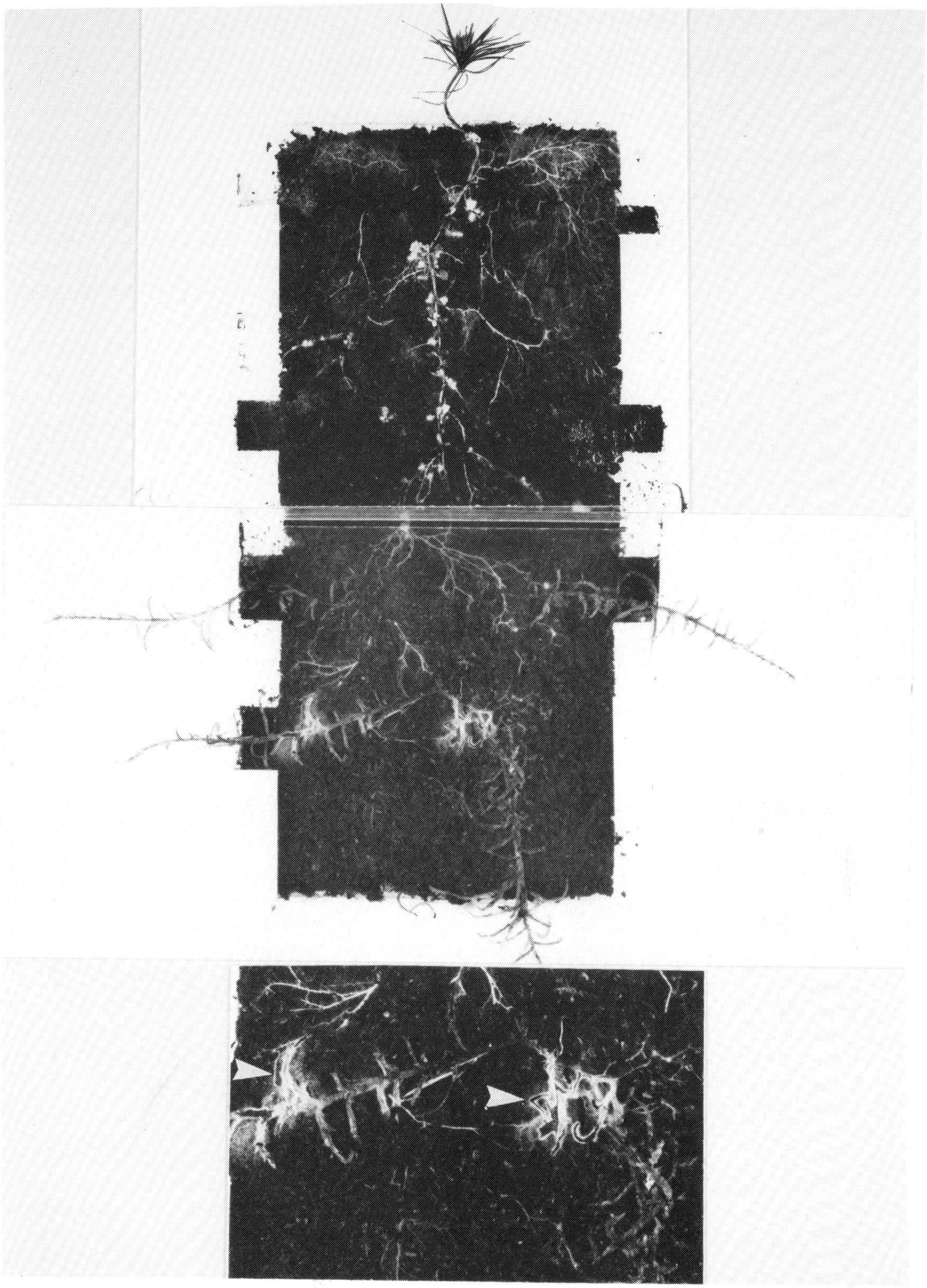

Plate 2a. Vertical section of 'L' shaped chamber supporting ectomycorrhizal seedling of *Pinus contorta* from which mycelium of *Suillus bovinus* has grown onto the lower horizontal section where it is selectively colonizing senescing, but not completely senescent, parts of shoots of the feather-moss *Pleurozium schreberi.* The enlarged view shows selective colonization of the moss shoot by *S. bovinus* showing sheath-like development of mycelium around the substrate (arrowed).

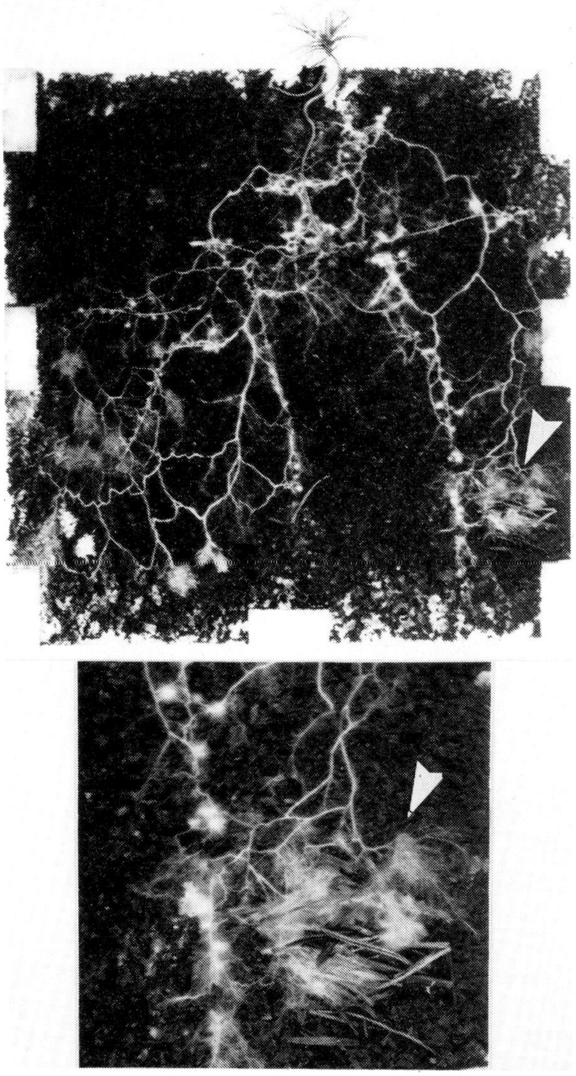

Plate 2b. Proliferation of mycelium of *S. bovinus* growing from *Pinus sylvestris* to form a 'patch' (arrowed) on substrate (pine litter) collected from the forest FH layer and introduced into otherwise homogenous unsterile peat. Detail of the 'patch' is shown in the enlargement below.

pectinolytic and cellulolytic activity are significantly higher than in adjacent areas not colonized by the fungi (Griffiths et al., 1987, 1991; Entry et al., 1991a, b). The result is that in addition to facilitating nutrient supply to seedlings, more rapid rates of litter decomposition are also recorded in association with mats (Entry et al., 1991a). Allocation of carbon to such host specific fungal systems ensures tight cycling of resources between the partners and one of its major consequences in the maintenance of a species poor stand.

These mats are formed in association with mature trees but have the potential to influence the pattern of recruitment into the community since they provide favourable sites for regeneration. Griffiths et al. (1991) observed that all seedlings of Douglas fir growing under the canopy of a maturing 60–75 year old stand of the same tree were associated with these mats.

Studies of patterns of regeneration of coniferous seedlings after fire strongly suggest that the presence of compatible mycorrhizal inoculum is of great importance. Amaranthus et al. (1990) reported that numbers of regenerating seedlings were five times greater beneath surviving plants of *Arubutus menziesii,* which is a host to ectomycorrhizal fungi, then it was beneath shrub species that were hosts to VA fungi. The seedlings regenerating under *Arbutus* also had a larger number of ectomycorrhizal root tips. Similarly, seedlings of Douglas fir planted into rhizosphere soil taken from immediately underneath hosts of ectomycorrhizal fungi such as *Lithocarpus densiflora, Quercus chrysolepis* and *Arbutus menziesii,* have been shown (Borchers and Perry, 1990) to have significantly greater biomass, height and numbers of ectomycorrhizal roots than those planted into non-rhizosphere soil collected at 4 metres distance from the nearest tree.

Attempts to simulate the natural process of regeneration by planting seedlings at the time of germination into undisturbed soil in the forest (Fleming, 1985; Newton and Pigott, 1991) have demonstrated the rapidity with which infection takes place in these situations and confirmed laboratory observations that the fungi already associated with established plants are largely responsible for the infection.

In contrast to the low and uniform C:N ratios found in the litter of most of the herbaceous species that are hosts to VA fungi, the woody hosts of ectomycorrhizal fungi produce litters of high C:N ratio and a wide range of qualities (Read, 1991). As a result, in the relatively species-poor communities typical of boreal and temperate forests the ectomycorrhizal fungi of the dominant trees proliferate most extensively in the distinctive organic substrates produced originally by their own host species. This may have been one of the key factors selecting in favour of the greater host specificity seen in ectomycorrhizal relative to VA dominated ecosystems. Such specificity will tend to maintain low diversity since it favours establishment and survival of a relatively small number of compatible hosts. However, despite the occurrence of genus level specificity for example between *Pseudotsuga* and *Rhizopogon,* or *Pinus* and *Suillus,* it is of considerable ecological importance that such hosts still share an overwhelming number, around 72% (Molina et al., 1991) of potentially compatible fungi of broad host range. This increases the likelihood that some emerging roots become integrated into a compatible network of absorptive hyphae (Read et al., 1985).

Conclusion

That the majority of plants co-evolved with mycorrhizal mutualists is now widely accepted but instead of adopting management practices which recognize the closeness of the relationship between autotroph and heterotroph most agriculturalists, horticulturalists, as well as foresters at the nursery stage, employ cultivation and fertilization techniques which reduce the extent and effectiveness of the mycorrhizal infection by damaging the fungal mycelium. For their part, by concentrating on laboratory or glasshouse experiments with individual roots or plants those researching on mycorrhiza have failed clearly to demonstrate the magnitude of the possible impacts of the symbiosis in nature. Increasingly, however, the evidence suggests that mycorrhizal fungi have the capability not only to influence the nutrient relations of plants but also to determine the nature and composition of the plant communities with which they are associated.

The fungal mycelia involved in these mutualisms exhibit low, but well defined levels of host specificity which, while optimizing the abilities of the heterotrophs to obtain carbon, also enables them to interlink and integrate those plant species with which they are compatible into assemblages or guilds. By influencing recruitment into and survivorship within these guilds they have a significant impact upon species composition and diversity of the plant community. Thus, while the high species diversity characteristic of phosphorus deficient grassland ecosystems dominated by plants with VA mycorrhizas may be attributable to a low level of 'host' specificity which enables a variety of 'hosts' to be incorporated into and supported by, a common mycelium, the exclusion of incompatible ruderals by direct antagonistic effects of the VA mycelium could be an equally important determinant of community structure.

The greater levels of 'host' specificity seen in ectomycorrhizal fungi and the possibility of a relationship between the specificity of the fungus for the 'host' and its affinities for the substrates produced by the 'host' are both likely to be significant features influencing the productivities and community structures of boreal, temperate and some tropical forests also of significance. Fungi equipped to flourish under the condition of tillage and fertilizer application routinely practiced in forest nurseries are most unlikely to be of great value to 'host' trees planted into raw humus or litter enriched soils of the forest.

It can be hoped that as awareness of the role of the mycorrhizal fungal mycelium in plant community processes increases so those involved in the management of ecosystems, be they conservationists, agriculturalists or foresters will adjust their practices so that they more sensitively reflect the functional attributes of the symbiosis. In so doing they will improve the quality of the environment as a whole.

References

Abbott L K and Robson A D 1985 Formation of external hyphae in soil by four species of vesicular-arbuscular mycorrhizal fungi. New Phytol. 99, 245–244.

Abbott L K, Robson A D and DeBoer G 1984 The effect of phosphorus on the formation of hyphae in soil by the vesicular-arbuscular mycorrhizal fungus *Glomus fasciculatum.* New Phytol. 97, 437–446.

Abuzinadah R A and Read D J 1986a The role of proteins in the nitrogen nutrition of ectomycorrhizal plants. I. Utilization of peptides and proteins by ectomycorrhizal fungi. New Phytol. 103, 481–493.

Abuzinadah R A and Read D J 1986b The role of proteins in the nitrogen nutrition of ectomycorrhizal plants. III. Protein utilization by *Betula, Picea* in mycorrhizal association with Hebeloma crustuliniforme. New Phytol. 103, 507–514.

Abuzinadah R A and Read D J 1989a The role of proteins in the nitrogen nutrition of ectomycorrhizal plants. IV. The utilization of peptides by birch (*Betula pendula* L.) infected with different mycorrhizal fungi. New Phytol. 112, 55–60.

Abuzinadah R A and Read D J 1989b The role of proteins in the nitrogen nutrition of ectomycorrhizal plants. V. Nitrogen transfer in birch (*Betula pendula*) grown in association with mycorrhizal and non-mycorrhizal fungi. New Phytol. 112, 61–68.

Allen E B and Allen M F 1980 Natural re-establishment of vesicular-arbuscular mycorrhizae following stripmine reclamation in Wyoming. J. Appl. Ecol. 17, 139–147.

Allen M F, Allen E B and Friese C G 1989 Responses of the non-mycotrophic plant *Salsola kali* to invasion by vesicular-arbuscular mycorrhizal fungi. New Phytol. 111, 45–49.

Amaranthus M P, Molina R and Perry D A 1990 Soil organisms, root growth and forest regeneration. Proc. Soc. Amer. for Spokane Washington. pp 89–93.

Borchers S L and Perry D A 1990 Growth and ectomycorrhiza formation of Douglas-fir seedlings grown in soils collected at different distances from pioneering hardwoods in southwest Oregon clear-cuts Can. J. For. Res. 20, 712–721.

Buwalda J G 1980 Growth of a clover–rye grass association with vesicular-arbuscular mycorrhizas. N. Z. J. Agric. Res. 23, 379–383.

Carleton T J and Read D J 1991 Ectomycorrhizas and nutrient transfer in conifer–feathermoss ecosystems. Can. J. Bot. 69, 778–785.

Chilvers G A and Pryor L D 1965 The structure of eucalypt mycorrhizas. Aust. J. Bot. 13, 245–261.

Coutts M P and Nicoll B C 1990a Growth and survival of shoots, roots and mycorrhizal mycelium in clonal Sitka spruce during the first growing season after planting. Can. J. For. Res. 20, 861–868.

Coutts M P and Nicoll B C 1990b Waterlogging tolerance of roots of Sitka spruce clones and of strands from *Thelephora terrestris* mycorrhizas. Can. J. For. Res. 20, 1896–1899.

Cromack K, Sollins P, Granstein W C, Speidel K, Todd A W, Spycher G, Ching Y-Li and Tod R L 1979 Calcium oxalate accumulation and soil weathering in mats of the hypogeous fungus *Hysterangium crassum*. Soil Biol. Biochem. 11, 463–468.

Dahlberg A and Stenlid J 1990 Population structure and dynamics in *Suillus bovinus* as indicated by spatial distribution of fungal clones. New Phytol. 115, 487–493.

Entry J A, Rose C L and Cromack Jr K 1991a Litter decomposition and nutrient release in ectomycorrhizal mat soils of a Douglas fir ecosystem. Soil Biol. Biochem. 23, 285–290.

Entry J A, Donnelly P K and Cromack Jr K 1991b Influence of ectomycorrhizal mat soils on lignin and cellulose degradation. Biol. Fert. Soils 11, 75–78.

Fenner M 1978 A comparison of the abilities of colonizers and closed turf species to establish from seed in artificial swards. J. Ecol. 66, 953–965.

Finlay R D and Read D J 1986a The structure and function of the vegetative mycelium of ectomycorrhizal plants. I. Translocation of ^{14}C-labelled carbon between plants interconnected by a common mycelium. New Phytol. 103, 143–156.

Finlay R D and Read D J 1986b The structure and function of the vegetative mycelium of ectomycorrhizal plants. II. The uptake and distribution of phosphorus by mycelial strands inter-connecting host plants. New Phytol. 103, 157–165.

Fitter A H 1977 Influence of mycorrhizal infection on competition for phosphorus and potassium by two grasses. New Phytol. 79, 19–25.

Fleming L V 1985 Experimental study of sequences of ectomycorrhizal fungi on birch (*Betula* sp.) seedling root systems. Soil Biol. Biochem. 17, 591–600.

Glenn M G, Chew F S and Williams P H 1985 Hyphal penetration of *Brassica* (Cruciferae) roots by a vesicular-arbuscular mycorrhizal fungus. New Phytol. 99, 463–472.

Griffiths R, Caldwell B A, Cromack Jr K, Castellano M A and Morita R Y 1987 A study of the chemical and microbial variables in forest soils colonized with *Hysterangium setchellii* rhizomorphs. *In* Mycorrhizae in the Next Decade: Practical Applications and Research Priorities, Proc. 7th NACOM. Eds. D M Silva, L L Hung and J H Graham. p 96. Institute of Food and Agricultural Science, University of Florida, Gainesville, Florida.

Griffiths R P, Castellano M A and Caldwell B A 1991 Hyphal mats formed by two ectomycorrhizal fungi and their association with Douglas-fir seedlings: A case study. Plant Soil 134, 255–259.

Grime J P 1979 Plant strategies and vegetation processes. John Wiley and Sons, New York. 222 p.

Grime J P, Mackey J M L, Sillier S H and Read D J 1987 Floristic diversity in a model system using experimental microcosms. Nature (London) 328, 420–422.

Grime J P, Hodgson J G and Hunt R 1988 Comparative plant ecology. A functional approach to common British species. Unwin and Hyman, London. 742 p.

Grubb P J 1986 The ecology of establishment. *In* Ecology and Design in Landscape. Symp. Br. Ecol. Soc. 24 . Eds. A D Bradshaw, D A Goode and E Thorpe. pp 83–97.

Hall I T 1978 Effects of endomycorrhizas on the competitive abilities of white clover. N. Z. J. Agric. Res. 21, 509–515.

Harley J L and Smith S E 1983 Mycorrhizal Symbiosis, Academic press, London. 483 p.

Harley J L and Harley E L 1987 Check list of mycorrhizas in the British flora. New Phytol. 105, 1–102.

Harvey A E, Larsen M J and Jurgensen M F 1976 Distribution of ectomycorrhizae in a mature Douglas fir/larch forest system in Western Montana. For. Sci. 22, 393–398.

Hetrick B A D, Kitt D G and Wilson G T 1988 Mycorrhizal dependence and growth habit of warm-season and cool-season tallgrass prairie plants. Can. J. Bot. 66, 1376–1380.

Hetrick B A D, Wilson G W T and Hartnett D C 1989 Relationship between mycorrhizal dependence and competitive ability of two tallgrass prairie grasses. Can. J. Bot. 67, 2608–2615.

Janos D P 1980 Mycorrhizae influence tropical succession. Biotropica 12, 56–64.

Koide R T and Lu X 1991 *Avena fatua* L seed and seedling nutrient dynamics as influence by mycorrhizal infection of the maternal generation. Plant Cell. Env. 14, 931-939.

Koide R T, Li M, Lewis J and Irby C 1988 Role of mycorrhizal infection in the growth and reproduction of wild US cultivated plants. I. Wild US cultivated oats. Oecologia 77, 537–549.

Lewis J D and Koide R T 1990 Phosphorus supply, mycorrhizal infection and plant offspring vigour. Functional Ecol. 4, 695–702.

Miller R M 1987 The ecology of vesicular-arbuscular mycorrhizal in grass and shrublands. *In* Ecophysiology of VA Mycorrhizal Plants. Ed. G R Safir. pp 135–170. CRC Press, Boca Raton, USA.

Molina R, Massicotte H and Trappe J M 1991 Specificity phenomena in mycorrhizal symbioses: Community-ecological consequences and practical implications. *In* Mycorrhizal Functioning: An Integrative Plant–Fungal Process. Ed. M F Allen. pp 357–423. Chapman and Hall, New York.

Newman E I 1969 Resistance to water flow in soil and plant. I. Soil resistance in relation to amounts of root : theoretical estimates. J. Appl. Ecol. 6, 1–12.

Newton A C and Pigott D 1991 Mineral nutrition and mycorrhizal infection of oak and birch. I. Nutrient uptake and the development of mycorrhizal infection during seedling establishment. New Phytol. 17, 37–44.

Ocampo J A Martin J and Hayman D S 1980 Influence of plant interactions on vesicular-arbuscular mycorrhizal infections. I. Host and non-host plants grown together. New Phytol. 84, 27–35.

Ocampo J A and Hayman D A 1981 Influence of plant interactions on vesicular-arbuscular mycorrhizal infections. II. Crop rotations and residual effects of non-host plants. New Phytol. 87, 333–334.

Parrish J A D and Bazzaz 1985 Nutrient content of *Abutilon theophrasti* seeds and competitive ability of the resulting plants. Oecologia 65, 247–251.

Persson H 1979 Fine-root production, mortality and decomposition in forest ecosystem. Vegetatio 41, 101–109.

Read D J 1991 Mycorrhizas in ecosystems. Experientia 47, 376–391.

Read D J 1992 The mycorrhizal mycelium. *In* Mycorrhizal Functioning. An integrative plant–fungal process. Ed. M Allen. pp 102–133. Chapman and Hall, New York.

Read D J and Birch C P D 1988 The effects and implications of disturbance of mycorrhizal mycelial systems. Proc. R. Soc. Edinburgh 94B, 13–24.

Read D J, Koucheki H K and Hodgson J 1976 Vesicular-arbuscular mycorrhiza in natural vegetation systems. I. The occurrence of infection. New Phytol. 77, 541–655.

Read D J, Franics R and Finlay R D 1985 Mycorrhizal mycelia and nutrient cycling in plant communities. *In* Ecological Interactions in Soil: Plants, Microbes and Animals. Eds. A H Fitter, D Atkinson, D J Read and M B Usher. pp 193–217. British Ecological Society special Publication 4, Blackwell, Oxford.

Reddell P and Malajczuk N 1984 Formation of ectomycorrhizae by jarrah (*Eucalyptus marginata* Donn ex Smith) in litter and soil. Aust. J. Bot. 32, 435–444.

Reeves F B, Wagner D W, Moorman T and Kiel J 1979 The role of endomycorrhizae in revegetation practises in the semi-arid west. I. A comparison of incidence of mycorrhizae in severely disturbed US natural environments. Am. J. Bot. 66, 1–13.

Sanders F E and Tinker P B 1973 Phosphate flow into mycorrhizal roots. Pestic. Sci. 4, 385–395.

Scherfose V 1990 Feinwurzelverterlung und Mycorrhizatypen von *Pinus sylvestris* in verschiedenen Bodentypen Berichte Forschungszentrum Wald ökosysteme. Univ. Göttingen, Reihe A Bd 62 166 p.

Sylvia D M 1990 Distribution, structure and function of external hyphae of vesicular-arbuscular mycorrhizal fungi. *In* Rhizosphere dynamics Eds. J E Box, L C Hammond. pp 144–167. Westview Press, Boulder, Co.

Tisdall J M and Oades J M 1979 Stabilization of soil aggregates by the root segments of rye grass. Aust. J. Soil Res. 17, 429–441.

Trappe J M 1987 Phylogenetic and ecological aspects of mycotrophy in the angiosperms from an evolutionary standpoint. *In* Ecophysiology of VA Mycorrhizal plants. Ed. G R Safir. pp 2–25. CRC Press, Boca Raton, USA.

Reprinted from *Plant and Soil*, Volume 159, No. 1 (1994), pp. 11–25.

An economic approach to evaluate the role of mycorrhizas in managed ecosystems

MURRAY MILLER, TERENCE MCGONIGLE and HEATHER ADDY
University of Guelph, Guelph, Ontario, Canada N1G 2W1

Key words: agriculture, arbuscular-mycorrhizas, economic analysis, management

Abstract

The benefits of management of mycorrhizas in agricultural and horticultural cropping systems remains problematic except where the indigenous fungal population is low. Most experiments have focused on the introduction of exotic fungal isolates. Promotion of plant growth by mycorrhizas can be enhanced by increasing the effectiveness of the indigenous fungi as well as by introducing more effective species. Lack of reliable methods for identification of fungal species colonizing roots is a major limitation to characterizing the change in mycorrhizal populations. Assessment of the role of mycorrhizas in commercial food production systems must include an economic analysis. To do so requires an evaluation of the response to increasing the effectiveness of the mycorrhizal symbiosis relative to increasing yield with addition of phosphorus fertilizer. Thus field experiments should be designed to measure the response to phosphorus addition with the existing mycorrhizal population as well as with the more effectively managed population.

In this paper we discuss changes that may be induced in mycorrhizal fungi by management to increase their effectiveness in promoting plant growth. We then suggest an economic analysis approach to assessing the potential benefits of this increase in effectiveness. We conclude with a discussion of research approaches needed to determine, in a more objective manner, the role of mycorrhizas in managed ecosystems.

Introduction

Mycorrhizas are widespread and abundant in natural ecosystems (Read et al., 1976; Malloch et al., 1980). Although the role of arbuscular mycorrhizas as mutualistic symbionts is generally accepted, questions remain regarding their role in ecosystems that are managed for production of crops for animal or human consumption. Mosse (1986) concluded her review of mycorrhizas in sustainable agriculture with the following: "Because of the range of mycorrhizal fungi, the inadequate understanding of soil–fungus relationships and the variability of plant response, it is difficult to predict the ultimate effects on crop production of modifications in mycorrhizal systems". A similar degree of uncertainty has been expressed by others (Menge, 1983; Hall, 1988). Menge (1983) suggested that, because of the widespread occurrence of mycorrhizal fungi in nearly all soils, "commercial use of mycorrhizal fungi is probably economically feasible at the present time in only three major agricultural areas; (1) disturbed sites, (2) fumigated soils, and (3) greenhouses".

Much of our understanding of mycorrhizal symbiosis has come from pot experiments, usually using mycorrhiza-free media or sterilized soil (Fitter, 1985; Hall, 1988). Several authors (Hayman, 1982; Powell, 1984; Fitter, 1985; Jeffries, 1987; Hall, 1988 and McGonigle, 1988) have summarized the field experiments that have been conducted. Although major increases in

A.D. Robson, L.K. Abbott & N. Malajczuk (eds.), Management of mycorrhizas in agriculture, horticulture and forestry, 27–35.

growth and yield have been reported, many of these can be attributed to inadequate experimental controls, or to soils in which the indigenous fungal population had been depleted by erosion, fumigation, disturbance or fallowing (Hall, 1988). McGonigle (1988) concluded that direct evidence for a mutualistic function of arbuscular mycorrhizal symbioses in the field is scant.

The reviews cited above focus on the potential for introduction of inoculum with a view to either increase the overall level of mycorrhizal activity or to replace indigenous with more effective fungal species or clones. Our ability to assess the role of mycorrhizas in field-scale managed ecosystems is still severely limited by our understanding of the behaviour of the arbuscular mycorrhizal symbiosis under field conditions, and by the lack of carefully conducted field trials in which the causes of responses to management of the symbiosis can be determined. Abbott and Robson (1991a) have presented some suggestions for overcoming these limitations.

In this paper, we extend the concepts presented by Abbott and Robson (1991a). We begin with a discussion of the ways in which the effectiveness of arbuscular mycorrhizas, defined as their ability to promote plant growth, may be increased. We then suggest an economic-analysis approach to assessing the potential benefit of increasing the effectiveness of mycorrhizas, and indicate possible mechanisms whereby these benefits may occur. We conclude with a discussion of research approaches needed to determine, in a more objective manner, the role of mycorrhizas in managed ecosystems.

Increasing the effectiveness of arbuscular mycorrhizas.

Most attempts to increase the effectiveness of arbuscular mycorrhizas have involved inoculation with exogenous species. However, soil and crop management practices can influence the indigenous mycorrhizal fungi in terms of the total quantity of mycorrhizal development. This has for example been shown for pesticides (Ocampo and Hayman, 1980), fertilizers (Strzemska, 1975), tillage (Mulligan et al., 1985; Miller and McGonigle, 1992) and crop sequence (Thompson, 1987). It might also be possible for changes to occur in the relative

Table 1. A representation of the ways in which management might change the mycorrhizal fungi at a given site. Upper case letters indicate greater relative abundance at the site. Different letters refer to different species or clones of mycorrhizal fungi. Fungus D is exotic to the site. The asterisks indicate a modification in the quantity or character of the fungi. Inoculation is used only in cases III, IV and V, not in I or II

Case	Without management	With management	Notes
I	A b c	a B c	To change the relative abundance of the fungi present such that fungus B replaces fungus A as the most abundant
II	A b c	A* b* c*	To modify the quantity or character of the fungi present without changing their relative abundance
III	A b c	a b c D	To inoculate with fungus D at a site where extensive mycorrhizal development already occurs
IV	A	a D	To inoculate with fungus D at a site where only small quantities of one mycorrhizal fungus remain
V	–	D	To inoculate with fungus D at a site where no mycorrhizal fungi remain

quantity of each of the species and clones of mycorrhizal fungi present, or in terms of their structural characteristics. All of these types of alterations, which are summarized in Table 1, might then lead to changes in the ability of the fungi to promote plant growth.

Choice of agricultural practice might influence the capacity of arbuscular mycorrhizal fungi to colonize roots (Abbott and Robson, 1991a), which could change the relative quantities of the different native fungi (Table 1, case I) if there are differential responses among species and clones. Although the proportion of root length colonized and the length of colonized root are readily determined, different mycorrhizal fungi in the roots are not easily distinguished from one another (Stribley, 1987). There are, however, a few examples of changes in the abundance of individual species and clones in response to imposed management conditions, such as liming (Wang et al., 1985), or cropping sequence (Johnson et al., 1991). Johnson et al. (1991) found continuous cropping with maize as compared to soybean resulted in significantly higher spore densities of two *Glomus* species at one site, but of different *Glomus* species at another site, even though all three species occurred at both sites. Differences in responses of this type from one site to the next might be because physiological characteristics differ among clones, even though they may have morphologically identical spores (Morton, 1990).

Although densities of spores in soil and extent of colonization have been widely used as measures of abundance of mycorrhizas, from a functional standpoint it is probably better to know the quantity of extraradical mycelium (Bethlenfalvay et al., 1982; McGonigle et al., 1990; Miller and McGonigle, 1992). However, this does present problems because mycorrhizal hyphae are not always easily distinguished from those of other fungi (Abbott and Robson, 1991b).

Management practices might also modify the effectiveness of the indigenous fungi (Table 1, case II) in ways quite separate from any effects on their relative abundance. As well as increased quantities overall, changes in structural features, such as the distribution of extraradical mycelium relative to individual roots, the frequency of connection of extraradical to intraradical mycelium, or the extent of production of plant–fungus interface per unit of colonized length, might conceivably occur. Differences certainly exist among arbuscular mycorrhizal fungi in terms of the pattern of hyphal length density of the extraradical mycelium as a function of distance from the rhizoplane (Jakobsen et al., 1992). The flexibility of this pattern of distribution for a given clone, in response to changed conditions, is not yet known. Studies on the detailed aspects of the structure of arbuscular mycorrhizas have recently begun to appear in the literature. Smith and Dickson (1991) found that the number of cortex cells containing arbuscules, relative to the total number of cortex cells, decreased with time, whereas Toth et al. (1990) found the percentage of root volume occupied by cells containing arbuscules was linearly related to colonization. It remains to be seen how parameters of these types respond to imposed conditions.

The effectiveness of the arbuscular mycorrhizal symbiosis in increasing crop yield may be determined to a large extent by the temporal pattern of colonization in relation to the phosphorus requirement of the crop. Miller and McGonigle (1992) observed greater phosphorus concentration in shoots and greater arbuscular colonization of roots of maize seedlings from a ridge till compared to a moldboard plow treatment (Fig. 1a). At a later stage there were no differences in either phosphorus concentration or arbuscular colonization. However, as shown by Barry and Miller (1989), increased shoot phosphorus concentration at these early growth stages may increase grain yield regardless of phosphorus concentration at later growth stages (Fig. 1b).

We recognize that there are limitations on what information can be obtained, using available methods, on changes in the abundance or the structural characteristics of the various clones and species of mycorrhizal fungi in a given field situation. In spite of this, the approach we adopt below is to discuss the possible outcomes of an improvement in the effectiveness of the mycorrhizal associations, due to one or more of these types of changes, in response to some aspect of management. In order to emphasize that management is primarily

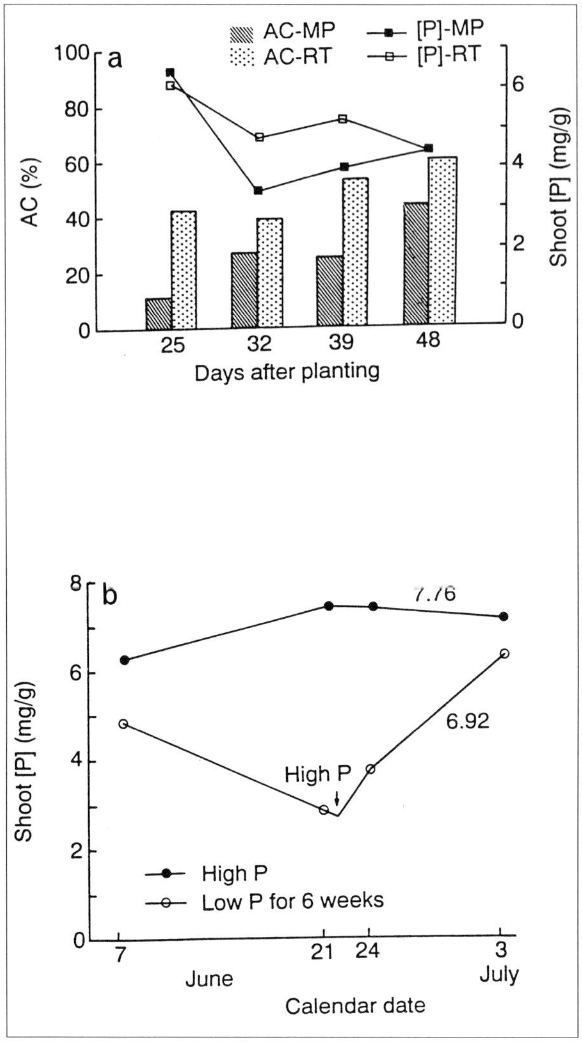

Fig. 1. a) Early season changes in arbuscular colonization (AC) and shoot phosphorus concentration of maize in field plots under moldboard plow (MP) or ridge tillage (RT) (Data from Miller and McGonigle, 1992) b) Early season changes in shoot phosphorus concentration for maize grown hydroponically in the field with 1.63 μM $H_2PO_4^-$ nutrient solution (High P) or 0.04 mM $H_2PO_4^-$ until 6 weeks after planting (Low P) followed by high P (arrow) until maturity. Numbers indicate final grain yield (t ha^{-1}). (Data from Barry and Miller, 1989).

being considered in cases where arbuscular mycorrhizal fungi are already present, in our analysis below the condition of the system without the management is denoted as mycorrhizal (M). The condition with management is then one of improved mycorrhizal effectiveness (M+). However, systems with only trace levels of mycorrhizas, or none at all, are not excluded from our classification M.

An economic-analysis approach to assessment of the role for management of mycorrhizas

Hall (1988) stated "whether VAM technology will ever be used in the production of field-sown crops and pastures will depend on the economics of employing it as compared with, for example, stimulating growth with fertilizer". A methodology for assessing the most economic rate of fertilizer based on the crop value and the fertilizer cost has been presented by Zebarth et al. (1991). A generalized relationship is presented in Fig. 2a. The maximum net return (MNR) occurs at the fertilizer rate at which the slope of the revenue and cost curves are the same. Fixed costs, in this analysis, include all costs that are not dependent on the rate of P applied.

Increasing the effectiveness of the mycorrhizal symbiosis may result in a number of different response curves. Although we realize there are many possible combinations, we have chosen to discuss three response patterns illustrated in Fig. 2b, c and d. In each, the curve labelled M is the response to phosphorus fertilizer with the prevailing mycorrhizal effectiveness, and the curve labelled M+ is that when the effectiveness has been increased through management (see Table 1).

The cost of increasing the effectiveness of the mycorrhizas must be included in any economic analysis. It has not been included in the fixed costs shown in any of the scenarios presented in Fig. 2 because it may vary from essentially zero with, for example, a change in tillage practice, to prohibitively high for field-scale inoculation using cultures of mixed propagules. Increasing the effectiveness of mycorrhizas may have benefits beyond the value of the crop produced. For example, mycorrhizas may improve soil structure (Tisdall, 1991) leading to reduced soil erosion (Reganold et al., 1987) which will affect yield in succeeding years (Larson et al., 1985). The present value of this improvement in soil structure is not easily determined. It depends not

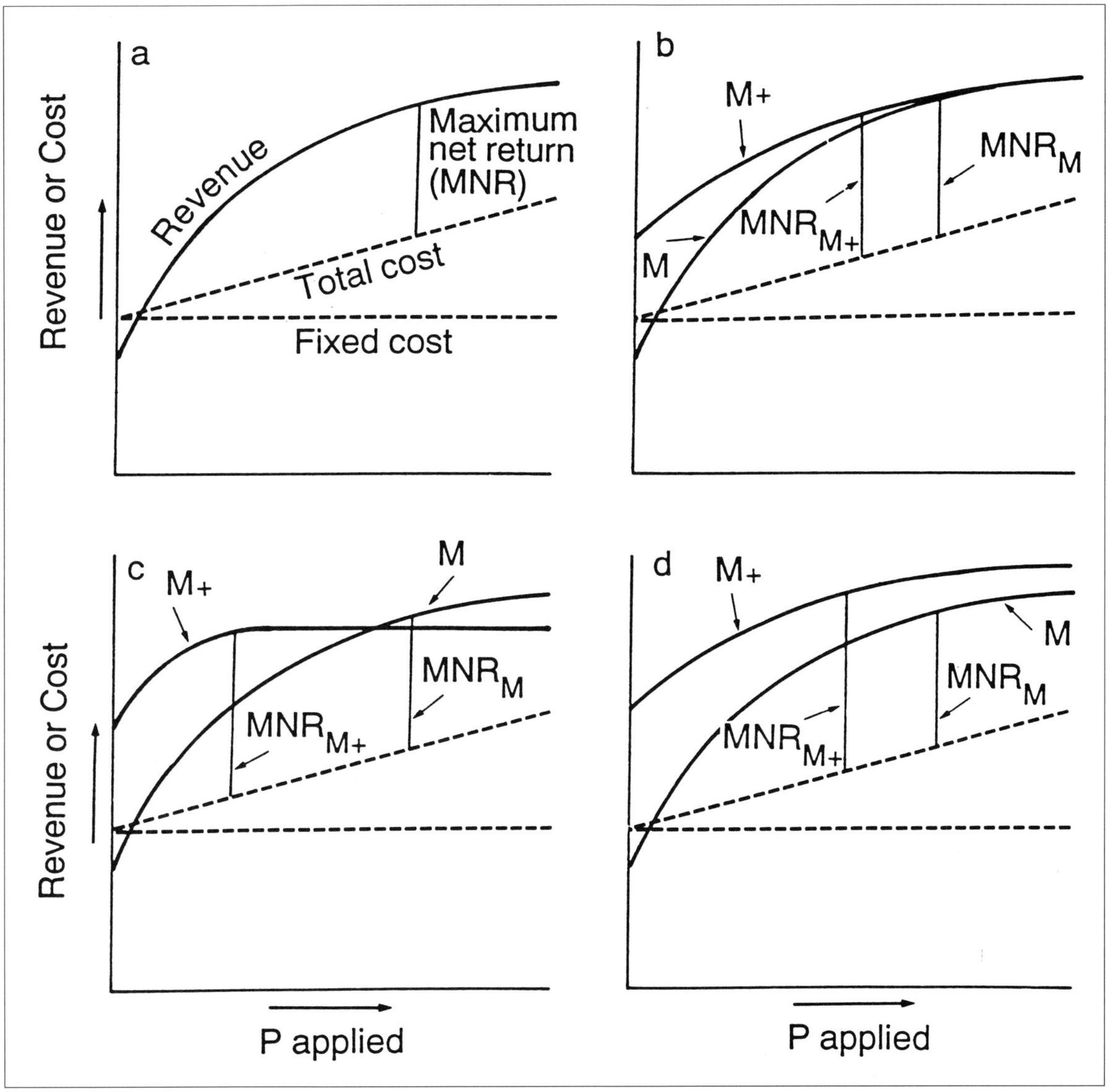

Fig. 2. Generalized relationship between effectiveness of increased mycorrhizal symbiosis and maximum net returns to added phosphorus with three possible response patterns.

only on the relation between soil erosion and crop yield (Battiston et al., 1987) but also on how much a farmer is prepared to spend today to gain a future return (Fox et al., 1991). A more effective mycorrhizal symbiosis may also reduce the sediment and phosphorus in runoff from agricultural land by reducing erosion due to improved soil structure, and by reducing the phosphorus concentration on the sediment because of the decreased available phosphorus required in the soil for most profitable yields. The contribution of phosphorus from agricultural land to surface water has been a major concern in the Great Lakes Region of North America for the past two decades (Miller et al., 1982). The effect of this environmental benefit is even more difficult to assess than the effect on long term crop yield. However, several studies have suggested that the off-farm impacts of soil erosion are greater than the on-farm impacts (Crosson,

1984). In addition, environmental effects benefit society rather than the farmer directly so can not as readily be factored into an economic analysis. If they are included, the net benefit from management of mycorrhizas would be:

$$\text{Net Benefit (NB)} = (MNR_{m+} - MNR_{m}) - MC + EB,$$

where MC = cost of management of mycorrhizas and EB = environmental benefit.

Response pattern No. 1

In the response pattern presented in Fig. 2b, although revenue is considerably greater for M+ at low rates of applied phosphorus, the maximum net return (MNR_{m+}) is similar to that for M. In this case, there would be a net benefit from management of the mycorrhizas only if the environmental benefit (EB) exceeded the cost of management (MC). If this were true *and* if the costs of management were shared between society and the farmer in proportion to the benefits each receive, management would be justified. Otherwise, the only justification for management would be an inability to apply the amount of phosphorus necessary to obtain MNR_m because of unavailability of either fertilizer or funds. Either may occur, most likely in developing countries.

Response pattern No. 2

A second response pattern is presented in Fig. 2c. In this case, yield, and hence revenue, with M+ and M increase at the same rate with low additions of phosphorus, but maximum yield and revenue is lower with M+. In spite of the lower maximum revenue, the net benefit from management of the mycorrhizas would be positive unless the cost of management exceeded the difference between MNR_{m+} and MNR_m plus EB.

The response patterns illustrated in Fig. 2b and 2c suggest that the only benefit from increased mycorrhizal effectiveness is from improved phosphorus absorption. Whether the net benefit is positive or negative depends on the phosphorus response relative to the carbon cost. The lower maximum yield in Fig. 2c could result if the increased effectiveness with M+ resulted in the carbon cost exceeding the phosphorus benefit. This may occur at high soil phosphorus values (Hall, 1978; Buwalda and Goh, 1982; Fairchild and Miller, 1990). Snellgrove et al. (1986), however, reported increased photosynthetic rate per unit leaf area in mycorrhizal plants which compensated for the increased carbon demand. Koide and Elliott (1989) and Fitter (1991) have recently presented theoretical frameworks within which the cost and benefits of mycorrhizal symbiosis can be considered. The degree to which maximum yield and revenue is reduced by increased mycorrhizal effectiveness will be determined, in part, by whether the carbon fixation by the plant is sink or source limited.

Response pattern No. 3

In a third response pattern (Fig. 2d) the maximum yield and revenue is increased by management of the mycorrhizal effectiveness. Such a response has been reported by Buwalda et al., 1985. The net benefit from an increase in effectiveness of the mycorrhizal symbiosis in this case would be positive unless the cost of increasing the effectiveness exceeded the difference between MNR_{m+} and MNR_m plus the environmental benefit.

In this case, the mycorrhizas have some benefit in addition to the increased phosphorus absorption. This may occur because of increased absorption of nutrients other than P from deficient soils (Kothari et al., 1990), decreased absorption of metals from soils with toxic levels (Heggo et al., 1990), improved soil structure (Tisdall, 1991), reduced pathogenesis (Dehne, 1982), an effect on hormonal relations (Gogala, 1991) or improved plant–water relations.

The most frequently reported changes in plant water relations that can be attributed to mycorrhizal colonization are increased transpiration and stomatal conductance (Levy et al., 1983; Koide, 1985; Allen and Allen, 1986). This effect has frequently been associated with improved shoot phosphorus nutrition (Fitter, 1988). In itself, however, higher transpiration in mycorrhizal plants would presumably place those plants in a more vulnerable position during drought stress. To be beneficial, a change in water relations must improve production by helping to maintain physiological activity during episodes of water stress. There are several mechanisms whereby this could occur, but there is little information available to assess their likelihood.

Research approaches for addressing the role of mycorrhizas

There are many questions that must be addressed if we are to assess adequately the role of arbuscular mycorrhizas in managed ecosystems. We recognize that some of these problems can be studied most effectively through research under controlled conditions. Our emphasis here however, is on the need for more detailed field work. The comments that follow apply equally to any method used to manage mycorrhizas.

Assuming that the major benefit of increased mycorrhizal effectiveness is from improved phosphorus nutrition of the crop, field experiments should provide a phosphorus response curve both with and without management of the mycorrhizas. Only then can the true economic benefit of such management be determined. Increased production at a single phosphorus fertilization level is not a sufficient basis for recommending management.

The methods used to manage the mycorrhizas must be practical within the production system being considered. For example, demonstrating a response to inoculation following soil fumigation may be useful when the system involves production of seedlings or cuttings for transplanting, but it is of little relevance for field scale crop production, where the cost of fumigation would be prohibitive.

Our ability to generalize from our experiments requires that we not only establish that plant responses are due to increased mycorrhizal effectiveness, but also that we elucidate the mechanisms involved. This will require changes in conceptual approach as well as improvements in techniques to monitor the behaviour of the fungi.

We need improvements in three key areas:

1. Determination of the species and clones of fungi responsible for colonization under different management systems. Immunological (Morton et al., 1987) and molecular genetical (Simon et al., 1992) techniques are likely candidates here.
2. Improved understanding of the relationship between intra- and extra-radical structures and effectiveness of the mycorrhizas. Arbuscular colonization may not be an adequate indicator of effectiveness, and assessment of extraradical hyphae has recently become more commonplace. Reliable methods to distinguish mycorrhizal hyphae in soil from those of other fungi are badly needed. In addition, estimation of hyphal length density alone may not be sufficient; it may be necessary to know the hyphal distribution in relation to roots (Jakobsen et al., 1992). All hyphae may not be equally effective in phosphorus absorption since there seem to be hyphae specialized for absorption (Friese and Allen, 1991).
3. Improved understanding of the inter-relationships between mycorrhizal fungi and other microbes in the rhizosphere. The biological interactions in the mycorhizosphere are complex (Garbaye, 1991), but if we are to improve our ability to manage the mycorrhizas, we must study them in the presence of the other rhizosphere organisms.

We recognize the immensity of the challenges that these suggestions present. We are convinced, however, that, until these challenges are met, the situation described by Mosse (1986) will continue to exist and we will remain unable to predict the consequences of modifications to mycorrhizal systems in crop production.

References

Abbott L K and Robson A D 1991a Field management of VA mycorrhizal fungi. *In* The Rhizosphere and Plant Growth. Eds. D L Keister and P B Cregan. pp 355–362. Kluwer Academic Publishers, Dordrecht, The Netherlands.

Abbott L K and Robson A D 1991b Factors influencing the occurrence of vesicular-arbuscular mycorrhizas. Agric., Ecos. Environ. 35, 121–150.

Allen E B and Allen M F 1986 Water relations of xeric grasses in the field: interaction of mycorrhizas and competition. New Phytol. 104, 559–571.

Barry D A J and Miller M H 1989 Phosphorus nutritional

requirements of maize seedlings for maximum yield. Agron. J. 81, 95–99.

Battiston L A, Miller M H and Shelton I J 1987 Soil erosion and corn yield in Ontario. I. Field evaluation. Can. J. Soil Sci. 67, 731–745.

Bethlenfalvay G J, Brown M S and Pacovsky R S 1982 Relationships between host and endophyte development in mycorrhizal soybeans. New Phytol. 90, 537–543.

Buwalda J G and Goh K M 1982 Host–fungus competition for carbon as a cause of growth depressions in vesicular-arbuscular mycorrhizal ryegrass. Soil Biol. Biochem. 14, 103–106.

Buwalda J G, Stribley D P and Tinker P B 1985 Effects of vesicular arbuscular mycorrhizal infection in first, second and third cereal crops. J. Agric. Sci. 105, 631–647.

Crosson P 1984 New perspectives on soil conservation policy. J. Soil Water Conserv. 39, 222–225.

Dehne H W 1982 Interaction between vesicular-arbuscular mycorrhizal fungi and plant pathogens. Am. Phytopathol. Soc. 72, 1115–1119.

Fairchild G L and Miller M H 1990 Vesicular-arbuscular mycorrhizas and the soil-disturbance induced reduction of nutrient absorption in maize. III Influence of P amendments to soil. New Phytol. 114, 641–650.

Fitter A H 1985 Functioning of vesicular-arbuscular mycorrhizas under field conditions. New Phytol. 99, 257–265.

Fitter A H 1988 Water relations of red clover *Trifolium repens* L. as affected by VA mycorrhizal infection and phosphorus supply before and during drought. J. Exp. Bot. 39, 595–603.

Fitter A H 1991 Costs and benefits of mycorrhizas: Implications for functioning under natural conditions. Experientia 47, 350–355.

Fox G, Weersink A, Sarwar G, Duff S and Deen B 1991. Comparative economics of alternative agricultural production systems: A review. Northeastern J. Agric. Resource Econ. 20, 124–142.

Friese C F and Allen M F 1991 The spread of VA mycorrhizal fungal hyphae in the soil: Inoculum types and external hyphal architecture. Mycologia 83, 404–418.

Garbaye J 1991 Biological interactions in the mycorhizosphere. Experientia 47, 370–375.

Gogala N 1991 Regulation of mycorrhizal infection by hormonal factors produced by hosts and fungi. Experientia 47, 331–340.

Hall I R 1978 Effects of endomycorrhizas on the competitive ability of white clover. N.Z.J. Agric. Res. 21, 509–515.

Hall I R 1988 Potential for exploiting vesicular-arbuscular mycorrhizas in agriculture. *In* Biotechnology in Agriculture. pp 141–174. Alan R. Liss, Inc.

Hayman D S 1982 Practical aspects of vesicular arbuscular mycorrhiza. *In* Advances in Agricultural Microbiology. Ed. N S Subba Pao. pp 325–373. Butterworth Scientific, London, U.K.

Heggo A, Angle J S and Chaney R L 1990 Effects of vesicular-arbuscular mycorrhizal fungi on heavy metal uptake by soybeans. Soil Biol. Biochem. 22, 865–869.

Jakobsen I, Abbott L K and Robson A D 1992. External hyphae of arbuscular mycorrhizal fungi associated with *Trifolium subterraneum* L. I. Spread of hyphae and phosphorus inflow into roots. New Phytol. 120, 372–380.

Jeffries P 1987 Use of mycorrhizae in agriculture. CRC Annu. Rev. Biotech. 5, 319–357.

Johnson N C, Pfleger F L, Crookston R K, Simmons S R and Copeland P J 1991 Vesicular arbuscular mycorrhizas respond to corn and soybean cropping history. New Phytol. 117, 657–663.

Koide R 1985 The effect of VA mycorrhizal infection and phosphorus status on sunflower hydraulic and stomatal properties. J. Exp. Bot. 36, 1087–1098.

Koide R and Elliott G 1989 Cost, benefit and efficiency of the vesicular-arbuscular mycorrhizal symbiosis. Functional. Ecol. 3, 252–255.

Kothari S K, Marschner H and Romheld V 1990 Direct and indirect effects of VA mycorrhizal fungi and rhizosphere microorganisms on acquisition of mineral nutrients by maize (*Zea mays* L.) in a calcareous soil. New Phytol. 116, 637–645.

Larson W E, Fenton T E, Skidmore E L and Benbrook C M 1985 Effects of soil erosion on soil properties as related to crop productivity and classification. *In* Soil Erosion and Crop Productivity. Eds. R F Follett and B.A. Stewart. pp 190–211. American Society of Agronomy, Inc., Crop Science Society of America, Inc. Soil Science Society of America, Inc., Madison, Wisc., U.S.A.

Levy Y, Syvertsem J P and Nemee S 1983 Effect of drought stress and vesicular-arbuscular mycorrhiza on *Citrus* transpiration and hydraulic conductivity of roots. New Phytol. 93, 61–66.

Malloch D, Pirozynski K A and Raven P H 1980 Ecological and evolutionary significance of mycorrhizal symbiosis in vascular plants. Proc. Natl. Aca. Sci. USA 77, 2113–2118.

McGonigle T P 1988 Numerical analysis of published field trials with vesicular-arbuscular mycorrhizal fungi. Functional. Ecol. 2, 473–478.

McGonigle T P, Evans D G and Miller M H 1990 Effect of degree of soil disturbance on mycorrhizal colonization and phosphorus absorption by maize in growth chamber and field experiments. New Phytol. 116, 629–636.

Menge J A 1983 Utilization of vesicular-arbuscular mycorrhizal fungi in agriculture. Can. J. Bot. 61, 1015–1024.

Miller M H and McGonigle T P 1992. Soil disturbance and the effectiveness of arbuscular mycorrhizae in an agricultural ecosystem. *In* Mycorrhizas in Ecosystems. Eds. D J Read, D H Lewis, A H Fitter, I J Alexander. pp 156–163. CAB International, Wallingford, Oxford.

Miller M H, Robinson J B, Coote D R, Spires A C and Draper D W 1982. Agriculture and water quality in the Canadian Great Lakes basin III. Phosphorus. J. Environ. Qual. 11, 487–493.

Morton J B 1990 Species and clones of arbuscular mycorrhizal fungi (Glomales, Zygomycetes): their role in macro- and micro-evolutionary processes. Mycotaxon 37, 493–515.

Morton J B, Wright S F and Sworobuk J E 1987 Sensitivity of monoclonal antibodies to detect propagules of *Glomus occultum*. *In* Mycorrhizae in the Next Decade: Practical Application and Research Priorities, Proc. 7th NACOM. Ed. D M Sylvia, L L Hung and J H Graham. p 317. Institute of Food and Agricultural Sciences, University of Florida, Gainesville, Florida.

Mosse B 1986 Mycorrhiza in a sustainable agriculture. Biol. Agric. Hort. 3, 191–209.

Mulligan M F, Smucker A J M and Safir G F 1985 Tillage modifications of dry edible bean root colonization by VAM fungi. Agron. J. 77, 140–144.

Ocampo J A and Hayman D S 1980 Effects of pesticides on mycorrhiza in field-grown barley, maize and potatoes. Trans. Br. Mycol. Soc. 74, 413–416.

Powell C L 1984 Field inoculation with VA mycorrhizal fungi. *In* VA mycorrhiza. Eds. C.L. Powell and D.J. Bagyaraj. pp 205–222. CRC Press, Inc., Boca Raton, Florida, U.S.A.

Read D J, Koucheki H K and Hodgson J 1976 Vesicular-arbuscular mycorrhiza in natural vegetation systems. New Phytol. 77, 641–653.

Reganold J P, Elliott L F, and Unger Y L 1987. Long-term effects of organic and conventional farming on soil erosion. Nature 330, 370–372.

Simon L, Lalonde M, and Burns T D 1992 Specific amplification of 18S fungal ribosomal genes from vesicular-arbuscular endomycorrhizal fungi colonizing roots. Appl. Environ. Microbiol. 58, 291–295.

Smith S E and Dickson S 1991 Quantification of active vesicular-arbuscular mycorrhizal infection using image analysis and other techniques. Aust. J. Plant Physiol. 18, 637–848.

Snellgrove R C, Stribley D P, Tinker P B and Lawlor D W 1986 The effect of vesicular-arbuscular mycorrhizal infection on photosynthesis and carbon distribution in leek plants. *In* Proc. 1st ESM, Dijon, France.

Stribley D P 1987 Mineral Nutrition. *In* Ecophysiology of VA Mycorrhizal Plants. Ed. G.R. Safir. pp 59–70. CRC Press, Boca Raton, Florida.

Strzemska J 1975 Mycorrhiza in farm crops grown in monoculture. *In* Endomycorrhizas Eds. F.E. Sanders, B. Mosse and P.B. Tinker. pp 529–535. Academic Press, London.

Thompson J P 1987 Decline of vesicular-arbuscular mycorrhizae in long fallow disorder of field crops, and its expression in phosphorus deficiency of sunflower. Aust. J. Agric. Res. 38, 847–867.

Tisdall J M 1991 Fungal hyphae and structural stability of soil. Aust. J. Soil Res. 29, 729–743.

Toth R, Doane C, Bennett H E and Alexander T 1990 Correlation between host–fungal surface areas and current colonization in VA mycorrhizae. Mycologia 82, 519–522.

Wang G M, Stribley D P, Tinker P B and Walker C 1985 Soil pH and vesicular-arbuscular mycorrhizas. *In* Ecological Interactions in Soil. Ed. A H Fitter. pp 219–224. Blackwell Scientific.

Zebarth B J, Sheard R W and Curnoe W E 1991 A soil test calibration method for potassium on alfalfa which allows for variation in crop yield and fertilizer cost. J. Prod. Agric. 4, 317–322.

Reprinted from *Plant and Soil*, Volume 159, No. 1 (1994), pp. 27–35.

Consideration of the taxonomy and biodiversity of Australian ectomycorrhizal fungi

MICHAEL A. CASTELLANO[1] and NEALE L. BOUGHER[2]
[1]*U.S. Department of Agriculture, Forest Service, Pacific Northwest Research Station, Forestry Sciences Laboratory, Corvallis, OR 97331, USA*
[2]*CSIRO Division of Forestry, Private Bag, P.O. Wembley, WA 6014, Australia*

Key words: ascomycotina, Australia, basidiomycotina, biodiversity, ectomycorrhizae

Abstract

Mycorrhiza management in forestry must be predicated on an understanding of fungal biology and ecology. A fundamental building block of the biology and ecology of any organism is accurate identification and an understanding of its relationship to other organisms.

The taxonomy of the larger fungi has been largely based on morphological classification of sexual structures but now Taxonomy routinely incorporates mating studies, and biochemical and molecular data. Taxonomy may not revolutionize theories on mycorrhiza but can clarify some of the inconsistencies due to misrepresentation or over-generalizations and inappropriate conclusions drawn from studies with inaccurately identified fungi. To illustrate this, we discuss an example where incorrect fungal names were repeatedly erroneously applied in morphological and physiological research reports on this fungus, e.g. *Laccaria laccata*. In this case subsequent taxonomic study revealed the reason for the conflicting research results reported for this fungus.

We discuss the status of identifying the ectomycorrhizal fungi in various forest communities in Australia and the relationships of this process to assessing their use in forestry. Recent intensive efforts to collect, isolate and identify Australian ectomycorrhizal fungi have revealed an enormous and unique species diversity, e.g., for truffle-like fungi, over 2000 collections from the last five years alone have yielded 2 new families, 24 new genera, and about 184 new species. Nearly 95% of the described and undescribed fungi from Australia are novel, with some 22 genera and 3 families endemic. In most cases the current systematic knowledge of mycorrhizal fungi is inadequate to support clear framework for Australian taxa. This reflects the traditional Northern Hemisphere view of the world, the uniqueness of the Australian fungal flora, and how poorly it is known. For example, the genus *Hymenogaster* had been widely acknowledged as the most reduced member of the Cortinariaceae. However recent work on *Hymenogaster* species from the Southern Hemisphere has offered a number of alternative affinities to various species.

We also discuss the role proper identification of the organisms involved plays in understanding the ecosystem. Emphasis should be placed on how species diversity equates with physiological and genetic diversity and how a sound taxonomic understanding of species and their systematic position is essential to properly manage them. Accurate taxonomic information will continue to be required as the basis for assessing the role of ectomycorrhizal fungi in sustained ecological development. Of particular significance is the role of ectomycorrhizal fungi in maintenance of plant diversity in natural ecosystems and those disturbed by management. In conclusion, we present some key research areas involving the use of taxonomy that need priority attention.

A.D. Robson, L.K. Abbott & N. Malajczuk (eds.), Management of mycorrhizas in agriculture, horticulture and forestry, 37–46.

Introduction

Foresters world-wide use many tools to aid afforestation and reforestation including mycorrhiza management. Over the past 20 years, in many disjunct areas of the world, mycorrhiza inoculation has received increased attention as a tool to improve seedling survival and growth in both the nursery and the field. Mycorrhiza management in nurseries has become more common as techniques have been developed to allow flexibility in crop inoculation and reliability in crop response (Castellano and Molina, 1989). The usefulness of mycorrhizal fungi in forest management must be predicated on an understanding of mycorrhiza biology and ecology. A fundamental building block of the biology and ecology of any organism is accurate identification and understanding systematic relationships.

Ectomycorrhizal fungi belong mainly to genera in the Basidiomycotina and Ascomycotina and to the genus *Endogone* in the Zygomycotina. Truffle-like mycorrhizal fungi (Table 1) form relatively large fruiting structures and are angiocarpic (spore mass enclosed) at maturity or have statismospores (spores not actively discharged). Compared to the Northern Hemisphere, the truffle-like fungi of Australia have received little attention (Castellano and Trappe, 1990). Because truffle-like fungi fruit beneath the soil surface, even at maturity, they are seldom encountered even by the most avid mushroom collector. Mushrooms, or toadstools, are above-ground fungi that are gymnocarpic (spore mass exposed) at maturity or have ballistospores (spores actively discharged).

The purpose of this paper is to present the status of identifying ectomycorrhizal fungi in various forest communities in Australia in relation to their potential in forest management. We also discuss how proper identification of organisms relates to our understanding of the ecosystem. We illustrate many of our points by using truffle-like fungi because the body of knowledge on the identity and ecology of mushroom species in Australia is still rather confused. We emphasize how species diversity relates to physiological and genetic diversity and how an accurate taxonomic understanding of species and their systematic position is essential if they are to be properly managed. Taxonomy is a major factor that can influence the contribution that mycorrhiza management can offer forest managers. Because there is insufficient data on fungi from the Southern Hemisphere, some examples are illustrated with Northern Hemisphere fungi. We expect that many, if not all, of these examples have analogies in the Southern Hemisphere. Finally, we highlight some research areas needing further attention.

Taxonomy

"A science can scarcely be said to exist before its material, whether it be species or elements or geologic and geographic features, has been organized and classified" (Wertenbaker, 1974). Mycorrhiza science is no exception; it requires description, names, and keys to taxonomic if not biological species to allow some coherent understanding of the biological and ecological role of mycorrhiza in forest ecosystems. Identification of the mycorrhizal roots themselves is not advanced enough to be used in a broad sense (Agerer, 1991), so we must rely on sporocarps.

The taxonomy of the larger fungi has been largely based on morphological classification of sexual fruiting structures and separation of discernible individuals into 'taxonomic species'. For the traditional mycologist, this includes both macroscopic (apparent to the unaided eye) and microscopic (visualized by use of the light and electron microscope) characters and is termed alpha taxonomy. In the narrow sense, taxonomy deals with 'biological species' that are discernible through the use of mating studies and by biochemical and molecular data and is termed beta taxonomy. The usefulness of both taxonomic and biological species concepts is determined in the context of the of the questions posed. If it is important to separate fungal populations as biological species, then DNA or mating studies might be used. If, however, an analysis of fungal community structure of a forested watershed is desired, then taxonomic species are most appropriate. It is a matter of expedience to deal with fungi as taxonomic

Table 1. Genera of presumptive ectomycorrhizal truffle-like fungi occurring with Southern Hemisphere hosts, Northern Hemisphere hosts or both [a]

Genera known only from Southern Hemisphere hosts		Genera known only from Northern Hemisphere hosts		Genera known from both Northern and Southern Hemisphere hosts	
Agrogaster	(B)	*Balsamia*	(A)	*Alpova*	(B)
Amylascus	(A)	*Barrsia*	(A)	*Arcangeliella*	(B)
Austrogaster	(B)	*Brauniellula*	(B)	*Chamonixia*	(B)
Austrogautieria	(B)	*Carbomyces*	(B)	*Choiromyces*	(A)
Boughera	(B)	*Cazia*	(A)	*Cortinarius*	(B) [b]
Castoreum	(B)	*Delastria*	(A)	*Cortinomyces*	(B)
Chondrogaster	(B)	*Destuntzia*	(B)	*Cystangium*	(B)
Claustula	(B)	*Fischerula*	(A)	*Elaphomyces*	(A)
Corditubera	(B)	*Gastroleccinum*	(B)	*Elasmomyces*	(B)
Cribbea	(B)	*Gastrosuillus*	(B)	*Endogone*	(Z)
Descomyces	(B)	*Genabea*	(A)	*Gastroboletus*	(B)
Dingleya	(A)	*Geopora*	(A)	*Gautieria*	(B)
Gallacea	(B)	*Gomphogaster*	(B)	*Genea*	(A)
Gelopellis	(B)	*Hydnobolities*	(A)	*Gymnomyces*	(B)
Gigasperma	(B)	*Hydnocystis*	(A)	*Hydnangium*	(B)
Gummiglobus	(B)	*Hydnotryopsis*	(A)	*Hydnotrya*	(A)
Gymnogaster	(B)	*Leucophelps*	(B)	*Hymenogaster*	(B)
Gymnohydnotrya	(A)	*Melanogaster*	(B)	*Hysterangium*	(B)
Gymnopaxillus	(B)	*Mycolevis*	(B)	*Leucogaster*	(B)
Horakiella	(B)	*Paradoxa*	(A)	*Macowanites*	(B)
Hybogaster	(B)	*Phlebogaster*	(B)	*Martellia*	(B)
Hysterogaster	(B)	*Picoa*	(A)	*Octavianina*	(B)
Labyrinthomyces	(A)	*Protogautieria*	(B)	*Peziza*	(A) [b]
Malajczukia	(B)	*Pyrenogaster*	(B)	*Protubera*	(B)
Mesophellia	(B)	*Rhizopogon*	(B)	*Radiigera*	(B)
Muciturbo	(A)	*Rhodactina*	(B)	*Richoniella*	(B)
Mycoamaranthus	(B)	*Seducula*	(B)	*Scleroderma*	(B) [b]
Mycoclelandia	(A)	*Tirmania*	(A)	*Sclerogaster*	(B)
Neogelopellis	(B)	*Truncocolumella*	(B)	*Secotium*	(B)
Nothocastoreum	(B)	*Tuber*	(A)	*Stephanospora*	(B)
Paurocotylis	(B)			*Stephensia*	(A)
Paxillogaster	(B)			*Terfezia*	(A)
Pirogaster	(B)			*Thaxterogaster*	(B)
Podohydnangium	(B)			*Torrendia*	(B)
Quadrispora	(B)			*Wakefieldia*	(B)
Reddellomyces	(A)			*Zelleromyces*	(B)
Rhodogaster	(B)				
Royoungia	(B)				
Ruhlandiella	(A)				
Setchelliogaster	(B)				
Singeromyces	(B)				
Smithiogaster	(B)				
Timgrovea	(B)				
Tremellogaster	(B)				

[a] A = Ascomycotina, B = Basidiomycotina, and Z = Zygomycotina
[b] hypogeous members only

species with some integration of their biology rather than as biological species.

Taxonomy may not revolutionize theories on mycorrhiza but can clarify some of the controversy in the literature which has arisen as a result of misrepresentation (Trappe and Molina, 1986). For example, *Laccaria laccata* has historically been taxonomically confused with many different morphological types attributed to it (Mueller, 1987). Mueller and Vellinga (1986) reduced 13 varieties and one form of *L. laccata* to 3 varieties and synonomized both *L. ohiensis* and *L. tetraspora* with *L. laccata*. Thus, where once 16 presumably recognizable entities existed, now there are 3. Prior to the work by Mueller, Molina (1982), Molina and Chamard (1983), and Hung and Molina (1986) had published on the use of *L. laccata* in forestry. From study of the herbarium voucher and the live isolate itself, the isolate they used proves to be *Laccaria bicolor* not *L. laccata*. Its ecology and biology may or may not be similar to that of correctly identified *L. laccata,* as reported in studies by Duponnois and Garbaye (1991) and Livingston (1991). Until recently, the generalized name *L. laccata* had also been inappropriately used in Western Australia. Tommerup et al. (1991), in a detailed taxonomic examination of *L. laccata* reported from Western Australia, revealed many of them to be more appropriately assigned to *L. fraterna*. This work included other genetic and observational data that partially explain the ecological success of *L. fraterna* as a pioneer in eucalypt plantations and allows us to understand some of its competitiveness against other ectomycorrhizal fungi.

The science of taxonomy is conducted by accumulating observations, then erecting and testing hypotheses (Trappe and Molina, 1986) that can change over time as more data become available. For example, the genus *Labyrinthomyces* first was described based on *L. steenisii* Boedijn (1939), a fungus found in Java. Subsequently, it was discovered that the same taxonomic species had been found previously in Tasmania but was described as *Stephensia varia* Rodway (1897). Trappe (1979) discovered that *S. varia* Rodway was untenable in the genus *Stephensia* and that it was the same as *L. steenisii* Boedijn. Because of priority, it was recombined as *Labyrinthomyces varius* (Rodway) Trappe with *Labyrinthomyces steenisii* Boedijn as a synonym. Thus the correct nomenclatural reference to the fungus as defined by published rules and regulations evolved (Voss et al., 1983). Such decisions require that taxonomists have access to enough data to draw conclusions. For example, recent studies, by the senior author and associates, of the 10 species previously assigned to the genus *Mesophellia,* in addition to abundant collections from our own field work, reveal that at least five different genera were erroneously grouped together in *Mesophellia*. One species is more correctly placed in a previously described genus while four other species require the coining of new generic names for segregation (Trappe et al., 1992; Trappe, Castellano and Malajczuk, unpublished results).

In some cases, the current systematic knowledge of mycorrhizal fungi is inadequate to support a taxonomic framework for Australian taxa. This reflects a Northern Hemisphere view of the world, the uniqueness of the Australian flora, and how poorly that flora is known. For example, the truffle-like genus *Hymenogaster* occurs worldwide and commonly associates with *Eucalyptus* and other Myrtaceae and Fagaceae in Australia (Bougher and Castellano, 1993). *Hymenogaster* as previously understood was poorly defined and had become a conglomeration of truffle-like Basidiomycotina with ornamented, brown spores and a loculate, non-gelatinised gleba. This was due to the confusion inherent in the original generic concept as circumscribed by Vittadini in 1831 (Bougher and Castellano, 1993). The only way to clarify the generic limits and align them with modern concepts was to start anew with re-examination of Vittadini's original eight species. The genus required narrower delimitation in light of comparison with the many Southern Hemisphere collections previously assigned to it. The eight original species are more appropriately assigned to three different genera. Traditionally, *Hymenogaster* has been acknowledged as the most reduced phylogenetic member of the Cortinariaceae with *Cortinarius–Thaxterogaster* ancestry (Fogel and Trappe, 1985). Our studies point to a different conclusion, with *Hymenogaster* sensu stricto not intimately aligned with *Cortinarius* on the basis of spore characters. In addition, all Southern Hemisphere species assigned to *Hymenogaster* sensu lato are excluded from *Hymenogaster* sensu stricto, and we assign them to five different genera (Bougher and Castellano, 1993). Each of the new genera have different phylogenetic links and hence relations to other fungi.

Table 2. Truffle-like fungal species numbers known to occur in Queensland, Australia

Fungal Class	Genus	Prior to 1988	Presently known
Ascomycotina	*Amylascus*	1	2
	Dingleya	–	1
	Elaphomyces	–	8
	Genea	1	1
	Gymnohydnotrya	1	1
	Labyrinthomyces	1	1
	Stephensia	–	1
	Total	4	15
Basidiomycotina	*Arcangeliella*	1	7
	Austrogautieria	–	2
	Boughera	–	2
	Castoreum	–	2
	Chamonixia	1	2
	Chondrogaster	–	3
	Cortinomyces	–	4
	Cribbea	3	4
	Cystangium	–	8
	Descomyces	–	2
	Gautieria	2	5
	Gelopellis	–	1
	Gummiglobus	–	2
	Gymnogaster	1	1
	Gymnomyces	–	4
	Hydnangium	1	2
	Hymenogaster	9	– [a]
	Hysterangium	5	15
	Hysterogaster	–	2
	Macowanites	–	8
	Malajczukia	–	2
	Martellia	2	10
	Mesophellia	3	4
	Mycoamaranthus	–	1
	Neogelopellis	–	1
	Nothocastoreum	–	2
	Octavianina	6	2 [a]
	Rhizopogon	1	3
	Richoniella	1	1
	Royoungia	–	1
	Scleroderma [b]	–	2
	Sclerogaster	–	3
	Secotium	1	1
	Stephanospora	1	2
	Thaxterogaster	–	7
	Timgrovea	–	2
	Wakefieldia	1	1
	Zelleromyces	2	14
	Total	41	134
	Grand total	45	149

[a] species originally placed here are now recognized as belonging to other genera
[b] hypogeous members only

Biodiversity

We define biodiversity as the collective genes, species, and ecosystems in a definable region (World Resources Institute 1992). Until 1988, relatively little was known of the ectomycorrhizal fungal diversity of the forests of Australia. Through support of the Australian Centre for International Agricultural Research and CSIRO, a project was begun to collect, isolate, and identify ectomycorrhizal fungi from a range of forest habitats throughout Australia. Ectomycorrhizal fungi were collected for identification and later study over a five year period. For truffle-like fungi, the more-than 2000 collections include about 1000 collections in Queensland and over 1000 collections in Tasmania. A preliminary assessment reveals 2 new families, 24 new genera, and perhaps 184 new species of truffle-like, presumptive ectomycorrhizal fungi (Tables 2 and 3). Nearly 95% of the described and undescribed truffle-like fungi from Australia are novel, with at least 22 genera and 3 families endemic (Table 1). This represents nearly a two-thirds increase in genera and nearly a threefold increase in the number of species known. It is increasingly apparent, as other regions of this extremely diverse continent are explored many more truffle-like fungal taxa will likely be uncovered.

Host specificity

Many ectomycorrhizal fungi express varying degrees of host specificity. Host range of the mycorrhizal fungus and host receptivity determine host specificity (Molina et al., 1992). Host range is the diversity of plant species with which a fungus can form mycorrhizae. Host range spans a spectrum from narrow, typically plant genus restricted, to intermediate, usually plant family restricted, to broad, usually encompassing two or more plant families. For example, a restricted association to a particular host genus may exist, such as *Descomyces albus* and *Hydnangium carneum* are found only in the presence of *Eucalyptus* spp. (Castellano and Bougher, unpublished results) Other ectomycorrhizal fungi are associated with a wider range of host genera and families, e.g., *Elaphomyces granulatus* Fries occurs with most genera in the Pinaceae and Fagaceae (Trappe, 1972). Host receptivity is a gauge of the diversity of mycorrhizal fungi that a particular plant species will form mycorrhizae with. Host receptivity is the number of fungi involved with the host and ranges from low to high. For example, some *Eucalyptus* species such as *E. grandis* and *E. tereticornis* have dozens, maybe even hundreds, of ectomycorrhizal fungal partners (Castellano and Bougher, unpublished results). In contrast Alnus has only a handful of fungal partners as compared to ectomycorrhizal conifers in the same plant communities (Miller et al., 1991). Foresters are interested not only in whether a particular plant forms mycorrhizae with a particular fungus but also in the physiological expression of this relationship in terms of seedling survival and growth in the field. *Pisolithus tinctorius* (Pers.) Coker and Couch will form mycorrhizae with numerous host genera including most in the Pinaceae and Fagaceae and at least some *Eucalyptus* species. In the Southeastern United States, *P. tinctorius* significantly stimulates field performance of many different pine species, especially those planted on degraded or otherwise stressful sites (Marx, 1980). On the other hand, although *P. tinctorius* forms mycorrhizae with *Pseudotsuga menziesii* when inoculated in Pacific Northwest nurseries, it does not confer any additional benefits to the host seedling in plantations and is replaced rapidly by indigenous mycorrhizal fungi (Castellano and Trappe, 1992).

Amplitude, resilience and duplication

Amplitude is the range over which a particular quality or property or function is expressed. As varied as ectomycorrhizal fungi are in morphological form, they are potentially equally varied in function within an ecosystem. Empirical data strongly suggest that some taxonomic species are not only host-specific but also habitat-dependent to some extent. For example, although *Tuber gibbosum* is apparently restricted to a host relationship with *Pseudotsuga menziesii,* it also exhibits a preference, as

Table 3. Truffle-like fungal species numbers known to occur in Tasmania, Australia

Fungal Class	Genus	Prior to 1988	Presently known
Ascomycotina	*Amylascus*	1	1
	Balsamia	1	1
	Dingleya	–	2
	Elaphomyces	1	5
	Genabea	1	1
	Genea	1	1
	Hydnocystis	1	1
	Labyrinthomyces	1	1
	Sphaerosoma	1	1
	Gen. nov.	–	3
	Total	8	17
Basidiomycotina	*Alpova*	1	2
	Arcangeliella	1	3
	Austrogautieria	2	5
	Boughera	–	2
	Castoreum	2	2
	Chamonixia	–	3
	Chondrogaster	–	2
	Cortinomyces	–	3
	Cystangium	2	4
	Descomyces	–	2
	Destuntzia	–	1
	Gautieria	2	2
	Gymnomyces	7	11
	Horakiella	1	1
	Hydnangium	3	3
	Hymenogaster	6	[a]
	Hysterangium	8	12
	Hysterogaster	1	3
	Leucogaster	–	1
	Macowanites	–	7
	Malajczukia	–	3
	Martellia	–	5
	Mesophellia	1	6
	Nothocastoreum	1	2
	Octavianina	2	2
	Protubera	–	2
	Quadrispora	–	2
	Rhizopogon	1	4
	Richoniella	–	1
	Scleroderma [b]	–	2
	Sclerogaster	–	2
	Secotium	1	1
	Setchelliogaster	–	1
	Stephanospora	–	1
	Thaxterogaster	–	5
	Zelleromyces	1	6
	Total	43	114
	Grand total	52	132

[a] species originally placed here are now recognized as belonging to other genera
[b] hypogeous members only

determined by sporocarp abundance, for low-elevation (30–770 m), young (10–80 years old) second growth stands in the Oregon Coast Ranges (J. Trappe, pers. comm.).

Resilience is the ability to recover from, or adjust to, change. Resilience is encoded in genetic material and through genetic recombination and drift will adjust from one generation to another. Genetic diversity is difficult to measure so we use the concept of ecological diversity to infer genetic diversity. What is the ecological diversity of an organism and what influences it? Ectomycorrhizal fungi are hypothesized to have appeared with the first Pinaceae, 100 million years ago and through the eons have evolved as have their hosts and concomitant habitats (Trappe, 1977). Major geologic events, such as uplifts and volcanoes, and climatic events such as, glaciation have greatly influenced the present complement of ectomycorrhizal fungi and their genetic make-up. In northern Europe, the major mountain ranges are oriented east-west so that glaciation forced many plants and their associated fungi into extinction as they literally 'ran out' of suitable habitat against the mountains (Barnowski, 1987). In contrast, the mountain ranges in North and South America and in Australia are oriented north-south, thereby allowing plants and fungi to migrate to the full extent of their ecological resilience (Barnowski, 1987).

Duplication is the repetition of a parameter—for our discussion we will address function. In some Northern Hemisphere habitats, a comprehensive list of the taxa of ectomycorrhizal fungi is available. Unfortunately, we have little understanding of the role or function each ectomycorrhizal species plays in these habitats. We lack an understanding of their basic biology, ecology, and physiology thereby making it difficult to predict distribution, except in a general way, or the effects human-caused or natural disturbance will have on fungal populations. On the other hand, we infer that ectomycorrhizal fungi that occur in unique or restrictive habitats have some competitive advantage in these habitats, for example, *Thaxterogaster pingue* is a dominant fungus (measured by sporocarp numbers and biomass) of the mid-high elevation Abies forests in the Pacific Northwest (J. Trappe, pers. comm.). What allows these fungi to flourish where other fungi do not? Is it because the habitat is exclusive or is one fungus more competitive than another under certain environmental and habitat constraints? How will this change with changes in global climate?

Some implications of rarity and abundance

Why are some fungi seldom encountered? Collecting truffle-like fungi is problematic because of their hypogeous habit, but even if we concentrate on areas, such as the Pacific Northwest U.S.A., that have been extensively searched we still have taxa that can be classified as uncommon or rare. For ectomycorrhizal fungi, host and habitat specificity as well as fitness determine rarity. If, for example, the fungus is host specific to a rare plant species, then it too would be considered rare; for example, there is an undescribed truffle-like genus found associated with *Noahdendron* in north Queensland (Castellano, unpublished results). If, however, the fungus is particularly adapted to a specific substrate, for example calcareous or serpentine soil, then the substrate determines distribution that is. Another influence is environment, for example it appears that fungi in the Mesopheliiaceae are stimulated to fruit after the stand has been subjected to fire, exclusion of fire would thus reduce abundance (C N Johnson, pers. comm.). Fungal incidence can be grouped as follows: (1) narrow distribution, low abundance or infrequent occurrence, for example, an unidentified *Leucogaster* sp. has only been found two or three times in a relatively small area (1 km^2) that has been intensely collected by the senior author and associates for over 30 years; (2) narrow distribution, high abundance, for example, *Cazia flexiascus* Trappe recently discovered and described from a single collection in southern Oregon (Trappe, 1989) is restricted in distribution but occurs gregariously every year in a second location not originally reported (Castellano, unpublished results); (3) wide distribution and low abundance, for example,

Brauniellula nancyae Smith occurs throughout Western North America with the genus *Pinus* but is not often encountered and is usually found in low numbers (Castellano et al,. 1989) and; (4) wide distribution and high abundance, for example, *Hysterangium coriaceum* Hesse is found widely distributed and in great abundance in north temperate Pinaceae forests (Castellano, unpublished results).

Priorities for additional research

We have attempted to illustrate some of the incompleteness and inaccuracies of the taxonomic literature on ectomycorrhizal fungi in Australia. Accurate taxonomic information is essential as the basis the assessing the role of ectomycorrhizal fungi in ecological development. Of particular significance is their role in maintenance of plant biodiversity in natural ecosystems and those disturbed by management.

There is much to learn, clarify, and discover about mycorrhizal fungi of Australia and the following priorities need to be addressed:

1. Provide improved accuracy of taxonomic knowledge of ectomycorrhizal fungi currently used in mycorrhizal research and commercial endeavours.
2. Modernize descriptions, keys etc. to mycorrhizal fungi to facilitate identification of mycorrhizal fungi.
3. Develop and maintain dried and live collections of ectomycorrhizal fungi, including voucher specimens in herbaria, aseptic culture collections, and germplasms for use by the science community.
4. Combine alpha and beta taxonomic techniques for a higher degree of resolution.
5. Continue knowledge accumulation on biodiversity in natural ecosystems emphasizing:
   a) ecosystems of high diversity or biological significance.
   b) ecosystems having particular ectomycorrhizal trees of current or potential economic benefit.
   c) ecosystems which are rare or under threat of destruction.

References

Agerer, R 1991 Characterization of ectomycorrhiza. *In* Methods In Microbiology. Vol. 23. Eds. J R Norris, D J Read, and A K Varma. pp 25–74. Academic Press, London.

Barnowski C W 1987 Response of vegetation to climate changes of different durations in the late Neogene. J Trends Ecol. Evol. 2, 247–250.

Boedijn K B 1939 The Tuberales of the Netherlands Indies. Bull Jard. Bot. Buitenzora Ser. 3. 16, 236–244.

Bougher N L and Castellano M A 1993 Delimitation of *Hymenogaster* sensu stricto and four new segregate genera. Mycologia 85, 273–293.

Castellano M A and Molina R 1989 The biological component: Mycorrhizae. *In* The Container Tree Nursery Manual, Volume 5. Agric. Handbk. 674. Eds. T D Landis, R W Tinus, S E McDonald and J P Barnett. pp 103–167. USDA, Forest Service, Washington, D.C.

Castellano M A and Trappe J M 1990 Australasian truffle-like fungi. I. Nomenclatural bibliography of type descriptions of Basidiomycotina. Aust. Syst. Bot. 3, 653–670.

Castellano M A and Trappe J M 1992 *Pisolithus tinctorius* fails to improve plantation performance of inoculated conifers in southwestern Oregon. New For. 5, 349–358.

Castellano M A, Trappe J M, Maser Z and Maser C 1989 Key to Spores of the Genera of Hypogeous Fungi of North Temperate Forests with Special Reference to Animal Mycophagy. Mad River Press, Arcata, California. 186 p.

Duponnois R and Garbaye J 1991 Mycorrhization helper bacteria associated with the Douglas-fir–*Laccaria laccata* symbiosis: effects in aseptic and in glasshouse conditions. Ann. Sci. For. 48, 239–251.

Fogel R and Trappe J M 1985 *Destuntzia*, a new genus in the Hymenogastraceae (Basidiomycotina). Mycologia 77, 732–742.

Hung L-L and Molina R 1986 Use of the ectomycorrhizal fungus *Laccaria laccata* in forestry. III. Effects of commercially produced inoculum on container-grown Douglas-fir and ponderosa pine seedlings. Can. J. For. Res. 16, 802–806.

Livingston W H 1991 Effect of methionine and 1-aminocyclopropane-1-carboxylic acid on ethylene production by *Laccaria bicolor* and *L. laccata.* Mycologia 83, 236–241.

Marx D H 1980 Ectomycorrhizal fungus inoculations: a tool for improving forestation practices. *In* Tropical Mycorrhiza Research. Ed. P Mikola. pp 11–71. Clarendon Press, Oxford.

Miller S L, Koo C D and Molina R 1991 Characterization of red alder ectomycorrhizae: A preface to monitoring below ground ecological responses. Can. J. Bot. 69, 516–531.

Molina R 1982 Use of the ectomycorrhizal fungus *Laccaria laccata* in forestry. I. Consistency between isolates in effective colonization of containerized conifer seedlings. Can. J. For. Res. 12, 469–473.

Molina R and Chamard J 1983 Use of ectomycorrhizal fungus *Laccaria laccata* in forestry. II. Effects of fertilizer forms and levels on ectomycorrhizal development and growth of container-grown Douglas-fir and ponderosa pine seedlings. Can. J. For. Res. 13:89–95.

Molina, R, Massicotte, H, and Trappe J M 1992 Specificity

phenomena in mycorrhizal symbioses: Community-ecological consequences and practical implications. *In* Mycorrhizal Functioning. An Integrative Plant–fungal Process. Ed. M F Allen. pp 357–423. Chapman and Hill, New York.

Mueller G M 1987 How does a taxonomist approach a problem like *Laccaria*? McIlvainea 8, 52–58.

Mueller G M and Vellinga E C 1986. Taxonomic and nomenclatural notes on *Laccaria B.* and Br. *Laccaria amethystea, L. fraterna, L. laccata, L. pumila,* and their synonyms. Persoonia 13, 27–43.

Rodway L 1897 Tasmanian fungi. Paps. Proc. Roy. Soc. Tasmania 1897, 129–144.

Tommerup I, Bougher N L and Malajczuk N 1991 *Laccaria fraterna* a common ectomycorrhizal fungus with mono- and bi-sporic basidia and multinucleate spores: comparison with the quadristerigmate, binucleate spored *L. laccata* and the hypogeous relative *Hydnangium carneum*. Mycol. Res. 95, 689–698.

Trappe J M 1972 Mycorrhiza-forming Ascomycetes. *In* Mycorrhizae. Proc. 1st NACOM April 1969, Illinois. Ed. E Hacskaylo. pp 19–37. USDA For. Serv. Misc. Publ.

Trappe J M 1977 Biogeography of hypogeous fungi: trees, mammals, and continental drift. *In* Proc. 2nd International Mycological Congress, Gainesville, Florida. p 675.

Trappe, J.M. 1979. The orders, families, and genera of hypogeous Ascomycotina (truffles, and their relatives). Mycotaxon 9, 297–340.

Trappe J M 1989 *Cazia flexiascus* gen et sp. nov., a hypogeous fungus in the Helvellaceae. Mem. New York Bot. Gard. 49, 336–338.

Trappe J M and Molina R J 1986 Taxonomy and genetics of mycorrhizal fungi: their interactions and relevance. *In* Mycorrhizae: Physiology and Genetics, Proc. 1st ESM. pp 133–146. INRA, Paris.

Trappe J M, Castellano M A and Trappe M J 1992 Australasian truffle-like fungi. IV. *Malajczukia* gen. nov. (Basidiomycotina, Mesophelliaceae). Aust. Syst. Bot. 5, 617–630.

Voss E G, Burdet H M, Chaloner W G, Demoulin V, Hiepko P, McNeill J, Meikle R D, Nicolson D H, Rollins R C, Silva P C and Greuter W. 1983. International Code of Botanical Nomenclature. W. Junk, The Hague. 472 p.

Wertenbaker W 1974 The Floor of the Sea: Maurice Ewing and the Search to Understand the Earth. Boston, Little and Brown, New York. 275 p.

World Resources Institute 1992 Global Biodiversity Strategy. A Policy-makers' Guide. WRI, Washington D.C. 35 p.

Levels of diversity in endomycorrhizal fungi (Glomales, Zygomycetes) and their role in defining taxonomic and non-taxonomic groups

JOSEPH B. MORTON and STEPHEN P. BENTIVENGA
Division of Plant and Soil Sciences, 401 Brooks Hall, P. 0. Box 6057, West Virginia University, Morgantown, WV 26506-6057, USA

Key words: developmental biology, ecology, evolution, molecular biology, morphology, ontogeny, phylogeny, vesicular-arbuscular mycorrhizal fungi

Abstract

Diversity in glomalean fungi is manifested at the molecular, morphological, and ecological levels. Characters at any of these levels can be ordered into hierarchical patterns defining taxonomic groups if they are conserved enough to be heritable through geologic time in all descendants of a common ancestor. At present, only morphological characters associated with mode of spore formation and in subcellular structure of spores are sufficiently stable and diverse to recognize at least 150 species. Ontogenetic comparisons indicate that species integrity, despite asexual reproduction, is the result of rigid internal constraints imposed on variation during the process of spore subcellular differentiation. Epigenetic factors dominate because the differentiation sequence is linear and each new stage is causally linked to preceding stages. Some morphological characters of the fungal mycelium also exist, but they define more inclusive groups at the family level and above. Most diversity in the mycorrhizae consists of life-history traits associated with abundance and architecture of fungal components, their rate of formation and longevity, and their cost in the symbiosis. These characters participate in processes at the molecular and ecological levels, so they are autonomous from morphological determinants. They often are labile or affected by external environmental conditions, so fewer stable taxonomic characters are likely to be discovered. Instead, molecular and ecological diversity has greater potential to define; (a) niche specificity of organisms/populations and (b) causal processes linked to host–fungus compatibility and mycorrhizal efficiency. Any taxonomic characters that relate to mycorrhizal functions will come only from comparative studies involving organisms from shared habitats rather than those having shared spore morphologies.

Introduction

Diversity is not an easy quantity to measure in biological systems because it embraces so many different processes. Most explanations of process centre around the species, since this taxonomic category links organisms/populations with higher taxa. Species have many different meanings, however, depending on the biological and ecological properties of their member organisms (Mishler and Brandon, 1987). The most universal species concepts are based largely on study of animals, but these organisms are taxonomically delimited more by somatic characters participating in life history strategies than reproductive parts (Stearns, 1976). Species concepts in plants are based on highly organized and complex characters of both somatic and reproductive structures (Stebbins, 1974). At the other end of the spectrum, fungal species are differentiated mainly by morphological characters in reproductive spores or fruiting bodies (Alexopoulos and Mims, 1979). This is particularly true of endomycorrhizal fungi in

A.D. Robson, L.K. Abbott & N. Malajczuk (eds.), Management of mycorrhizas in agriculture, horticulture and forestry, 47–59.

Glomales, Zygomycetes (Morton et al., 1992).

Glomalean fungi have a number of unique biological and ecological properties that warrant more complete explanations, particularly within a taxonomic framework. First, they originated in an ancient obligate symbiosis at least 250 million years ago that was conserved concomitant with evolution and radiation of land plants (Malloch et al., 1980; Morton, 1990a). Second, most of the fungi appear to be obligatorily asexual, with only one unconfirmed exception reported (Tommerup and Sivasithamparam, 1990). Third, subcellular morphological diversity of spores is unparalleled among fungi in the kingdom Mycota in organization, complexity, and stability (Morton and Benny, 1990). Fourth, organisms of a species are so widely dispersed that they can be found on all major continents. Fifth, taxonomic groups do not correlate with ecological niches (Morton, 1990a). These aspects of glomalean fungi, when taken together, indicate that molecular, morphological and ecological levels of diversity are produced by different, and even independent, processes. The daunting task for systematists and other biologists is to make sense of this diversity. In this paper, we will show the extent to which different levels of diversity can be incorporated into taxonomy based on the causal processes involved and how that taxonomy relates to the mycorrhizal symbiosis. We also discuss diversity in non-taxonomic entities such as fungal organisms or populations and its importance in understanding mycorrhizal interactions.

Morphological diversity

Taxonomic groups are recognized and ordered by morphological characters that are stable and discrete. When these characters are used to define phylogenetic relationships by common ancestry, taxa are recognized as products of evolutionary processes (Wiley, 1981). Phylogenetic relationships are of a historical nature and thus cannot be defined by direct experimentation. They must be reconstructed from the distribution of characters found in fossils and present-day organisms. For the present, only morphological characters of the mycelium and spores of glomalean organisms contain enough information to hierarchically order species in a phylogenetic tree (Morton, 1990b) and define taxonomic groups (Morton and Benny, 1990). Enough discrete variation is expressed in multinucleate spores to recognize over 150 species (Morton, 1988; Schenck and Pérez, 1990). This diversity is a remarkable evolutionary achievement because it is confined to subcellular levels of organization.

Morphological characters in glomalean fungi are not easily interpreted, regardless if they are used for diagnostic or classification purposes (Morton, 1992). Variation in characters able to group and rank asexual organisms into a species is a reflection of either developmental constraints within the organism or selective constraints imposed by the external environment (Morton, 1990a). To understand the magnitude of these constraints, information is needed on both developmental patterns and identifiable external factors (Maynard Smith et al., 1985). However, individual characters and their boundary conditions in glomalean fungi have no empirical basis and thus remain vague constructs. Developmental studies provide the only empirical means to recognize discrete characters, their origin, their transformational states during differentiation, and the sequence in which they are formed (Gould, 1977; Lovtrup, 1974). Phylogenetic patterns also can be deduced through the study of comparative developmental sequences in different taxa (Kluge and Strauss, 1985). For these reasons, work in our laboratory has focused on developmental comparisons of fungal taxa in the International Culture Collection of Arbuscular and Vesicular-arbuscular Mycorrhizal Fungi (INVAM). Some results are reported in this paper as the basis for new explanations of the nature of morphological characters, their origin and their persistence.

Fungal mycelium and its parts

The fungal mycelium of a mycorrhiza cannot be separated into discrete ontogenetic stages because there is no alternation of generations. It grows without finite beginnings (except from a spore) or endings. Friese and Allen (1991) were able to recognize morphological and architectural differences in extraradical mycelia grouped functionally as infective, absorptive and runner hyphae. These hyphal types had no known taxonomic value, however, because none were linked to named organisms. Abbott and Robson (1979) and Abbott (1982) provided evidence that morphological, architectural, and histochemical properties of intraradical hyphae may be taxonomically informative at the genus level or above. Fungi in INVAM corroborate some of the reported diagnostic hyphal characters: long infection units with 'H' connections between parallel strands in *Glomus*; pale staining by trypan blue and constrictions near branch points in *Acaulospora* and *Entrophospora*; and coiled, irregularly swollen hyphae with lateral projections or knobs in *Gigaspora* and *Scutellospora* (Morton, unpublished results). These characters will have increased taxonomic importance if they are shown to be consistent in a wider range of fungi propagated in a standardized host-soil environment.

The mycelia of endomycorrhizae also differentiate structures with specialized or unknown functions. Even though these structures are meristic (repeating), they are discrete and conserved enough to define higher taxonomic groups (Morton and Benny, 1990; Fig. 1). Arbuscules branch from intra- and intercellular mycelium in the root cortex (Bonfante-Fasolo, 1984). Arbuscule-like structures were discovered in fossils from the Triassic (Stubblefield et al., 1987), suggesting that arbuscule design and general morphology has been conserved for at least 250 million years. The most probable mechanisms of this stability are morphogenetic limitations and epigenetic factors that prevent disruption of the critical processes associated with mycorrhiza establishment, growth, and interface activity with the host plant (Bonfante-Fasolo, 1984; Gianinazzi, 1991). Morton (1990b) uses arbuscules as the only character to unite Glomineae and Gigasporineae in Glomales. However, a more detailed study of arbuscule morphology by Brundrett and Kendrick (1990) reveals distinct differences in isolates of *Gigaspora* and *Glomus* species; abrupt narrowing of branch hyphae in the former, and incremental reductions in hyphal width in the latter.

Intraradical vesicles are formed inclusively in Glomaceae and Acaulosporaceae, but distinct differences are evident in each family (Abbott, 1982). Vesicles in Glomaceae usually are subglobose to elliptical whereas those in Acaulosporaceae are pleiomorphic, knobby, and stain lightly in trypan blue. McGee (1986) also reports differences in vesicle morphology among sporocarpic species in *Glomus*. However, these differences are not discrete enough to differentiate many species.

Auxiliary cells are found only in genera of Gigasporineae (Fig. 1). They are unique to this suborder because they fail tests of similarity and homology with vesicles of fungi in Glominae (Morton, 1990b). Their function is strictly speculative, and cladistic analysis provides no clues of relationships between form and function. The shape and length of cell wall ornamentations differ enough to separate genera (Walker and Sanders, 1986), but a continuum of variation in each genus makes these characters taxonomically uninformative at the species level.

Spores and subcellular structure

Families and genera of fungi in Glomales differ in the structures from which spores originate (Morton and Benny, 1990; Fig. 1). Spores in *Glomus* are formed on cylindrical to flared sporogenous hyphae. The sporogenous hyphae may form individually, in loose to tight aggregates with little or no organization, or in highly organized aggregations that result in compact sporocarps (Morton, 1988). It is our view that the only species now in *Sclerocystis, S. coremioides* Berk and Broome, is just a more advanced step in the sporocarp specialization series found in *Glomus*. This transition series is difficult to interpret taxonomically because many sporocarpic species have not been shown or verified as mycorrhizal. There is a high probability of convergent evolution in these species, making it difficult to ascertain whether some of them even belong in Glomales.

In *Acaulospora* and *Entrophospora,* a saccule forms terminally on a sporogenous hypha, after which spores are initiated laterally from or within this hypha (Fig. 1). Origin of the sporiferous saccule cannot be explained from

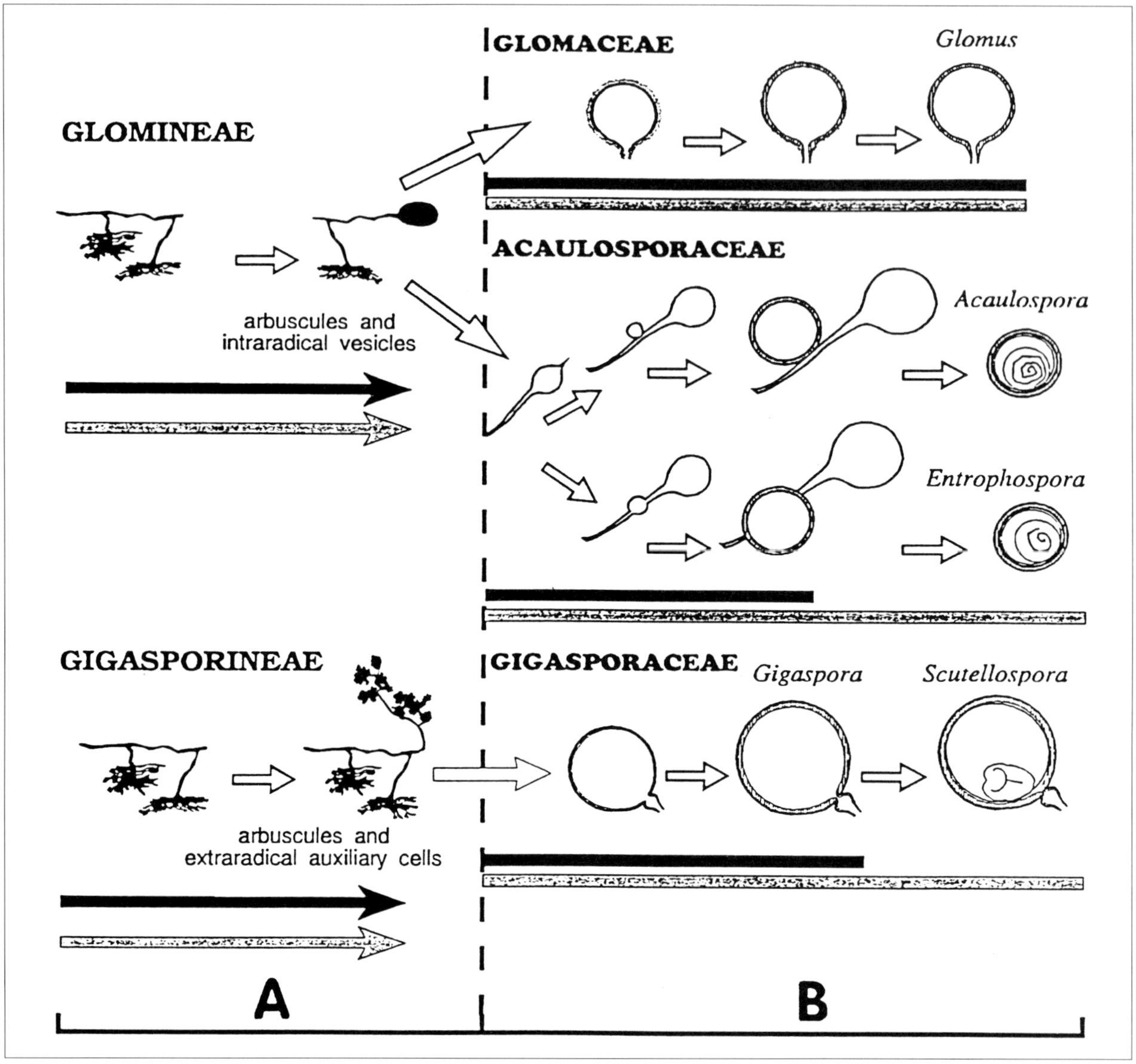

Fig. 1. Taxonomic groups in Glomales, Zygomycetes based on growth and differentiation of parts in a fungal organism. Arrows and bars signify indeterminate and determinate growth and differentiation, respectively. Black and grey fills indicate growth and differentiation processes, respectively. (A) Mycelial growth; differentiation of arbuscules, vesicles, and auxiliary cells. (B) Growth (expansion) of spores; differentiation of the spore wall and inner flexible walls (when present).

developmental patterns because no intermediate phenotypes have been found. Some evidence of linkage between *Acaulospora* and *Glomus* is found in *A. appendicula* Spain, Sieverding and Schenck. This fungus forms a saccule, spores from the saccule with glomus-like hyphal attachments, as well as a separate *Glomus* morph (Schenck et al., 1984; Morton and Benny, 1990). Dimorphism is a rarity among known glomalean species, and its appearance in *A. appendicula* may have been due to an improbable fusion event that led to mixing and retention of different developmental and structural genes.

Spores of *Gigaspora* and *Scutellospora* species expand from a bulbous sporogenous cell (Fig. 1) and can grow much larger than spores of other genera (up to 700 μm in *S. scutata* Walker and Diederichs). Because of equivalent hyphal attachments and mode of spore formation, spores of fungi in each genus are not easily distinguished under a stereomicroscope.

Spores of all glomalean fungi develop in two separate phases; (1) spore expansion coupled with differentiation of a spore wall at the subcellular level and (2) synthesis of one or more flexible inner walls that originate independently of the spore wall. Recognition of these discontinuities in developmental events conceptually alters how morphological diversity in the spore should be interpreted.

Traditionally, characters in spores have included different structural elements named as various types of walls (Walker, 1983; Morton, 1988) and the various properties of those walls (e.g. colour, ornamentation pattern, thickness, histochemical reactivity). All were considered individual characters and given equal taxonomic weight. Berch (1986) argued that empirical character analysis was needed, particularly in defining origin of individual characters. Some characters appeared to be correlated based on their distribution among taxa (Morton, 1990b), suggesting some were subordinate parts of more complex characters. In this paper, we will show that morphological diversity in spores can be partitioned into three hierarchical levels of organization. The first level encompasses spore structures identified by separate origin during both two phases of development. They include the spore wall and any inner flexible walls synthesized. The second level consists of phenotypically distinct characters sharing a common origin within each structure. These characters correspond to the wall terminology of

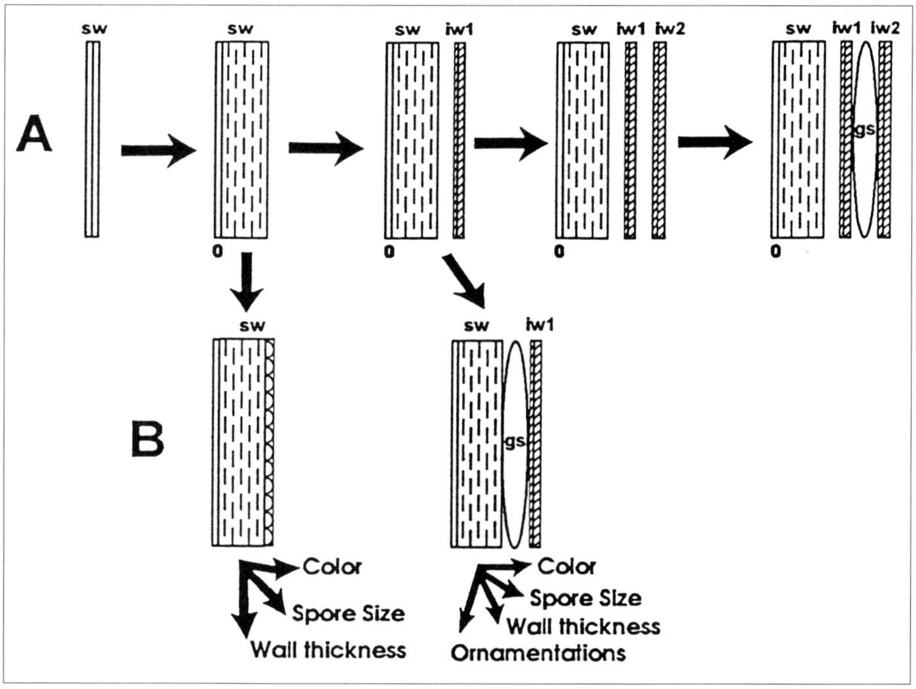

Fig. 2. Discrete stages in differentiation of subcellular structures, characters, and character states in *Gigaspora* and selected *Scutellospora* species. (A) Murographic representation of five stages in differentiation of *S. heterogama* spores. Subcellular structures; spore wall (sw), first inner wall (iwl), second inner wall (iw2), germination shield (gs). Fill patterns signify characters of each structure; none = outer layer, vertical dashed lines = laminae, angled lines = flexible layers; o = ornamentation character state. (B) Murographs illustrating phenotypes of adult spores in: left = *Gigaspora* species, each species separated by different character states of the spore wall (arrows); right = *S. coralloidea, S. fulgida, S. gregaria, S. persica,* and *S. verrucosa,* each species separated by character states of the spore wall (arrows). Morphology of the inner walls in all *Scutellospora* species shown here differs from that in species descriptions (unpublished results).

Walker and others, except that they no longer are separate entities. Instead, they correspond to wall layers suggested by Berch (1986). The third, and lowest, level embodies all states of each character in the level above, such as ornamentations, colour, thickness, plasticity, histochemical reactions in a particular wall layer.

The spore wall is the only structure to differentiate concurrently with growth of the spore. Differentiation can be partitioned into discrete stages in all genera, but we will discuss only *Scutellospora heterogama* (Nicol. and Gerd.) Walker and Sanders for brevity. The spore wall begins as two thin and rigid adherent layers of near-equal thickness (Fig. 2A). The outer layer thickens slightly and differentiates crowded warts of various lengths (unit wall sensu Walker). The inner layer is the first of multiple layers that become pigmented and more rigid (laminated wall sensu Walker). A unique transitory stage between the two stages of spore wall differentiation is not shown to keep the discussion simplified. Juvenile pores at the end of the second stage cannot be distinguished from adult spores with fully differentiated inner walls.

Variation in layers of the spore wall, such as colour, thickness, ornamentations, and spore size (expansion of all layers) cannot be linked to functional properties (e.g. survival, dispersal) because morphology, ecology, and biogeography show no evidence of being correlated. The spore wall has a clear-cut role in germination only in *Gigaspora,* where the innermost layer (germinal wall sensu Spain et al., 1989) differentiates rounded protuberances and is the origin of germ tubes (Maia, 1991).

The second phase of spore development proceeds with de novo synthesis of one or more inner flexible walls. These subcellular structures are found in species of *Acaulospora, Entrophospora, Glomus,* and *Scutellospora.* In *S. heterogama* spores (Fig. 2A), the differentiation sequence is causally linked; the first inner wall must be formed and differentiated before the second one appears. Within each inner wall structure, differentiation results in formation of two phenotypically distinct layers (membranous walls sensu Walker). Termination of differentiation in the second inner wall is marked by a chemical transformation of the inner layer so that it produces a dextrinoid reaction in Melzer's reagent. The germination shield is part of the causal sequence, because it is synthesized on (or from?) the second inner wall after appearance of the dextrinoid reaction.

Causally-linked sequences such as the one in *S. heterogama* spores indicate epigenetic mechanisms are operating within the organism to channel and constrain new variation. These constraints are powerful, as indicated by identical stages of spore differentiation in varied host and soil conditions (Franke, 1992). Similar mechanisms are active in controlling origin of variation in higher eukaryotic organisms (Lovtrup, 1974; Riedl, 1978), but diversification occurs at the subcellular level in spores of glomalean fungi where the magnitude of the constraints are expected to be greater.

Differentiation of de novo inner walls are not unique to glomalean fungi. In spores of some mucoraceous fungi, a single inner wall is synthesized which is essential for germ tube formation (references in Grove, 1976). The two flexible inner walls of *S. heterogama* also can be linked directly to spore germination, but only when defined as two ontogenetically indivisible structures with two phenotypically divisible characters in each. Formation of the germination shield in *S. heterogama* spores is linked to the second inner wall as a result of the linear causal sequence in differentiation. The first inner wall of *S. heterogama* spores has no obvious function until other *Scutellospora* species are added to the comparison (Fig. 2B). Spores of *S. coralloidea* (Trappe and Gerd.) Walker and Sanders, *S. fulgida* Koske and Walker, *S. gregaria* (Schenck and Nicol.) Walker and Sanders, *S. persica* (Koske and Walker) Walker and Sanders, and *S. verrucosa* (Koske and Walker) Walker and Sanders have the same differentiation sequence and form the same subcellular structures and characters as those of *S. heterogama* through the third stage (Fig. 2A—B). The only inner wall in spores in *S. fulgida* and sister species has identical function to the second inner wall of mature *S. heterogama* spores. Within a phylogenetic framework, then, the third stage inner wall of juvenile *S. heterogama* spores is homologous and can be interpreted as a historically conserved structure that had a

Fig. 3. Accessions in the International Culture Collection of Arbuscular and VA Mycorrhizal Fungi (circles) grouped and ranked into three *Glomus* species (each with the same fill pattern). (A) Geographic origin of different isolates. (B) Hypothesized phylogenetic relationship among species based on stable spore phenotypes.

germination-associated function in spores of an ancestral fungus

The epigenetic outcome of events in spore differentiation is stable spore morphologies that can unite geographic isolates into cohesive species (Fig. 3A). At the same time, it must be remembered that these species are based on morphological characters confined to spores, which constitute only one cell lineage of the whole organism. It follows that genetic, biochemical, behavioural, or physiological changes in fungal parts participating in mycorrhizal interactions will not correlate with differences in spore subcellular diversity and a species concept based on that morphological variation (Morton et al., 1992). This does not mean that species-level diversity in spores does not have considerable utility in the study of mycorrhizal processes. Organisms producing similar spore phenotypes can be identified within natural or managed plant communities, in different environments or geographic locations, or at the beginning and end of a particular experiment. Without these taxonomic boundaries, large collections such as INVAM would be impossible to manage.

Further developmental studies will improve definition of characters and weight them according to the taxonomic information they contain. At the present time, taxonomic information is not complete enough for even the most experienced individuals to make accurate identifications on all isolates they encounter (Morton, 1992). The following steps are recommended to circumvent this problem; (a) provide a culture the experimental organism(s) alone or with other members of the native population to a living culture collection such as INVAM so that they can be studied by others, (b) worry less about species identification than careful characterization of the spore phenotype to insure that the organism going into the experiment is the same one existing at the end, (c) prepare vouchers of spores and deposit them in a safe place so that identification may be determined or verified at some time (e.g., herbaria, INVAM), and (d) substitute an isolate name in place of the specific epithet when identifications are impossible or difficult—anything is better than educated guesses that mislead rather than inform.

Relationship between ontogeny and phylogeny

Ontogenetic sequences recapitulate phylogeny, but only when new characters are added by terminal addition (Gould, 1977). When stages in different taxa are considered homologous, then they can be arranged hierarchically by cladistic analyses (Wiley, 1981). Evolutionary directionality is assigned to each homologized stage by its temporal precedence; for any given stage, the preceding stage is ancestral. In Fig. 2, for example, the first two stages of spore differentiation in *S. dipapillosa, Gigaspora,* and a subgroup of *Scutellospora* consist of similar structures and characters (NOT character states). *Gigaspora* is ancestral to the *Scutellospora* subgroup phylogenetically because adult spores of the former are equivalent to the penultimate stage of the latter ontogenetically. The *Scutellospora* subgroup is ancestral to *S. dipapillosa* because of the same recapitulatory pattern; adult spores of the former are equivalent to the third stage of the latter. Morphological variation in character states of the spore wall layers (Fig. 2B) are not causally-linked in the linear sequence (being unique innovations that are non-disruptive) and thus cannot be assigned evolutionary directionality from ontogenetic comparisons alone. Cladistic methodology applicable to these character states are beyond the scope of this paper and will not be discussed. The important point is that this hierarchy of patterns permits recognition of characters differing in their phylogenetic value (or weight). For example, spore subcellular structures receive the most weight because they are highly conserved and shared by a more inclusive group of species. Character states have more limited distributions and thus may only group organisms into species if they are stable enough.

Spore wall structure, as defined ontogenetically in this paper, corroborates morphological characters of the fungal mycelium in separating Glomales into two suborders (Fig. 1). In Glomineae, the spore wall consists of rigid laminae with one or more outer layers that slough (evanescent wall sensu Walker). The spore wall in Gigasporineae, alternatively, consists of laminae with a permanent outer layer. When each layer was interpreted as separate walls, this distinction between suborders was not so obvious. The morphological gap between suborders will grow even wider if differences in arbuscule branching (Brundrett and Kendrick, 1990) prove to be congruent.

Characters and character states of spore wall structures group and rank geographic isolates into a species, despite the potentially distinct molecular and ecological properties of each (see next sections). Species can be ordered phylogenetically (Fig. 3B) when homologous characters (indicative of shared ancestry) are distinguished from those which are analogous (independent ancestry) (Morton et al., 1992). Without these distinctions, characters and the species they may define are merely aggregates with little biological meaning or value.

Molecular diversity

Molecular characters include base sequences or restriction fragment-length polymorphisms of DNA regions, isoenzymes, glycoproteins, and fatty acids. They are taxonomically informative only if they are conserved enough to reconstruct phylogenetic relationships between ancestors and their descendants. Various regions of the nuclear rDNA gene have been used most widely in systematic studies of eukaryotic groups, including the fungi (Bruns et al., 1991). Nucleotide sequences of a highly conserved region of the 18S ribosomal gene have been compared in isolates of selected glomalean species (Simon et al., 1992). Even though the data are is too limited in scope for broad phylogenetic interpretations, the evidence suggests great potential for defining relationships at higher taxonomic levels (e.g. family level and above). Comparative sequences also may lead to discovery of non-mycorrhizal relatives. More variable fragments have the potential to elucidate non-taxonomic relationships among organisms within a species, but their usefulness will depend on the magnitude of sequence variation.

Proteins or glycoproteins identified by immunological procedures appear to have some diagnostic value in distinguishing isolates of species (Aldwell et al., 1983; Wright et al., 1987), but there is no evidence they are informative in defining relationships. The problem is that there are too few characters to determine whether the antigenic molecules are homologues or analogues. Sanders et al. (1992) suggest that affinities between antigens of *A. laevis* and *G. margarita* might be taxonomically informative, but they ignore the homology criterion. Fungal antigens that distinguish organisms within a species (non-taxonomic diversity) are an elusive goal, in part because molecules with potential variability appear to have low immunogenicity. A solution to discovering such target antigens is to separate soluble proteins or glycoproteins electrophoretically and then immunize with purified extracts.

Isoenzymes have been used to extract information on taxonomic relationships in Glomales (Sen and Hepper, 1986), but the level of comparison (intraspecific, interspecific) was not verifiable. These molecules are not conserved enough in higher organisms to recognize taxonomic groups (Murphy et al., 1990), a trend borne out in other comparisons among glomalean organisms (Rosendahl, 1989). Important questions remain: Is there enough variation among organisms to differentiate them by isozyme banding patterns? Is there so much variation within an organism that progeny of different spores or hyphal fragments produce different banding patterns?

Fatty acid analysis has not been applied extensively to fungi, so very little is known about diversity or stability of molecules in organisms grown on different substrates (or host plants), exposed to varying environmental conditions or of different age. Jabaji-Hare (1988) observed different fatty acids in suborganismal parts of one *Gigaspora* and three *Glomus* species. However, the data were of no taxonomic value because the distribution of specific fatty acids was based on too few comparisons. In more recent studies, fatty acids such as 18:1 cis 9, 16:0, and 20:4 were found in spores of species in all genera (Graham, Hodge and Morton, unpublished results). This distribution indicates convergent evolution of these molecules. Alternatively, 20:0 is shared only by spores of all *Gigaspora* species tested.

Jabaji-Hare (1988) attempted to use fatty acid distribution patterns to identify non-mycorrhizal relatives. However, the nutrition source of mycorrhizal and non-mycorrhizal fungi is so different that it may obscure genotypic differences affecting fatty acid composition. Standardization is not possible, since glomalean fungi cannot be cultured in in vitro systems and many non-mycorrhizal fungi cannot be propagated on plant hosts.

Very careful species identification is needed in studies directed toward reconstruction of genetic phylogenies or to develop specific probes for various assays. Only then can a researcher know with confidence if the comparisons are within a species or between selected isolates of different species. Confirmations should be obtained by experienced taxonomists as well as adherence to the procedures outlined in the section on morphological diversity. Voucher specimens of the organism are essential.

Ecological diversity

Ecological studies with the goal of measuring species-level diversity currently rely on production of spores. However, measurements of diversity based on extracted spores are problematical for two reasons. First, the absence of spores does not necessarily indicate the absence of organisms forming mycorrhizae. Evidence is accumulating that threshold levels of colonization are required before sporulation can occur (Gazey et al., 1992). Second, most glomalean fungi cannot be identified accurately from field-collected spores (Morton, 1992). These spores rarely are identical to those in culture, the latter being newly-formed and with all morphological features intact. The solution is to obtain cultured spores of indigenous species, compare their morphology to those in the field, determine a 'mind's-eye' view of differences and similarities, and then proceed with a more extensive field-sampling regime. Trap (or bait) cultures of indigenous fungi can be established by mixing rhizosphere soil with a sterile growth medium in pots, and then seeding each pot with a selected host plant in a greenhouse. The benefits and inherent limitations of this methodology are discussed by Morton (1992).

Species-level diversity provides useful information on dispersal patterns, biogeographic ranges of species, and indirect evidence of competitive interactions by member organisms in plant communities. However, the species is not the relevant unit of comparison in studying mycorrhizal phenomena in individual plants or local plant communities because all members are not unified by any interactive process such as gene flow. Even though spore characters are the basis of glomalean taxonomy at the species level, they provide no information on life-history traits (Stearns, 1976) such as fungal biomass in a mycorrhiza, amount and architecture of external hyphae, number and size of spores, and the timing and cost of vesicle/mycelium/auxiliary

Table 1. Fungal properties potentially able to influence mycorrhiza occurrence and function (from Brundrett, 1991; Smith and Gianinazzi-Pearson, 1988; Gianinazzi-Pearson and Gianinazzi, 1989)

	Spore properties	Vegetative properties
Inoculum potential	Viability, dormancy factors, germination frequency, number and architecture of germ tubes, tolerance of germ tubes to edaphic factors, host recognition by germ tubes.	Abundance of hyphae and spores produced from the parent organism; hyphae viability, branch architecture, longevity, tolerance to edaphic factors, host recognition phenomena.
Arbuscules	Not applicable, except to control quantity and quality of carbon transported to spore initials.	Longevity, volume, surface area, rate and efficiency of nutrient exchange with host cell.
Intraradical hyphae	Not applicable, except as origin of spores in some species, transport mechanisms control allocation of carbon and other compounds to extraradical hyphae.	Rate of formation, length, volume and surface area, branch architecture, volume of root colonized, nutrient partitioning mechanisms, nutrient rate and efficiency of transport, allocation of resources to growth, differentiation of arbuscules versus vesicles.
Extraradical hyphae	Not applicable, except as origin of spores in most species; source of nuclei, intracellular organelles, carbon substrates for lipid synthesis.	Rate of formation, length, volume and surface area, branch architecture, volume of soil colonized, nutrient uptake and transport mechanisms, nutrient partitioning mechanisms, rate and efficiency of transport; allocation of resources to growth, uptake, or sporulation / auxiliary cell formation.

cell/spore production to the plant host (Table 1). Conclusions by some workers (e.g. Sieverding, 1991) that plant growth responses are a species-level property have no experimental or theoretical support. Most processes occurring in mycorrhizas are caused by interactions at the organism or population level.

Virtually all conclusions about mycorrhizal phenomena are generalized to include all glomalean fungi, from host–fungus cost–benefit relationships (Fitter, 1991; Koide, 1991) to phosphate metabolism (Bolan, 1991; Gianinazzi-Pearson and Gianinazzi, 1989) to lower level cellular and molecular effects (Gianinazzi, 1991). This 'top-down' approach masks considerable diversity in fungal organisms. If a researcher's main concern is the outcomes of mycorrhizal symbioses at the ecosystem or landscape levels, then knowledge of organismal diversity is not crucial (O'Neill et al., 1991). However, patterns and processes at lower levels must be sufficiently well understood to make generalizations with confidence. For example, mycorrhizal associations cannot be assumed to always have beneficial consequences. In some host–fungus-environment combinations, for example, the relationship can border on pathogenesis (e.g. Mojo and Hendrix, 1986). If prediction is a goal, then observations must start at the level of the individual organism, followed by comparisons among organisms to determine if an effect is species-specific. The more inclusive an effect, the more likely causal mechanisms will be found. Many equivalent effects, such as enhanced growth responses, can be caused by many different factors affecting the mycorrhizal association (Table 1). Generalizations derived from single isolate comparisons to all glomalean fungi are useful for propaganda purposes, but otherwise lose all value in generating testable hypotheses.

The fungal mycelium of an endomycorrhizal fungus has considerable potential to differentially respond to varied niches in the root or soil environment (Table 1). Nuclei migrate, the mycelium can be partitioned morphologically and metabolically, and germ tubes start new individuals that must adapt to host-soil conditions at the same or different locations. Little is known of genetic variability in a fungal organism or its influence on fungal responsiveness to environmental fluxes. Most certainly, asexuality does not have the same consequences as that hypothesized in higher organisms. Any somatic mutations would be heritable simply because almost all cells of a fungal organism are totipotent and can start new individuals. Somatic conflicts arising from accumulation of harmful mutations (Muller's ratchet—see Maynard Smith, 1989) are averted if deleterious nuclei are partitioned in non-germinating structures (e.g. senescing parts of the hyphal network or possibly even in auxiliary cells, vesicles).

Brundrett (1991) provides the most comprehensive and balanced look at diversity in both partners in the mycorrhizal symbiosis. Thirty-eight citations on page 187 of his article document that fungal organisms (NOT species) exhibit considerable variation in their fitness, adaptability to edaphic conditions, and ability to influence host response to mycorrhiza formation. Because glomalean fungi are able to grow only in mycorrhizas, niche breadth (focusing on range of living conditions for the fungus) and symbiotic effectiveness (ability of the fungus to produce growth benefits in the host) are often difficult to separate. When each was addressed as a separate phenomenon (Stahl and Christensen, 1991; Boerner, 1990), ecological and symbiotic interactions were autonomous in different isolates. Different causal mechanisms most probably were active.

Organisms within species must be grouped by at least two sets of criteria if the goal is to discover conserved properties that might have taxonomic value. The first set consists of edaphic conditions surrounding an organism/population (as emphasized by Bethlenfalvay et al., 1989). Organisms must be grouped by shared environments, regardless of their species name. Important parameters include species composition in both the plant and fungal community sampled, climatological and seasonal averages, and chemical and physical properties of the rhizosphere sample. Second are the functional attributes of organisms (as emphasized by Sylvia and Burks, 1988). By grouping organisms according to various kinds of mycorrhizal responses, comparisons can be made to elucidate causal mechanisms. For example, measurement of mycorrhizal colonization, external hyphal production, and growth response together may provide indices of efficiency of different fungal isolates. Other parameters, such as rate of colonization and rate of plant growth or nutrient uptake, provide additional useful information.

Acknowledgements

This work is supported by the West Virginia Agricultural and Forestry Experiment Station as scientific paper no. 2373 and by National Science Foundation grant no. DIR9015519.

References

Abbott L K 1982 Comparative anatomy of vesicular-arbuscular mycorrhizas formed on subterranean clover. Aust. J. Bot. 30, 485–499.

Abbott L K and Robson A D 1979 A quantitative study of the spores and mycorrhizas formed by a species of *Glomus* with reference to its taxonomy. Aust. J. Bot. 27, 363–375.

Aldwell F E B, Hall I R, and Smith J M B 1983 The identification of vesicular-arbuscular mycorrhizal fungi using immunofluorescence. Soil Biol. Biochem. 15, 439–445.

Alexopolous C J and Mims C W 1979 Introductory Mycology. Third Ed. Wiley and Sons, New York. 632 p.

Berch S M 1986 Endogonaceae: Taxonomy, specificity, fossil record, phylogeny. Front. Appl Microbiol. 2, 161–188.

Bethlenfalvay G J, Franson R L, Brown M S and Mihara K L 1989 The *Glycine–Glomus–Bradyrhizobium* symbiosis. IX. Nutritional, morphological and physiological responses of nodulated soybean to geographic isolates of the mycorrhizal fungus *Glomus mosseae*. Physiol. Plant. 76, 226–232.

Boerner R E J 1990 Role of mycorrhizal fungus origin in growth and nutrient uptake by *Geranium robertianum*. Am. J. Bot. 77, 483–489.

Bolan N S 1991 A critical review on the role of mycorrhizal fungi in the uptake of phosphorus by plants. Plant Soil 134, 189–207.

Bonfante-Fasolo P 1984 Anatomy and morphology. *In* VA Mycorrhizas. Eds. C L Powell and D J Bagyaraj. pp 5–33. CRC Press, Boca Raton, Florida.

Brundrett M 1991 Mycorrhizas in natural ecosystems. Adv. Ecol. Res. 21, 171–313.

Brundrett M and Kendrick B 1990 The roots and mycorrhizas

of herbaceous woodland plants. II. Structural aspects of morphology. New Phytol. 114, 469–479.

Bruns T D, White T J and Taylor J W 1991 Fungal molecular systematics. Annu. Rev. Ecol. Syst. 22, 525–564.

Fitter A H 1991 Cost and benefits of mycorrhizas: Implications for functioning under natural conditions. Experientia 47, 350–355.

Franke M 1992 Relationships among species and isolates of arbuscular mycorrhizal fungi based on morphology, ontogeny, and plant–fungus interactions. M. Sc. Thesis, West Virginia University, Morgantown, West Virginia.

Friese C F and Allen M F 1991 The spread of VA mycorrhizal fungal hyphae in the soil: inoculum types and external hyphal architecture. Mycologia 83, 409–418.

Gazey C, Abbott L K and Robson A D 1992 The rate of development of mycorrhizas affects the onset of sporulation and production of external hyphae by two species of Acaulospora. Mycol. Res. 96, 643–650.

Gianinazzi S 1991 Vesicular-arbuscular (endo-) mycorrhizas: cellular, biochemical and genetic aspects. Agric. Ecos. Environ. 35, 105–119.

Gianinazzi-Pearson V and Gianinazzi S 1989 Phosphorus metabolism in mycorrhizas. *In* Nitrogen, Phosphorus and Sulfur utilization by Fungi. Eds. L Boddy , R Marchant and D J Read. pp 227–241. Cambridge University Press, Cambridge.

Gould S J 1977 Ontogeny and Phylogeny. Belknap Press, Cambridge, MA.

Grove S N 1976 Form and function in zygomycete spores. *In* The Fungal Spore, Form and Function. Eds. D J Weber and W M Hess. pp 559–592. Wiley and Sons, New York.

Jabaji-Hare S 1988 Lipid and fatty acid profiles of some vesicular-arbuscular mycorrhizal fungi: Contribution to taxonomy. Mycologia 80, 622–629.

Kluge A J and Strauss R E 1985 Ontogeny and systematics. Annu. Rev. Ecol. Syst. 16, 247–268.

Koide R T 1991 Nutrient supply, nutrient demand and plant response to mycorrhizal infection. New Phytol. 117, 365–386.

Lovtrup S 1974 Epigenetics—A Treatise on Theoretical Biology. John Wiley and Sons, New York.

Maia L C 1991 Morphological and ultrastructural studies of spores and germ tubes of selected arbuscular mycorrhizal fungi (Glomales). Ph.D. Dissertation, University of Florida, Gainesville, Florida.

Malloch D W, Pirozynski K A and Raven P H 1980 Ecological and evolutionary significance of mycorrhizal symbioses in vascular plants (*a review*). Proc Nat. Acad. Sci. 77, 2113–2118.

Maynard Smith J 1989 Evolutionary Genetics. Oxford University Press, Oxford, UK. 325 p.

Maynard Smith J, Burlan R, Kauffman S, Alberch P, Campbell J, Goodwin B, Lande R, Raup D and Wolper L 1985 Developmental constraints and evolution. Q. Rev. Biol. 60, 265–287.

McGee P A 1986 Further sporocarpic species of *Glomus* (Endogonaceae) from South Australia. Trans. Br. Mycol. Soc. 87, 123–129.

Mishler B D and Brandon R N 1987 Individuality, pluralism, and the phylogenetic species concept. Biol. Phil. 2, 397–414.

Mojo H S and Hendrix V W 1986 The mycorrhizal fungus *Glomus macrocarpum* as a cause of Tobacco Stunt Disease. Phytopathology 76, 688–691.

Morton J B 1988 Taxonomy of VA mycorrhizal fungi: classification, nomenclature, and identification. Mycotaxon 32, 267–324.

Morton J B 1990a Species and clones of arbuscular mycorrhizal fungi (Glomales, Zygomycetes): their role in macro- and micro-evolutionary processes. Mycotaxon 37, 493–515.

Morton J B 1990b Evolutionary relationships among arbuscular mycorrhizal fungi in the Endogonaceae. Mycologia 82,192–207.

Morton J B 1993 Problems and solutions for integration of glomalean taxonomy, systematic biology, and the study of endomycorrhizal phenomena. Mycorrhiza 2, 97–109.

Morton J B and Benny G L 1990 Revised classification of arbuscular mycorrhizal fungi (Zygomycetes): a new order, Glomales, two new suborders, Glomineae and Gigasporineae, and two new families, Acaulosporaceae and Gigasporaceae, with an emendation of Glomaceae. Mycotaxon 37, 471–491.

Morton J, Franke M and Cloud G 1992 The nature of fungal species in Glomales (Zygomycetes). *In* Mycorrhiza in Ecosystems. Eds. D J Read, D H Lewis, A H Fitter and I J Alexander. pp 65–73. CAB International, Oxon, UK.

Murphy R W, Sites J W Jr, Buth D G and Haufler C H 1990 Proteins I: Isozyme electrophoresis. *In* Molecular Systematics. Eds. D M Hillis and C Moritz. pp 45–126. Sinauer Assoc., Inc. Sunderland, Mass.

O'Neill E G, O'Neill R V and Norby R J 1991 Hierarchy theory as a guide to mycorrhizal research on large-scale problems. Environ. Pollut. 73, 271–284.

Riedl R 1978 Order in Living Organisms. Wiley-Interscience, New York.

Rosendahl S 1989 Comparisons of spore-cluster forming *Glomus* species (Endogonaceae) based on morphological characteristics and isoenzyme banding patterns. Opera Bot. 100, 215–223.

Sanders I R, Ravolanirina F, Gianinazzi-Pearson V, Gianinazzi S and Lemoine M C 1992 Detection of specific antigens in the vesicular arbuscular mycorrhizal fungi *Gigaspora margarita* and *Acaulospora laevis* using polyclonal antibodies to soluble spore fractions. Mycol. Res. 96, 477–480.

Schenck N C and Pérez Y 1990 Manual for the Identification of VA Mycorrhizal Fungi. 3rd Ed. Synergistic Publ., Gainesville, Forida.

Schenck N C, Spain J L, Sieverding E and Howeler R H 1984 Several new and unreported vesicular-arbuscular mycorrhizal fungi (Endogonaceae) from Colombia. Mycologia 76, 685–699.

Sen R and Hepper C M 1986 Characterization of vesicular-arbuscular mycorrhizal fungi (*Glomus* spp.) by selective enzyme staining following polyacrylamide gel electrophoresis. Soil Biol. Biochem. 18, 29–34.

Sieverding E 1991 Vesicular-Arbuscular Mycorrhiza Management in Tropical Agrosystems. Deutche Gesellschaft fur Technische Zusammenarbeit, Eschborn. 371 p.

Simon L, Lalonde M and Bruns T D 1992 Specific amplification of 18S fungal ribosomal genes from vesicular-arbuscular endomycorrhizal fungi colonizing roots. Appl. Environ. Microbiol. 58, 291–295.

Smith S E and Gianinazzi-Pearson V 1988 Physiological interactions between symbionts in vesicular-arbuscular mycorrhizal plants. Annu. Rev. Plant Physiol. Plant Mol. Biol. 39, 221–244.

Spain J L, Sieverding E and Schenck N C 1989 *Gigaspora ramisporophora*: a new species with novel sporophores from Brazil. Mycotaxon 34, 667–677.

Stahl P D and Christensen M 1991 Population variation in the mycorrhizal fungus *Glomus* mosseae: breadth of environmental tolerance. Mycol. Res. 95, 300–307.

Stearns S C 1976 Life-history tactics: A review of the ideas. Q. Rev. Biol. 51, 3–47.

Stebbins G L 1974 Flowering Plants. Evolution Above the Species Level. Belknap Press Cambridge, Mass.

Stubblefield S P, Taylor T N and Trappe J M 1987 Fossil mycorrhizae: a case for symbiosis. Science 237, 59–60.

Sylvia D M and Burks J N 1988 Selection of a vesicular-arbuscular mycorrhizal fungus for practical inoculation of *Uniola paniculata*. Mycologia 80, 565–568.

Tommerup I C and Sivasithamparam K 1990 Zygospores and asexual spores of *Gigaspora decipiens*, an arbuscular mycorrhizal fungus. Mycol. Res. 94, 897–900.

Walker C 1983 Taxonomic concepts in the Endogonaceae: Spore wall characteristics in species descriptions. Mycotaxon 18, 443–455.

Walker C and Sanders F E 1986 Taxonomic concepts in the Endogonaceae. III. The separation of *Scutellospora* gen. nov. from *Gigaspora* Gerd. and Trappe. Mycotaxon 27, 169–182.

Wiley E 0 1981 Phylogenetics: The Theory and Practice of Phylogenetic Systematics. Wiley and Sons, New York.

Wright S F, Morton J B and Sworobuk J E 1987 Identification of a vesicular-arbuscular mycorrhizal fungus by using monoclonal antibodies in an enzyme-linked immunosorbent assay. Appl. Environ. Microbiol. 53, 2222–2225.

Reprinted from *Plant and Soil*, Volume 159, No. 1 (1994), pp. 47–59.

The ecology of ectomycorrhiza formation and functioning

G.D. BOWEN
CSIRO Division of Soils, Private Mail Bag No. 2, Glen Osmond, SA 5064, Australia

Key words: basidiospores, fungus succession, infection dynamics, mycelial growth, mycelial strands, mycorrhizal responses, rhizosphere, sclerotia

Abstract

Factors in inoculum potential/infection levels/plant growth response are analysed and experimental approaches to propagule germination, growth through soil and rhizosphere growth are indicated. It is suggested that seedling rhizosphere germination of basidiospores occurs particularly with early stage fungi, while late stage fungi may be advantaged by germination/growth on exudates from older parts of roots or on litter (and its associated microorganisms) and their tolerance of antimicrobial substances in litter. Relative growth in the rhizosphere is likely to dominate the mycorrhizal species composition and this may be a good selection method for organisms tolerant of stress and pollution conditions. As mycorrhizal function in nutrient uptake is largely determined by fungus growth into soil, there is need for much more experimental study of factors affecting this, and of the potential photosynthate drain this could represent. The possibility of selection/breeding mycorrhizal fungi with both 'early stage' and 'late stage' attributes is raised.

Introduction

The last 10–15 years have witnessed an increasing interest in the ecology/physiological-ecology of ectomycorrhizal fungi. Here I examine some ecological aspects of ectomycorrhizas (ECM), stressing the need for conceptual and experimental approaches to such questions as the relations between inoculum potential and response to inoculation, the fungus succession with age of the stand, and differing effects of mycorrhizal fungi on plant growth, a negative response to inoculation sometimes being found in the glasshouse and field (e.g. Stenström et al., 1990). I believe some rationale of these matters can be obtained by a better understanding of infection dynamics and microbial ecology. The important area of microbial interactions will not be addressed as this is the topic of another paper (Fitter and Garbaye, Chapter 12).

Populations and infection

One of the keys to understanding responses to inoculation and to the sufficiency of naturally occurring inoculum for 'adequate' mycorrhiza formation is a quantitative understanding of the relations between population, infection, and plant growth. This is also at the core of effects of clear felling, for while fewer mycorrhizas are often (not always) found on highly disturbed sites, particularly where logging slash has been burned (Perry et al., 1987), the real question is whether this represents a mycorrhizal deficiency (Parke et al., 1984).

Experimental studies on populations of propagules of various types in relation to infection levels/plant response and on the survival of propagules of various types during forest growth and following clear cutting are remarkable for their sparsity, yet quantitative evaluation of these (and of the speed of

A.D. Robson, L.K. Abbott & N. Malajczuk (eds.), Management of mycorrhizas in agriculture, horticulture and forestry, 61–67.

infection) in relation to the soil environment are necessary if we are to manage mycorrhizal populations for maximum effect. In their particular test soil, Theodorou and Bowen (1973) found 4 month old seedlings of *Pinus radiata* with 16% mycorrhiza infection to be 20% smaller than those with 58% mycorrhiza infection (significance P = 0.01); plants with 41% mycorrhiza infection were intermediate. In many instances plant stimulation is related to the level of infection but in other instances this is not so (Le Tacon and Bouchard, 1986). Such is not unexpected, for fungi will differ in their effectiveness and also the 'need' for high infection will depend on the fertility status of the soil (assuming a nutrient uptake mode of function) and populations of fine root pathogens and parasites.

The different distances hyphae from spores, sclerotia and mycelial strands can grow through soil will markedly affect infection dynamics. Small propagules such as basidiospores ('r' strategy, Bowen, 1980) are not capable of extensive growth through soil, so infection is more a function of their having sufficient numbers for a root to make contact with them. Much larger propagules, i.e. sclerotia and mycelial strands, are capable of growth through soil for considerable distances and high infection levels will usually be obtained with fewer propagules (Bowen, 1980). There have been no studies of dose-response curves using sclerotia of ECM fungi. Mycelial strands (in fungi which produce them) associated with live roots have been shown to be effective inoculum sources over large distances (Fleming, 1983; Finlay and Read, 1986). Because of the very different nature of different types of inoculum, it is extremely difficult to compare inoculum potentials quantitatively.

In the reafforestation context, the longevity of different types of propagules, is particularly important but has received little definitive experimental study. It is assumed that basidiospores and sclerotia have considerable longevity. However, mycelial strands separated from their energy source—the plant root—appear to be very fragile. For example, Fleming (1983) found mycelial strands were poorly effective inocula in soil cores severed from roots.

Inoculum potential/infection relations for any one fungus/plant combination will vary with many environmental factors, e.g. soil chemical status, soil pH, presence of heavy metals, soil moisture, soil temperature, organic matter and the soil microflora. The quantitative relations for each mycorrhizal fungus in a mixture will vary and these will determine the mycorrhizal composition to a large extent and be important in the mycorrhizal succession phenomenon as a stand ages. For example, *Pisolithus tinctorius* appears to be specially adapted to form mycorrhiza under high soil temperature (e.g. 30°C and higher, Marx et al., 1970) but appears not to be a good competitor in the presence of other microorganisms (Bowen and Theodorou, 1979). Stenström (1991) found mycorrhiza formation by *Thelephora terrestris, Laccaria laccata* and *Hebeloma crustuliniforme* was not sensitive to flooding but that by *Suillus flavatus* and *S. bovinus* was. It is to be expected that pollutants and acidification of soils can strongly affect mycorrhizal formation and species percentages on roots. Erland (1992) recorded different mycorrhizal fungi dominated at different pH's and Kasuya et al. (1990) recorded different sensitivities to aluminium by ECM fungi forming mycorrhiza on *Pinus caribaea*.

The numbers of ECM formed under various conditions result from effects of environment on fungus factors in soil and the rhizosphere and on root susceptibility to infection and spread. Bowen (1987) has suggested that the slope of the inoculum/infection curve (experimentally determined) will largely reflect the effects of environment on pre-infection phases such as spore germination and fungus growth through soil while changes in maximum infection points with an abundance of inoculum will reflect effects on root susceptibility (or fungus virulence). The physiology of susceptibility to ECM infection is still a matter of debate and will not be discussed further here, except to point out that host plant physiology may well affect root exudation and the growth of fungi in the rhizosphere e.g. P effects on loss of amino acids, organic acids and carbohydrates (Schwab et al., 1983).

Fungal succession

It is commonly observed that as a forest ages, fungal species which dominated in the early stages may be replaced by other fungi, referred to as 'late stage'. The studies of Mason et al. (1983) on birch and those of Gardner and Malajczuk (1988) on eucalypts are examples of the phenomenon. The separation into 'early stage' and 'late stage' is by no means always clear-cut, for some fungus species can persist, at least to some extent, throughout the life of the plantation, and this is especially marked where fungi have been introduced to an exotic tree species in an environment lacking ECM fungi. Where a natural situation pertains, as the environment changes with stand age, various fungi will be advantaged and will replace some of the prominent early stage fungi.

Phenomena such as replacement, competition and persistence must be definable in terms of growth and spread rates in a changing environment which has both positive and negative attributes, e.g. possible antibiosis of litter to some fungi.

In very few instances, the separate phases of infection/spread: spore germination, fungus growth through soil and growth along roots have been studied experimentally. However, very simple methods are available for the study of spore germination (e.g. Theodorou and Bowen, 1987), for fungus growth through soil to roots (e.g. Henis and Ben Yephet, 1970), for growth of the fungi along roots (e.g. Theodorou and Bowen, 1971) and for fungus growth into soil (see below). Any of these factors may be critical in mycorrhiza formation or in changing the mycorrhizal composition in different environments.

Spore germination

Melin (1962) suggested the stimulation of spore germination by tree roots. Fries and Birraux (1980) showed that the presence of roots of host species in laboratory media increased basidiospore germination of some ECM fungi from the occasional 0.1% in laboratory media alone to 1%, sometimes 10%, while Theodorou and Bowen (1987) demonstrated that *Rhizopogon luteolus* spores coated onto the root had up to 69% germination; with *Suillus luteus* it was 34%. Fries and Swedjmark (1986) and Theodorou and Bowen (1987) indicated great specificity in this stimulation and Fries et al. (1987) found abietic acid to be an activator for Suillus. Specificity in spore stimulation is an important inoculum conservation mechanism (Theodorou and Bowen, 1987). Theodorou (unpublished results) has shown high phosphate nutrition of seedlings markedly reduced spore germination on the root.

Almost all of the basidiospores responding to seedling roots are of species generally regarded as 'early stage' fungi, e.g. various species of *Hebeloma, Paxillus, Suillus, Thelephora, Laccaria, Pisolithus, Rhizopogon, Scleroderma* and *Inocybe* (Fries, 1987 and references therein; Theodorou and Bowen, 1987; Fox, 1983, 1986; Fries and Birraux, 1980 and references therein). Negative results have usually been obtained with putative late stage fungi such as species of *Leccinum, Cortinarius, Russula, Lactarius* and *Amanita* (Fries and Swedjmark, 1986; Fox, 1983, 1986). *Boletus* and *Paxillus* have also been recorded as negative, but Melin (1962) claimed stimulation of *Boletus, Amanita muscaria, Paxillus involutus* and some species of *Cortinarius, Russula* and *Lactarius*. With several species of *Russula* he was completely unsuccessful. If the observation made here is confirmed, it may explain why many 'late stage' fungi do not usually form mycorrhizas in early stages—the spores don't germinate. Infection of seedlings with 'late stage' fungi can occur with mycelial strands in the forest (Fleming, 1983) and with mycelial inoculum in pure culture. It would be worth experimentally studying if litter, or microorganisms associated with litter, stimulate spore germination of late stage fungi or whether exudates from older parts of roots stimulate them.

Growth through soil to the roots

There appears to be little study with ECM fungi on germination of sclerotia with regard to distance from the root, nor of growth of hyphae toward the root and the effects of soil physical and chemical factors. Horan and Chilvers (1990) have indicated a selective chemotrophic attraction of mycorrhizal fungi to substances diffusing from compatible host root apices, suggesting this could be the signal which initiates the ECM infection process. Some of the factors influencing growth of hyphae from roots into soil (see below) will be important. Can fungi growing through mineral soil grow equally well through litter, with its different chemical and microbiological composition?

Growth along roots, the ectotrophic phase

There have been few experimental studies on the growth of ECM fungi in the rhizosphere in soil, yet this may be a key to understanding the mycorrhizal composition on a root as it ages, and to selection of fungi for tolerance of deleterious factors such as acid precipitation and pollution of other kinds. It has been recognised increasingly (e.g. Theodorou and Bowen, 1971; Bowen and Theodorou, 1979; Erland et al., 1990) that studies in laboratory media may have little relevance to growth in the rhizosphere, where the physico-chemical environment is different and where there are indirect effects via effects on root exudation and changed rhizosphere pH as well as direct effects. Bowen and Theodorou (1979) showed 3-fold differences between early stage fungi in their colonization of the rhizosphere of *P. radiata*, large differences between ECM fungi in their tolerance of, and stimulation by, other bacteria and large differences between isolates of *R. luteolus* in their growth in the rhizosphere at 16°C soil but few differences at 20°C (Theodorou and Bowen, 1971).

Slow growth of 'late stage' fungi at least on young roots, (e.g. Mason et al., 1983) as well as poor spore germination may explain their failure to dominate the rhizosphere of young roots. Their success in later stages may be that they are able to use unique substrates coming from other parts of the root or that they can locally increase exudation by production of hormones. They may be tolerant of soil microorganisms colonizing older parts of roots, or of organisms associated with litter, tolerant of the antimicrobial effect of litter (e.g. Rose et al., 1983) or they may obtain extra sources of nutrient from litter (e.g. organic N, Abuzinadah et al., 1986). All of these hypotheses are amenable to relatively simple experimental study.

More research is needed on selection of mycorrhizal fungi not only for speed of infection but also for competitive ability and for persistence (e.g. Villeneuve et al., 1991) particularly on a background of indigenous mycorrhizal fungi and under a range of conditions. Mason et al. (1986) indicated particular fungi are more persistent in some soils. Tolerance of adverse conditions will also be important e.g. tolerance of *Cenoccum geophilum* mycorrhizas to drying (Pigott, 1982).

Mycelial growth into soil

A quantitative understanding of factors affecting growth of extramatrical hyphae is important, not only to uptake of nutrient and water, but also to other mycorrhizal functions e.g. the potential for accumulating toxins such as lead and cadmium (and keeping them out of the plant) and to spread of the infection but there has been little quantitative study of mycelial growth into soil. The development of extramatrical hyphae varies considerably between fungi (and hosts) and environment. A great many ECM fungi e.g. *R. luteolus* (Skinner and Bowen, 1974), *S. bovinus* (Finlay and Read, 1986) and *H. crustuliniforme* (Dosskey et al., 1990) form extensive, highly branched mycelial strands penetrating soil and litter for up to 40 cm (and more) (Finlay and Read, 1986). Others e.g. *Gautieria monticola* and *Hysterangium setchellii,* form dense mats in litter in Douglas fir stands of Pacific North-Western USA (Griffiths et al., 1991) and greatly change the micro-environment and microbial activities. At the other extreme, some ECM produce little hyphal growth into soil under most conditions.

Differences in hyphal growth into soil from mycorrhizas may partly explain differences in plant response in nutrient limiting conditions e.g. a lower response by *Salix viminalis* to *T. terrestris* than to *Laccaria proxima* (Jones et al., 1990). Tinker et al. (1991) found ECM fungi infecting *Eucalyptus cuccifera* produced more external hyphae and stimulated plant growth and P inflow more than VAM infection of the same species. Perhaps this is relevant to the observation that ECM species of Caesalpinaceae (and other tree species in tropical Africa) occur in soils poorest in phosphate (Högberg, 1986). The effect of low P levels on infection processes for VAM compared with ECM would be an interesting study.

Profuse development of hyphae in soil sometimes may present an undesirably large sink for assimilate, and this could be one reason for a negative effect of ECM on plant growth sometimes recorded e.g. studies of Stenström et al. (1990) and of Dosskey et al. (1990). On the other hand, an increase in the sink for photosynthate with ECM (Bevege et al., 1975) may reduce feedback inhibition of photosynthesis, thus stimulating photosynthesis (Dosskey et al., 1991).

The temporal aspect of extramatrical hyphae has received little attention but is likely to be important in mycorrhizal functioning. Cairney and Alexander (1992) found less transfer of ^{32}P from older mycorrhizas to the plant and Downes et al. (1992) found less extramatrical hyphae from ageing than from young ECM of *Picea sitchensis*. Another aspect is the effect of changing environment during the life of an ECM—Coutts and Nicoll (1990) found waterlogging caused the death of mycelial strands of *T. terrestris* which subsequently regenerated.

The few investigations of ECM hyphal growth into soil have used different methods. Finlay and Read (1986) observed mycelial strand growth and connection between plants growing in 4 cm wide plexiglass chambers in a soil/peat mixture. Rygiewicz et al. (1988) grew hyphae onto glass-fibre paper in a 'mycocosm' chamber, separated from roots in soil by a small air gap. While useful for some questions, the method does not examine soil factors affecting mycelial growth. Jones et al. (1990), using a recovery method for hyphae in soil similar to that used by Abbott and Robson (1985) for VAM hyphae (viz. blending a soil sample and measuring hyphal length by the line intersect method) measured 313 m hyphae m^{-1} mycorrhizal root of *Salix viminalis* and *L. proxima* but only 106 m for *T. terrestris,* at 10 mg P kg^{-1} soil and lesser differences at other P levels. While this approach is valuable for comparing different fungi and overall effects of soil variables it cannot separate the direct effects of soil factors on mycelial growth from indirect effects via effects on plant growth and assimilate supply. Skinner and Bowen (1974) approached the direct soil factors quantitatively by growing inoculated *P. radiata* in plexiglass sided boxes 1.2 cm wide. When mycelial growth into soil occurred, cores of the soil of 2.4 cm diameter were replaced by (unsterilized) cores of other soils or of the test soil with different treatments. After 4–6 weeks cores were removed and fungal hyphae (mainly mycelial strands) were removed and weighed. The amount of the test soil was small compared to that of the rest of the soil, thus assuring that effects measured were direct effects on mycelial growth, not indirect via plant vigour. Some soils inhibited mycelial strand growth completely; soil compaction reduced growth from 20.4 mg/core to 4.3 mg/core in unsterilized soils; added phosphate in non-sterile soil quadrupled strand growth. Soil sterilization increased mycelial strand growth considerably.

The approaches of Jones et al. (1990) and Skinner and Bowen (1974) could be used to advantage in examining soil/plant/fungus factors in mycorrhizal functioning. This should be a high priority area for future research.

Conclusion

An understanding of factors discussed in this paper is essential to selection/breeding and management of fungi which can be introduced successfully, are effective in a wide range of conditions and will persist. However, while there are many records of stimulation of young plants by inoculation—and we have *some* understanding of this—we know very little, if anything, of stimulation of older trees by mycorrhizal fungi and the scope for selection/breeding of fungi benefiting older trees. The greatest chance for control/management of the mycorrhizal suite on a plant is by inoculation in the nursery phase and one must ask if it will be possible to select/breed highly efficient fungi which combine the characteristics both of 'early stage' and 'late stage' fungi. This focuses attention on the reasons for success of particular fungi, their growth in the rhizosphere (and domination of it) and their growth into soil (and spread), on young and on older trees.

Experimental approaches to many aspects of infection and spread have now been developed. Happily, we now also have the emergence of several powerful tools in microbial ecology to follow introduced organisms. As well as immunofluorescence, ELISA and monoclonal antibodies, we also have isozyme analysis, hydrolysis—gas chromatography, AFCP and DNA probes to assist in identifying fungi, sometimes at the individual mycorrhiza level. The use of monokaryons and bikaryons with selected modified characteristics will also be valuable. These are all powerful tools in assessing some of the key ecology questions of ectomycorrhizas.

References

Abuzinadah R A, Finlay R D and Read D J 1986 The role of proteins in the nitrogen nutrition of ectomycorrhizal plants. II. Utilization of protein by mycorrhizal plants of *Pinus contorta*. New Phytol. 103, 495–506.

Abbott L K and Robson A D 1985 Formation of external hyphae in soil by four species of vesicular-arbuscular mycorrhizal fungi. New Phytol. 99, 245–255.

Bevege D I, Bowen G D and Skinner M F 1975 Comparative carbohydrate physiology of ecto- and endo-mycorrhizas. *In* Endomycorrhizas. Eds. F E Sanders, B Mosse and P B Tinker. pp. 149–174. Academic Press, London.

Bowen G D 1980 Misconceptions, concepts and approaches in rhizosphere biology. *In* Contemporary Microbial Ecology. Ed. D G Ellwood. pp. 283–304. Academic Press, London.

Bowen G D 1987 The biology and physiology of infection and its development. *In* Ecophysiology of VA Mycorrhizal Plants. Ed. G R Safir. pp. 28–57. CRC Press, Boca Raton, Florida.

Bowen G D and Theodorou C T 1979 Interactions between bacteria and ectomycorrhizal fungi. Soil Biol. Biochem. 11, 119–126.

Cairney J W G and Alexander I J 1992 A study of ageing of spruce (*Picea sitchensis*) ectomycorrhizas. III. Phosphate absorption and transfer in ageing *Pisea sitchensis*/*Tylospora fibrillosa* ectomycorrhizas. New Phytol. 122, 159–164.

Coutts M P and Nicoll B C 1990 Waterlogging tolerance of roots of Sitka spruce clones and of strands from *Thelephora terrestris* mycorrhizas. Can. J. For. Res. 20, 1894–1899.

Dosskey M G, Linderman R G and Boersma L 1990 Carbon-sink stimulation of photosynthesis in Douglas fir seedlings by some ectomycorrhizas. New Phytol. 115, 269–274.

Dosskey M G, Boersma L and Linderman R G 1991 Role for the photosynthate demand of ectomycorrhizas in the response of Douglas fir seedlings to drying soil. New Phytol. 117, 327–334.

Downes G M, Alexander I J and Cairney J W G 1992 A study of ageing of spruce (*Picea sitchensis*) ectomycorrhizas. I. Morphological and cellular changes in mycorrhizas formed by *Tylospora fibrillosa* and *Paxillus involutus*. New Phytol. 122, 141–152.

Erland S 1992 Effects of liming on pine ectomycorrhiza. *In* Mycorrhizas in Ecosystems. Eds. D J Read, D H Lewis, A H Fitter and I J Alexander. pp 113-118. CAB International, Wallingford, UK.

Erland S, Söderström B and Anderson S 1990 Effects of liming on ectomycorrhizal fungi infecting *Pinus sylvestris*. II. Growth rates in pure culture at different pH values compared to growth rates in symbiosis with the host plant. New Phytol. 115, 683–688.

Finlay R D and Read D J 1986 The structure and function of the vegetative mycelium of mycorrhizal plants. II. The uptake and distribution of photosynthate by mycelial strands interconnecting host plants. New Phytol. 103, 157–165.

Fleming J 1983 Succession of mycorrhizal fungi on birch: infection of seedlings planted around mature trees. Plant Soil 71, 263–267.

Fox F M 1983 Role of basidiospores as inocula of mycorrhizal fungi of birch. Plant Soil 41, 269–273.

Fox F M 1986 Groupings of ectomycorrhizal fungi of birch and pine based on establishment of mycorrhizas on seedlings from spores in unsterile soils. Trans. Br. Mycol. Soc. 87, 371–380.

Fries N 1987 Ecological and evolutionary aspects of spore germination in the higher basidiospores. Trans. Br. Mycol. Soc. 88, 1–7.

Fries N and Birraux D 1980 Spore germination in *Hebeloma* stimulated by living plant roots. Experienta 36, 1056–1057.

Fries N and Swedjemark G 1986 Specific effects of tree roots on spore germination in the ectomycorrhizal fungus *Hebeloma mesophaeum* (Agaricales). *In* Physiological and Genetical Aspects of Mycorrhizae. Eds. V Gianinazzi-Pearson and S Gianinazzi. pp. 725–731. INRA Press, Paris.

Fries N, Serck-Hanssen K, Dimberg L H and Theander O 1987 Abietic acid, an activator of basidiospore germination in ectomycorrhizal species of the genus *Suillus* (Boletaceae). Exp. Mycol. 11, 360–363.

Gardner J H and Malajczuk N 1988 Recolonization of rehabilitated bauxite mine sites in Western Australia by mycorrhizal fungi. For. Ecol. Manag. 24, 27–42.

Griffiths R P, Castellano M A and Caldwell B A 1991 Hyphal mats formed by two ectomycorrhizal fungi and their association with Douglas fir seedlings: A case study. Plant Soil 134, 255–259.

Henis Y and Ben-Yephet Y 1970 Effect of propagule size of *Rhizoctonia solani* on saprophytic growth, infectivity and virulence on bean seedlings. Phytopathology 60, 1351–1356.

Högberg P 1986 Soil nutrient availability, root symbiosis and tree species composition in tropical Africa (*a review*). *In* Mycorrhizas and Nitrogen-Fixing Root Nodules in Trees in East and South-Central Africa. Ph.D Thesis. Swedish University of Agricultural Science, Umea, Sweden.

Horan D P and Chilvers G A 1990 Chemotropism—the key to ectomycorrhizal formation. New Phytol. 116, 297–301.

Jones M D, Durall D M and Tinker P B 1990 Phosphorus

relations and production of extramatrical hyphae by two types of willow ectomycorrhizas at different soil phosphorus levels. New Phytol. 115, 259–267.

Kasuya M C M, Muchovej R M C and Muchovej J J 1990 Influence of aluminium on in vitro formation of *Pinus caribaea* mycorrhiza. Plant Soil 124, 73–79.

Le Tacon F and Bouchard D 1986 Effects of different ectomycorrhizal fungi on growth of larch, Douglas fir, Scots pine and Norway spruce seedlings in fumigated nursery soil. Acta Oecologia/Oecologia Applicate 7, 389–402.

Marx D A, Bryan W C and Davey C B 1970 Influence of temperature on aseptic synthesis of ectomycorrhizae by *Thelephora terrestris* and *Pisolithus tinctorius* on loblolly pine. For. Sci. 16, 424–431.

Mason P A, Wilson J, Last F T and Walker C 1983 The concept of succession in relation to the spread of sheathing mycorrhizal fungi on inoculated tree seedlings growing in unsterile soils. Plant Soil 71, 247–286.

Mason P A, Last F T and Wilson J 1986 Effects of different soils on the establishment and influence of sheathing mycorrhizas. *In* Physiological and Genetical Aspects of Mycorrhizae. Eds. V. Gianinazzi-Pearson and S Gianinazzi. pp. 767–772. INRA Press, Paris.

Melin E 1962 Physiological aspects of mycorrhizae of forest trees. *In* Tree Growth, Ed. Th T Kozlowski. pp. 247–263. Ronald Press, New York.

Parke J L, Lindeman R G and Trappe J M 1984 Inoculum potential of ectomycorrhizal fungi in forest soils of south west Oregon and northern California. For. Sci. 30, 300–304.

Perry D A, Molina R and Amaranthsu M P 1987 Mycorrhizae, mycorhizospheres and reforestation: current knowledge and research needs. Can. J. For. Res. 17, 929–940.

Pigott C D 1982 Survival of mycorrhiza formed by *Cenococcum geophilum* in dry soils. New Phytol. 92, 513–517.

Rose S L, Perry D A, Pilz D and Schoeneberger M M 1983 Alklopathic effects of litter on the growth and colonization of mycorrhizal fungi. J. Chem. Ecol. 9, 1153–1162.

Rygiewicz P T, Miller S L and Durall D M 1988 A root microcosm for growing ectomycorrhizal hyphae apart from host roots while maintaining symbiotic integrity. Plant Soil 109, 281–284.

Schwab S M, Menge J A and Leonard R T 1983 Quantitative and qualitative effects of phosphorus on extracts and exudates of sudan-grass roots in relation to vesicular arbuscular mycorrhiza formation. Plant Physiol. 73, 761–765.

Skinner M F and Bowen G D 1974 The penetration of soil by mycelial strands of ectomycorrhizal fungi. Soil Biol. Biochem. 6, 57–61.

Stenström E 1991 The effects of flooding on the formation of ectomycorrhizae in *Pinus radiata* seedlings. Plant Soil 131, 247–250.

Stenström E, Ek M and Unestam T 1990 Variation in field response of *Pinus sylvestris* to nursery inoculation with four different ectomycorrhizal fungi. Can. J. For. Res. 20, 1796–1803.

Theodorou C and Bowen G D 1971 Influence of temperature on the mycorrhizal association of *Pinus radiata*. Aust. J. Bot. 19, 13–20.

Theodorou C and Bowen G D 1973 Inoculation of seeds and soil with basidiospores of mycorrhizal fungi. Soil Biol. Biochem. 5, 763–771.

Theodorou C and Bowen G D 1987 Germination of basidiospores of mycorrhizal fungi in the rhizosphere of *Pinus radiata*. New Phytol. 106, 217–223.

Tinker P B, Jones M D and Durall D 1991 A functional comparison of ecto- and endo-mycorrhizas. *In* Mycorrhizas in Ecosystems—Structure and Function, Abstracts 3rd ESM, Sheffield.

Villeneuve N, Le Tacon F and Bouchard D 1991 Survival of inoculated *Laccaria bicolor* in competition with native ectomycorrhizal fungi and effects on the growth of out-planted Douglas fir seedlings. Plant Soil 135, 95–107.

Reprinted from *Plant and Soil*, Volume 159, No. 1 (1994), pp. 61–67.

An ecological view of the formation of VA mycorrhizas

L.K. ABBOTT and C. GAZEY
Soil Science and Plant Nutrition, School of Agriculture, The University of Western Australia, Nedlands, WA 6009, Australia

Key words: competition, diversity, infection-morphology, modelling, natural/disturbed ecosystems, vesicular-arbuscular mycorrhizas

Abstract

In spite of the major advances in understanding the functioning of symbioses between plants and arbuscular mycorrhizal fungi, details of the ecology of mycorrhizal fungi are not well documented. The benefits of the association are related to the timing and extent of colonization of roots, and fungi differ in their contribution to plant growth and presumably to soil aggregation. Knowledge of the processes that lead to successful colonization of roots by beneficial fungi at appropriate times for the host plants will form the basis of guidelines for soil management to maximize the benefits from the symbiosis.

Fungi differ in the manner and extent to which they colonize roots. They also differ in their capacity to form propagules. The importance of hyphae, spores and propagules within living or dead mycorrhizal roots also differs among species and for the same species in different habitats. The relationships between colonization of roots and propagule formation, and between propagule distribution and abundance and subsequent mycorrhiza formation, for different fungi in field environments, are not well understood.

Methods for quantifying mycorrhizal fungi are not especially suitable for distinguishing among different fungi within roots. Consequently, the dynamics of colonization of roots by different fungi, within and between seasons, have been little studied. Research is required that focuses on the dynamics of fungi within roots as well as on changes in the abundance of propagules of different fungi within soil. Interactions between fungi during the colonization of roots, the colonization of soil by hyphae and sporulation are all poorly understood. Without knowledge of these processes, it will by difficult to predict the likely success of inoculation with introduced fungi. Such knowledge is also required for selecting soil management procedures to enhance growth and survival of key species within the population.

The relative tolerance of various fungi to perturbations in their surroundings will provide a basis for identifying those fungi that are likely to persist in specific environments. The processes that influence mycorrhizal fungi in field soils can be identified in controlled studies. However, greater emphasis is required on studying these processes with mixed populations of fungi. The role played by diversity within populations of mycorrhizal fungi is virtually unexplored.

Introduction

VA mycorrhizal fungi are ubiquitous soil microorganisms. These fungi have great potential to enhance plant growth and soil aggregation. However, the extent to which they do this in the field is unclear (Fitter, 1989; Read, 1991). The benefit of mycorrhizal fungi in experimental conditions with individual fungi is generally related to the rate and extent of mycorrhiza formation (Abbott and Robson, 1991; Graham et al., 1991). However, there are

A.D. Robson, L.K. Abbott & N. Malajczuk (eds.), Management of mycorrhizas in agriculture, horticulture and forestry, 69–78.

indications that this may not be so for all fungi (e.g. Sylvia and Burks, 1988). In the field, the timing of mycorrhiza formation is also likely to be important for benefits to plant growth (Sanders and Fitter, 1992).

All VA mycorrhizal fungi do not contribute equally to nutrient uptake and plant growth (Abbott and Robson, 1984) and roots in field soils may become colonized by a range of fungi (e.g. Blaschke, 1991). Roots that are extensively colonized by ineffective fungi may be scored at a similar level to effective fungi using most current methods of estimation of mycorrhizas which do not distinguish among fungi. Thus the usual approach of assessing the mycorrhizal root as a whole, without considering the diversity of fungi within the colonized roots, limits our ability to assess contributions of these fungi in field soils. Where mycorrhizal roots are formed predominantly by one fungus there may be a clearer relationship between the colonization of roots and mycorrhiza benefit, as may be the case in the study by Thompson (this volume).

In this review we will examine the relevance of gaining a greater understanding of the dynamics of colonization of root systems by different fungi in natural and disturbed ecosystems. An understanding of how individual fungi colonize roots from mixed populations will be valuable for assessing the contribution of VA mycorrhizas in all environments. The role of diversity within mycorrhizal populations will be addressed in the context of how natural and intentional disturbance of field soil changes the relative abundance of these organisms within roots.

Mycorrhiza formation in field soils

Read (1991) recently re-emphasized the concept of the mycorrhizal symbiosis as a community of organisms, with many hosts and fungi involved. Although there are more than 150 species of VA mycorrhizal fungi (Morton, 1988), relatively little is known about the way that fungi within a population co-exist to form mycorrhizas on root systems in field soils. Most information about the dynamics and diversity of mycorrhizal fungi in field soils is derived from studies of the abundance and types of spores or of the total length of mycorrhizal root (e.g. Brundrett and Kendrick, 1990; Collins Johnson et al., 1991; Gemma et al., 1989). Although the presence of fungi within the mycorrhizal root at appropriate times during the season is essential for the effective functioning of the symbiosis, we are generally ignorant of the extent to which mycorrhizas are formed by different fungi in field soils. It is not possible to assume that the abundance of spores of a fungus in soil will accurately indicate the quantity of root colonized by that fungus in a field soil (Scheltema et al., 1987). For individual combinations of fungi and host this may occur, but it is unlikely in soils containing diverse populations of fungi and host plants (Giovannetti et al., 1988).

The quantity of mycorrhizal root colonized by each fungus within a soil can change throughout the season (Rosendahl et al., 1989). The dynamics of mycorrhiza formation by individual fungi will differ because of differences in (i) the growth of hyphae from propagules (e.g. Tommerup, 1983; Nasarajah and Nawawi, 1987), (ii) the intrinsic rates of infection from propagules (e.g. Wilson, 1984) and (iii) the capacity of fungi to use carbon substrates from host roots (e.g. Pearson and Jakobsen, 1993). Although it is likely that changes in the dynamics of propagule numbers, including spores, will reflect changes in the dynamics of colonization of roots, most of the data that supports this is for fungi that have been studied individually. The nature of relationships between colonization of roots and propagule formation (including sporulation) is known for relatively few fungi grown alone (Douds and Schenck, 1990; Gazey et al., 1992). It is not known whether these relationships hold for fungi growing in the presence of other fungi. Differences in carbon use by different fungi are likely to affect their interactions in roots (Pearson et al., 1993). It is possible that such interactions will also influence the abundance of spores and hyphae produced by each fungus relative to the rate and length of root that they colonize.

The processes involved in the colonization of roots by mycorrhizal fungi in the absence of competing mycorrhizal fungi have been well documented (see reviews by Bowen, 1987;

Table 1. Characteristics of VA mycorrhizal fungi and plants which may influence processes involved in mycorrhiza formation

Process	Characteristic [a]
Germination of propagules	Length of time required for spores to become mature and to overcome dormancy
	Rate of germination of spores
	Length and distribution of hyphae from spores, old hyphae and root pieces in disturbed and undisturbed situations
Root colonization	Susceptibility of roots of different ages and different host species to infection
	Rate of growth of hyphae and characteristics of spread of hyphae within the roots
	Rate of growth of roots
Hyphal spread in soil	Length and distribution of hyphae in the soil
Formation of propagules	Number and time of propagules produced in relation to the state of colonization

[a] Many of the above characteristics will be affected by edaphic factors such as: the phosphorus status of the soil, pH, salinity, temperature and moisture. These factors may alter the fungus, the plant or both partners of the symbiosis.

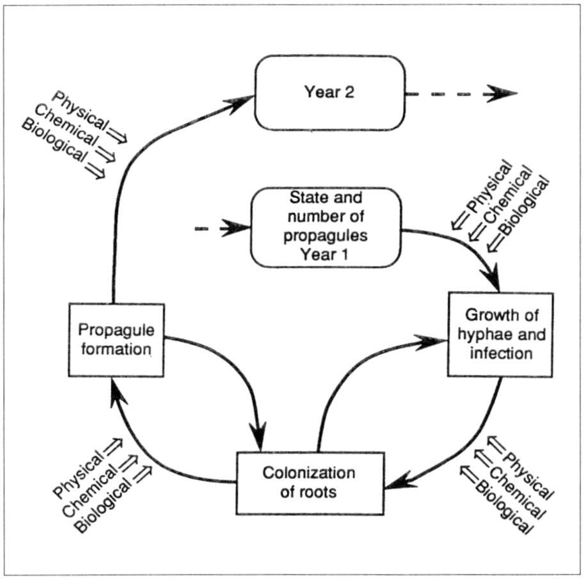

Fig. 1. The number and physiological state of propagules of each VA mycorrhizal fungus present in soil at one time (Year 1) will be reflected in the magnitude of the processes involved in the life cycle of VA mycorrhizal fungi. These processes will be influenced by physical, chemical and biological soil conditions which will affect subsequent numbers and states of propagules of each fungus present in soil in the following year.

Brundrett, 1991; Tommerup, 1992). These processes are summarized in Table 1 and Fig. 1. There is potential for interactions to occur among fungi during mycorrhiza formation but competitive interactions are not well understood (Wilson and Tommerup, 1992). An understanding of competitive interactions is required to predict the likely outcome of management practices, including inoculation, on colonization of roots by various fungi within a soil. The major obstacle to understanding competitive interactions among mycorrhizal fungi in field soils has been the technical difficulty in distinguishing these fungi within roots.

Distinguishing fungi within roots

Several methods have been used to distinguish different mycorrhizal fungi within roots to demonstrate seasonal changes in mycorrhiza formation, competitive interactions and survival of inoculant fungi in field soils. Morphological, immunological and electrophoretic approaches have all been applied, and the recent development of DNA/RNA technologies may also prove valuable. The approaches are complementary and can be used together in studies of the ecology of VA mycorrhizal fungi. The choice of technique depends on the level of specificity required.

Not all mycorrhizal fungi can be seen within roots using the usual staining methods (Morton,

1985; Merryweather and Fitter, 1991). There is potential for the length of root colonized by different species of fungi to be estimated in other ways. Furthermore, greater precision in the identification of specific fungi will sometimes be required.

Morphology of fungi within roots

It is possible to discriminate among some VA mycorrhizal fungi using visual differences in morphology of fungal hyphae and vesicles within roots (Abbott, 1982; Lopez-Aguillon and Mosse, 1987). Before morphological criteria can be used to distinguish fungi, careful measurement and observations of single isolates of fungi need to be made for defined host plants and soil conditions. The morphology of fungi within roots may change with host (e.g. Daniels-Hetrick et al., 1985; Boyetchko and Tewari, 1990). Even roots of different age on the same plant may lead to some alteration in the morphology of the fungi (Hepper, 1985). Furthermore, edaphic factors such as phosphorus supply can affect the morphology of the fungi within roots (Mosse, 1973; Amijee et al., 1989). However, in a quantitative study of the morphology of an isolate resembling *Glomus monosporum*, there was little effect of phosphorus supply for mycorrhizas on subterranean clover, although vesicles were not formed at the highest level of P supply (Abbott and Robson, 1979). Similarly this fungus was morphologically similar in three hosts grown at two levels of nitrogen supply.

Closely related fungi may have similar morphologies and therefore discrimination among fungi at the level of species will not generally be possible using morphological criteria. However, some species within genera (e.g. *Glomus* and *Acaulospora*) do have markedly different vesicle morphology (Morton, 1988), and hyphal diameter and growth pattern within roots (Abbott, 1982). Therefore, if soils contain fungi which display these distinct morphologies in roots of certain hosts, and if these morphologies have been defined, quantification of mycorrhizas formed by different species will be possible.

Although all VA mycorrhizal fungi cannot be distinguished on the basis of their morphology, the morphological approach has been valuable in several ways. First, it has been used to assess the success of inoculation of fungi in field soils. For example, Powell (1979) was able to recognize the inoculant *Glomus tenuis* in roots of plants grown in 10 out of 23 soils used. *G. tenuis* was absent prior to inoculation in these ten soils. In another study, the inoculant *Glomus* sp. was clearly present in roots of subterranean clover up to six weeks after field inoculation at two field sites (Abbott et al., 1983). At a third site, a fungus (or fungi) with similar morphology to the inoculant fungus was already present and inoculation did not change the proportion of roots colonized by hyphae resembling the inoculant fungus. In numerous glasshouse studies, the outcome of competition between fungi during colonization of roots has been assessed using morphological criteria (e.g. Wilson, 1984; Lopez-Aguillon and Mosse, 1987; Pearson et al., 1993). Finally, the infectivity of groups of fungi with different morphologies has been used to quantify the activities of mycorrhizal fungi within field soils (Abbott and Robson, 1982; Jasper et al., 1987; Scheltema et al., 1987). These studies show the potential of propagules of different fungi to colonize roots and provide information that can be used to predict sites where colonization by effective mycorrhizal fungi is likely to be limited.

In some studies of field soils, colonization of roots by different fungi or by groups of fungi with similar morphologies has been estimated according to differences in the diameter of hyphae in roots (e.g. Sparling and Tinker, 1978a; Sainz et al., 1990; Rosendahl et al., 1989; Wang et al., 1985). Although the mycorrhizas in these cases cannot be attributed to particular species or genera, this approach gives more information about the types of fungi colonizing the roots than does an assessment of colonization as a whole.

Further studies are required to delineate the morphology of species and isolates growing in association with one host. The approach provides a readily available means for discriminating among mycorrhizas formed by some fungi. The need to further distinguish between fungi for the purpose of understanding the function of mycorrhizas is not known. The extent to which fungi with similar morphologies have similar function is an important area for further study.

Gel electrophoresis

Electrophoretic separation of isozymes was used to assess the competitive interactions between three species of *Glomus* during colonization of leek roots (e.g. Hepper et al., 1988). Isozyme activity was linked to a hierarchy of competitiveness among the fungi. This approach provides an opportunity to assess the level of colonization of roots by fungi that appear morphologically similar.

Serological techniques

Several immunological techniques have been useful for identifying fungi within roots. Wright and Morton (1989) identified *Glomus occultum* in roots with a dot-immunoblot technique and a monoclonal antibody prepared for this fungus. As this fungus does not stain with conventional stains, the serological approach is of great value. Fluorescent antibody techniques have been successfully used to trace the hyphae of *Gigaspora margarita* in the field after inoculation (Fries and Allen, 1991). Hyphae and spores of several species of *Glomus* were labelled using an indirect immunofluorescent technique in a study by Kough et al. (1983). An antiserum was prepared for *Glomus epigaeum*, however the serum was only genus specific rather than species specific (Kough et al., 1983). This technique may not be suitable for use with all fungi because the reactive sites are not always on the surface of the hyphae (Wilson et al., 1983). An ELISA method was used to estimate the spread of *Glomus mosseae* through soil containing *Acaulospora laevis* (Aldwell and Hall, 1986). All of these serological methods give a higher level of specificity than the morphological approach.

Molecular techniques

The application of molecular techniques to VA mycorrhizal fungi (e.g. Simon et al., 1992) has the potential to allow precise identification (Morton, 1988) and quantification of mycorrhizal fungi within roots of field grown plants. It is most likely that these techniques will complement other less specific techniques in field studies aimed at expanding knowledge of the diversity and dynamics of colonization of roots by field populations of mycorrhizal fungi.

Modelling mycorrhiza formation in mixed communities

The colonization of roots by mycorrhizal fungi is a dynamic process that may include host control via the supply of carbon substrates for fungal growth. The quantity and types of propagules within the soil will affect the dynamics of colonization (see Bowen, 1987). The information required to model colonization of mycorrhizal fungi includes the effects of soil and environmental conditions on (i) growth of hyphae from propagules, (ii) colonization of roots, (iii) sporulation and other propagule formation, and (iv) hyphal growth in soil (see Table 1). Other parameters include the intricacies of spore dormancy and maturation, susceptibility of roots to infection, and carbon/phosphorus relationships in the plant that influence colonization and sporulation.

Models of colonization of root systems have enhanced our understanding of the processes involved for several species of VA mycorrhizal fungi (e.g. Smith and Walker, 1985; Sanders and Sheikh, 1983). Parameters measured include the rate of initiation of infection points and their subsequent growth in axial and lateral roots, and effects of varying environmental conditions. These models now need to be extended to other fungi. With the availability of substantial knowledge of the biology of a range of VA mycorrhizal fungi from different genera (see Table 4, in Abbott et al., 1992), there is potential to develop a more complex model describing simultaneous colonization of roots by several fungi. The effect of one fungus on the colonization and propagule formation of another fungus can be determined experimentally, but the outcome should be predictable from an understanding of the biology of each fungus. A systems modelling approach will be useful for identifying major gaps in understanding of the processes of colonization of roots by different fungi in field soils.

Seasonal patterns in the formation of mycorrhizas can vary considerably from year to year (Allen et al., 1989; Sanders, 1990) or can remain relatively stable (Sparling and Tinker, 1978b). These effects are not easily understood. Both Sparling and Tinker (1978b) and Sanders (1990) concluded that it was possible that some of the variation could be related to selective colonization of roots by different types of fungi and that uniform levels of total colonization could mask peaks in colonization by different fungi. These comments support the need to probe further the diversity and dynamics of colonization of root systems by populations of VA mycorrhizal fungi.

In spite of the complexities of the system, there will continue to be major blocks to successful management of VA mycorrhizal fungi in agriculture, horticulture, forestry and revegetation unless further progress is made towards predicting the likely outcome of an imposed management program on the fungi involved. By attempting to model the whole system, limitations to successful colonization of symbiotically effective fungi within a population will be identified.

Disturbance effects on the diversity of fungi

Changes in soil conditions will modify the dominance of particular fungi during mycorrhizal formation in field soils. Soil disturbance, either intentional (e.g. by mining or cultivation) or unintentional (e.g. by erosion or natural tree felling), will be a major source of changes in the physical, chemical and biological properties of the soil. Therefore, studies of the effects of disturbance (via whatever cause) will play a useful role in assessing likely changes in fungal dominance during colonization of roots. This understanding will also be useful for predicting propagule levels from season to season (Fig. 1).

According to the hypothesis of Connell (1978), the diversity of species should be lower for minimal and maximal disturbance, for recent and distant disturbance and for minor and major levels of disturbance than it is for intermediate levels of disturbance (Fig. 2a). The extent to which this hypothesis applies to the diversity of VA mycorrhizal fungi is not known. The hypothesis can be used to predict that the diversity of fungi within roots of plants in an undisturbed ecosystem may increase, decrease or remain the same with low to moderate disturbance (Fig. 2b). As the severity of disturbance increases, the diversity of fungi should decrease. An initial increase in diversity may arise from the input of surrounding soil (e.g.

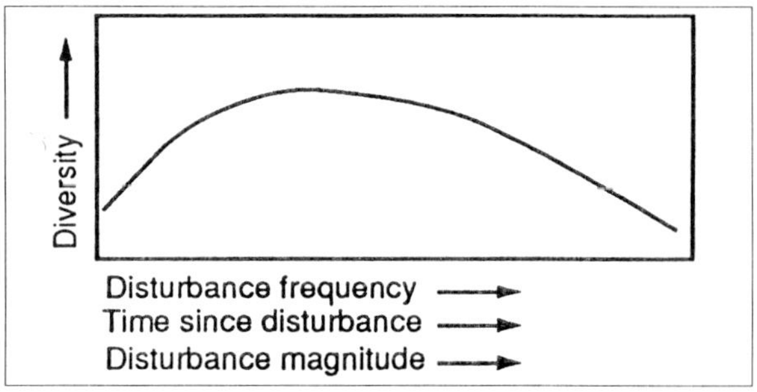

Fig. 2a. 'Intermediate disturbance' hypothesis of Connell (1978) predicts that the diversity of VA mycorrhizal fungi would be highest in soils of 'moderate' disturbance (in terms of frequency and magnitude) and at 'intermediate' times since disturbance. A major consideration in the testing of this hypothesis lies in the definition of these intermediate states.

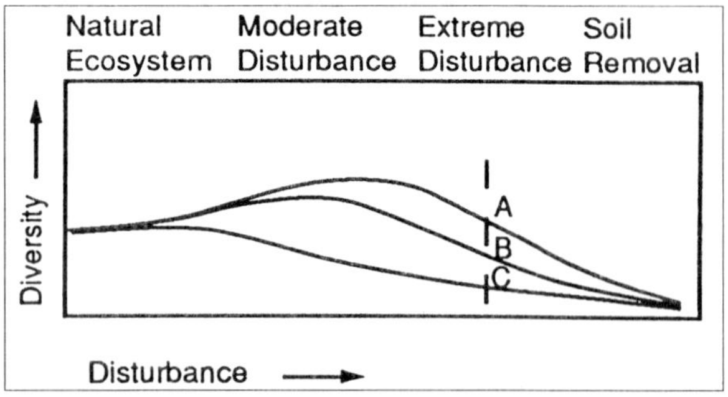

Fig. 2b. Based on this hypothesis, the diversity of mycorrhizal fungi is likely to change with the severity of disturbance. A, B, C represent possible changes in the diversity of three populations which had similar diversities in their natural (relatively undisturbed) state as the severity of disturbance increases. The subsequent changes in diversity could reflect differences in the the effects of (i) changed soil conditions on each fungus within each population or (ii) differences in the way that the disturbances alter soil conditions in different habitats.

soil deposition after wind erosion). A decrease in initial diversity may be due to soil loss, a change in soil conditions to those unsuitable for all fungi originally present, or a loss of species due to the competitive interactions of other species in the population that may be favoured in the disturbed conditions. Depending on the initial species composition of the population, the extent of changes will also differ (Fig. 2b). Natural or intentional disturbance is a common feature of the habitat of all VA mycorrhizal fungi, and an understanding of the effects of disturbance at the population level will complement existing studies of the effects on individual fungi (Jasper et al., 1989a; 1989b; 1993). The intermediate disturbance hypothesis provides a starting point for addressing the effects of disturbance on the diversity of VA mycorrhizal fungi.

Although some VA mycorrhizal fungi may occur in soil without contributing substantially to the total length of root that is colonized by all mycorrhizal fungi, they may dominate when the soil conditions are altered, such as by liming (e.g. Wang et al., 1985). In a long term liming experiment, colonization at the low pH was dominated by fungi with fine hyphae, in contrast, this fungus was not detected in the alkaline soil (Wang et al., 1985). It may be an advantage to maintain high levels of diversity of VA mycorrhizal fungi within soil, irrespective of their individual contribution to mycorrhiza formation. This would ensure maximum opportunity for suitable fungi to dominate when soil conditions are changed with agricultural, horticultural or forest practice.

A major consideration in the development of a model of mycorrhiza formation at the population level is the effect that natural or intentional soil disturbance may have on the individual fungi in soil within and between seasons. The extent to which the most suitable fungi would dominate as soil conditions change is not known. If soils are to be successfully managed to enhance the capacity of effective fungi to colonize a major proportion of the mycorrhizal root, it will be essential to be able to predict which fungi are most likely to be favoured in a particular soil environment and in the presence of particular hosts plants.

Conclusion

The objective of *identifying procedures for managing mycorrhizal fungi* is more appropriately restated as *managing soil conditions to suit the growth and activity of beneficial populations of mycorrhizal fungi.* To achieve this it will be necessary to understand the patterns and processes involved in establishing, maintaining and changing populations of mycorrhizal fungi in soil. It will also require an understanding of the effects of soil disturbances on the stability and dynamics of populations in relation to associated changes in root growth and host plant species. For agricultural and horticultural soils it will require a knowledge of the effects of practices that alter the chemical, physical and biological characteristics of a soil on the formation of mycorrhizas by individuals within the original population. The extent to which populations need to be diverse is not known.

Decisions regarding the management of soil to enhance the colonization of roots by preferred VA mycorrhizal fungi from season to season need to be based on a sound knowledge of the biology of individual species within populations (Abbott and Robson, 1991). Ecological studies are required to understand the processes that lead to the establishment and maintenance of requisite populations of fungi, with emphasis on mycorrhiza formation rather than on quantities of propagules (including spores). A modelling approach that takes an overview of the formation of mycorrhizas by separate species within populations would help to identify key components of the system that are poorly understood. Models of this type can provide new insights into the behaviour of complex systems (Thornley, 1976). The complexities of interactions between species of VA mycorrhizal fungi during root colonization, the processes of propagule formation, distribution, infectivity and survival for each fungus, and the effects of changing soil conditions on all these processes require a level of integration not easily accomplished without the aid of a systems-modelling approach.

Two important questions that need to be considered in the context of selecting soil management procedures that lead to maximum benefits from mycorrhizal fungi are: Do fungi that colonize roots in similar ways and have similar morphologies within roots function similarly? and What is the relevance of species diversity to the functioning of mycorrhizas in field soils? If the answer to the first question is 'yes' there will be a reduced need for identifying fungi to high levels of specificity. Faster progress will subsequently be possible towards understanding the dynamics of populations of fungi within mycorrhizal roots. The question of diversity of fungi needs further study in relation to the symbiotic effectiveness of individuals within diverse populations. Unfortunately the effectiveness of many of the species of fungi already described are unknown, as is the effectiveness of isolates of most species (Morton, 1988).

Acknowledgments

We especially thank Alan Robson for many valued discussions and for his encouragement. We are also grateful to Ian Abbott, Robert Black, Peter Braunberger, Mark Brundrett, David Jasper, Sato Juniper, Nick Pearson and Peter Schweiger for their comments and support. Financial contributions are acknowledged from the Australian Wool Research and Development Corporation, the Grain Research and Development Corporation, and the Australian Mineral Industries Research Association.

References

Abbott L K 1982 Comparative anatomy of vesicular-arbuscular mycorrhizas formed on subterranean clover. Aust. J. Bot. 30, 485–499.

Abbott L K and Robson A D 1979 A quantitative study of the spores and anatomy of mycorrhizas formed by a species of *Glomus* with reference to its taxonomy. Aust. J. Bot. 27, 363–375.

Abbott L K and Robson A D 1982 Infectivity of vesicular arbuscular mycorrhizal fungi in agricultural soils. Aust. J. Agric. Res. 33, 1049–1059.

Abbott L K and Robson A D 1984 The effect of mycorrhizae on plant growth. *In*: VA Mycorrhizae. Eds. C Ll Powell and D J Bagyaraj pp 113–130. CRC Press, Boca Raton, Florida.

Abbott L K and Robson A D 1991 Factors influencing the occurrence of VA mycorrhizal fungi. Agric. Ecos. Environ. 35, 121–150.

Abbott L K, Robson A D and Hall I R 1983 Introduction of vesicular arbuscular mycorrhizal fungi into agricultural soils. Aust. J. Agric. Res. 34, 741–749.

Abbott L K, Robson A D and Gazey C 1992 Selection of inoculant vesicular-arbuscular mycorrhizal fungi *In:* Methods in Microbiology Vol 24: Techniques for the Study of Mycorrhiza. Eds. J R Norris D J Read and A K Varma. pp 1–22. Academic Press, London.

Aldwell F E B and Hall I R 1986 Monitoring the spread of *Glomus mosseae* through soil infected with *Acaulospora laevis* using serological and morphological techniques. Trans. Br. Mycol. Soc. 87, 131–134.

Allen M F, Richards J H and Busso C A 1989 Influence of clipping and soil water status on vesicular-arbuscular mycorrhizae of two semi-arid tussock grasses. Biol. Fert. Soils 8, 285–289.

Amijee F, Tinker P B and Stribley D P 1989 The development of endomycorrhizal root systems VII A detailed study of effects of soil phosphorus on colonization. New Phytol. 111, 435–446.

Blaschke H 1991 Multiple mycorrhizal associations of individual calcicole host plants in the alpine grass-heath zone. Mycorrhiza 1, 31–34.

Bowen G D 1987 Infection processes in plants—Development of vesicular arbuscular mycorrhizae *In:* Ecophysiology of VA mycorrhizal plants. Ed. G R Safir. pp 27–58. CRC Press, Boca Raton, Florida.

Boyetchko S M and Tewari J P 1990 Root colonization of different hosts by the vesicular-arbuscular mycorrhizal fungus *Glomus dimorphicum* Plant Soil 129, 131–136.

Brundrett M 1991 Mycorrhizas in natural ecosystems *In:* Advances in Ecological Research Vol 21. Eds. M Begon A H Fitter and A Macfadyen. pp 171–313. Academic Press, London.

Brundrett M and Kendrick B 1990 The roots and mycorrhizas of herbaceous woodland plants 1 Quantitative aspects of morphology. New Phytol. 114, 457–468.

Collins Johnson N, Zak D R, Tilman D and Pfleger F L 1991 Dynamics of vesicular-arbuscular mycorrhizae during old field succession. Oecologia 86, 349–358.

Connell J H 1978 Diversity in tropical rainforests and coral reefs. Science 199, 1302–1310.

Daniels-Hetrick B A, Bloom J and Feyerherm S M 1985 Root colonization pattern of *Glomus epigaeum* in 9 host species. Mycologia 77, 825–829.

Douds D D and Schenck N C 1990 Relationship of colonization and sporulation by VA mycorrhizal fungi to plant nutrient and carbohydrate contents. New Phytol. 116, 621–627.

Fitter A H 1989 The role and ecological significance of vesicular-arbuscular mycorrhizas in temperate ecosystems. Agric. Ecos. Environ. 29, 137–151.

Friese C F and Allen M F 1991 The spread of VA mycorrhizal fungal hyphae in the soil—inoculum types and external hyphal architecture. Mycologia 834, 409–418.

Gazey C, Abbott L K and Robson A D 1992 The Rate of development of mycorrhizas affects the onset of sporulation and production of external hyphae by 2 species of *Acaulospora*. Mycol. Res. 96, 643–650.

Gemma J N, Koske R E and Carreiro M 1989 Seasonal

dynamics of selected species of VA mycorrhizal fungi in a sand dune. Mycorrhizal Res. 92, 317–321.

Giovannetti M, Schubert A, Cravero M C and Salutini L 1988 Spore production by the vesicular arbuscular mycorrhizal fungus *Glomus monosporum* as related to host species, root colonization and plant growth enhancement. Biol. Fert. Soils. 6, 120–124.

Graham J H, Eissenstat D M and Drouillard D L 1991 On the relationship between a plants mycorrhizal dependency and rate of vesicular-arbuscular mycorrhizal colonization. Functional Ecol. 56, 773–779

Hepper C M 1985 Influence of age of roots on the pattern of vesicular arbuscular mycorrhizal infection in leek and clover. New Phytol. 101, 685–695.

Hepper C M, Azcon-Aguilar C, Rosendhal S and Sen R 1988 Competition between three species of *Glomus* used as spatially separated introduced and indigenous mycorrhizal inocula for leek *Allium porrum* L. New Phytol. 110, 207–215

Jasper D A, Robson A D and Abbott L K 1987 The effect of surface mining on the infectivity of vesicular arbuscular mycorrhizal fungi. Aust. J. Bot. 35, 641–652.

Jasper D A, Abbott L K and Robson A D 1989a Soil disturbance reduces the infectivity of external hyphae of vesicular-arbuscular mycorrhizal fungi New Phytol. 112 93–99

Jasper D A, Abbott L K and Robson A D 1989b Hyphae of a vesicular-arbuscular mycorrhizal fungus maintain infectivity in dry soil except when the soil is disturbed. New Phytol. 112, 101–107.

Jasper D A, Abbott L K and Robson A D 1993 The survival of infective hyphae of vesicular-arbuscular mycorrhizal fungi in dry soil: an interaction with sporulation. New Phytol. 124, 473–479.

Kough J, Malajczuk N and Linderman R G 1983 Use if the indirect immunofluorescent technique to study the vesicular-arbuscular fungus *Glomus epigaeum* and other Glomus species. New Phytol. 94, 57–62.

Lopez-Aguillon R and Mosse B 1987 Experiments on the competitiveness of three endomycorrhizal fungi. Plant Soil 97, 155–170.

Merryweather J W and Fitter A H 1991 A modified method for elucidating the structure of the fungal partner in a vesicular-arbuscular mycorrhiza. Mycol. Res. 95, 1435–1437.

Morton J B 1985 Underestimation of most probable numbers of vesicular-arbuscular endophytes because of non-staining mycorrhizae. Soil Biol. Biochem. 17, 383–384.

Morton J B 1988 Taxonomy of VA mycorrhizal fungi: classification nomenclature and identification. Mycotaxon. 32, 267–324.

Mosse B 1973 Plant growth responses to vesicular-arbuscular mycorrhiza IV In soil given additional phosphate. New Phytol. 72, 127–136.

Nadarajah P and Nawawi A 1987 Effect of temperature on germination and growth of vesicular arbuscular mycorrhizal fungi *In:* Mycorrhizae in the Next Decade, Practical Applications and Research Priorities, 7th NACOM. Eds. D M Sylvia L L Hung and J H Graham. p 214. Institute of Food Agriculture Science, University of Florida, Gainsville, Florida.

Pearson J N and Jakobsen I 1993 Intrasymbiotic exchange of carbon and phosphorus between cucumber and three arbuscular mycorrhizal fungi. New Phytol. 124, 481–488.

Pearson J N, Abbott L K and Jasper D A 1993 Mediation of competition between two colonizing VA mycorrhizal fungi by the host plant. New Phytol. 123, 93–98.

Powell C L 1979 Inoculation of white clover and ryegrass seed with mycorrhizal fungi. New Phytol. 83, 81–85

Read D J 1991 Mycorrhizas in ecosystems. Experientia 474, 376–390.

Rosendahl S, Sen R, Hepper C M and Azcon-Aguilar C 1989 Quantification of three vesicular-arbuscular mycorrhizal fungi *Glomus* spp. in roots of leek *Allium porrum* on the basis of activity of diagnostic enzymes after polyacrylamide gel electrophoresis. Soil Biol. Biochem. 21, 519–522.

Sainz M J, Vilarino A and Arines J 1990 Competition between *Glomus tenue* and some coarse fungi for colonizing red clover roots in acid soils. Agric. Ecos. Environ. 29, 337–340.

Sanders F E and Sheikh N A 1983 The development of vesicular-arbuscular mycorrhizal infection in plant root systems. Plant Soil. 71, 223–246.

Sanders I R 1990 Seasonal patterns of vesicular-arbuscular mycorrhizal occurrence in grasslands. Symbiosis 9, 315–320.

Sanders I R and Fitter A H 1992 The ecology and functioning of vesicular arbuscular mycorrhizas in co-existing grassland species 1 seasonal patterns of mycorrhizal occurrence and morphology. New Phytol. 1204, 517–524.

Scheltema M A, Abbott L K and Robson A D 1987 Seasonal variation in the infectivity of VA mycorrhizal fungi in annual pastures in a Mediterranean environment Aust. J. Agric. Res. 38, 707–715.

Simon L, Lalonde M and Bruns T D 1992 Specific amplification of 18S fungal ribosomal genes from vesicular-arbuscular endomycorrhizal fungi colonizing roots. Appl. Environ. Microbiol. 581, 291–295.

Smith S E and Walker N A 1985 A mathematical model of vesicular-arbuscular mycorrhizal infection in roots of *Trifolium subterraneum. In:* Ecology and Management of Soil-borne Plant Pathogens. Eds. C A Parker A D Rovira K J Moore and P T W Wong. pp 88–91. Am. Phytopathol. Soc.

Sparling G P and Tinker P B 1978a Mycorrhizal infection in Pennine grassland III Effects of mycorrhizal infection on the growth of white clover. J. Appl. Ecol. 15, 959–964.

Sparling G P and Tinker P B 1978b Mycorrhizal infection in Pennine grassland I Levels of infection in the field. J Appl. Ecol. 15, 943–950.

Sylvia D M and Burks J N 1988 Selection of a vesicular arbuscular mycorrhizal fungus for practical inoculation of *Uniola paniculta*. Mycologia 80, 565–568.

Thornley J H M 1976 Mathematical models in plant physiology. Academic Press, London. 313 p.

Tommerup I C 1983 Temperature relations of spore germination and hyphal growth of vesicular-arbuscular mycorrhizal fungi in soil. Trans. Br. Mycol. Soc. 81, 381–387.

Tommerup I C 1992 Methods for the study of the population biology of vesicular arbuscular mycorrhizal fungi *In:* Methods in Microbiology Vol 24: Techniques for the Study of Mycorrhiza. Eds. J R Norris D J Read and A K Varma. pp 23–52. Academic Press, London.

Wang G M A, Stribley D P, Tinker P B and Walker C 1985

Soil pH and vesicular-arbuscular mycorrhizas. *In:* Ecological Interactions in Soil. Ed. A H Fitter. pp 219–224. Blackwell Sci. Publ., Oxford.

Wilson J M 1984 Competition for infection between three vesicular-arbuscular mycorrhizal fungi. New Phytol. 97, 427–435.

Wilson J M and Tommerup I C 1992 Interactions between fungal symbionts: VA mycorrhizae. *In:* Mycorrhizal Functioning: An Integrative Plant Fungal Process. Ed. M F Allen. pp 199–248. Chapman and Hall, London.

Wilson J M, Trinick M J and Parker C A 1983 The identification of vesicular-arbuscular mycorrhizal fungi using immunofluorescence. Soil Biol. Biochem. 15, 439–447.

Wright S F and Morton J B 1989 Detection of vesicular arbuscular mycorrhizal fungus colonization of roots by using a dot-immunoblot assay. Appl. Environ. Microbiol. 55, 761–763.

Reprinted from *Plant and Soil*, Volume 159, No. 1 (1994), pp. 69–78.

Comparative structure of vesicular-arbuscular mycorrhizas and ectomycorrhizas

R. LARRY PETERSON[1] and PAOLA BONFANTE[2]
[1]*Department of Botany, University of Guelph, Guelph, Ontario, Canada N1G 2W1*
[2]*Dipartimento Di Biologia Vegetale dell' Universita e Centro di Studio sulla Micologia del Terreno, viale Mattioli 25, 10125, Torino, Italy*

Key words: ectomycorrhizas, function, interface, structure, vesicular-arbuscular mycorrhizas

Abstract

During the establishment of vesicular-arbuscular mycorrhizas, fungal hyphae contact the root surface, form appressoria and initiate the internal colonization phase. Structural changes occur in the cell wall, the cytoplasm and the nucleus as the fungus progresses from a presymbiotic to a symbiotic phase. Nuclei in spores are in G_1 whereas in intraradical hyphae they are in G_1 and G_2. Changes in nuclear organization are evident in various stages in the colonization process. Dramatic changes in both symbionts occur as the nutrient exchange interface is established between arbuscules and root cortical cells. An interfacial matrix, consisting of molecules common to the primary wall of the cortical cell, separates the cortical cell plasma membrane from the fungal cell wall.

Ectomycorrhizas are characterized structurally by the presence of a mantle of fungal hyphae enclosing the root and usually an Hartig net of intercellular hyphae characterized by labyrinthine branching. As hyphae contact the root surface, they may respond by increasing their diameter and switching from apical growth to precocious branching. The site of initial contact of hyphae may be either the root cap or the 'mycorrhiza infection zone'. The mantle varies considerably in structure depending on both the plant and fungus genome. In some ectomycorrhizas, the mantle may be a barrier to apoplastic transport, and in most it may store polyphosphate, glycogen, lipids and perhaps protein.

Introduction

Structural characteristics of the various types of mycorrhizas have been documented extensively so that a basis now exists for correlating the anatomy with functional aspects of each mycorrhizal type. The symbionts in these mutualistic symbioses establish, by marked structural changes, an interface for nutrient exchange (Smith and Smith, 1990; Bonfante and Scannerini, 1992). In the resulting integrated system, the mycorrhiza, the host benefits ultimately by showing enhanced growth while the fungus may require the mycorrhizal association to complete its life cycle.

In this article the two major mycorrhizal types, vesicular-arbuscular mycorrhizas (VAMs) and ectomycorrhizas will be considered in terms of structure-function relationships. Recent reviews have summarized the anatomical aspects of VAMs (Bonfante, 1984) and ectomycorrhizas (Kottke and Oberwinkler, 1986).

Morphology to understand functioning in VAMs

VAM fungi cannot be maintained in axenic culture, do not produce sexual structures, and are obligate symbionts which usually limits studies of the morphology of VAMs to the mature colonization phase. The development of new

A.D. Robson, L.K. Abbott & N. Malajczuk (eds.), Management of mycorrhizas in agriculture, horticulture and forestry, 79–88.

techniques, i.e. an in vitro dual culture system with *Agrobacterium rhizogenes* T-DNA-transformed carrot roots (Bécard and Piché, 1989), germination of VAM spores and production of a limited quantity of axenic mycelium, use of host mutants and non-host genotypes of normal myc$^+$ plants (Gianinazzi-Pearson et al., 1991; Bradbury et al., 1991), now allow morphological studies of the presymbiotic phase of VAM fungi. Moreover, the improvement of some techniques (cryo-fixation, freeze-substitution and affinity techniques) has offered important tools to study the symbiotic phase in more detail.

The presymbiotic phase of VAM fungi

Spores of many VAM fungal species may germinate and form limited hyphal growth in nutrient-poor media or even in deionized water but swollen or collapsed hyphae and frequent septation begin to appear after 15–20 days of culture, leading to the arrest of hyphal growth (Hepper, 1984). This saprophytic growth may involve flavonols which enhance independent growth of VAM fungi in vitro (Bécard et al., 1992).

Growing hyphae of in vitro produced mycelium are characterized by a subconical apex, whereas apices from arrested hyphae appear rounded or swollen. Ultrastructurally, they show a thin fibrillar wall, a cytoplasm rich in organelles, abundant glycogen and only a limited number of vacuoles (Figs. 1, 2). In *Gigaspora margarita,* the germinating hyphae lead to the development of larger and yellow hyphae (10–12 µm). From these, a special cell type, the auxiliary cell, is initiated which is rich in nuclei, organelles, membranes and glycogen particles (Figs. 3, 4).

The distribution of nuclei within spores and germinating hyphae has been of particular interest, since a hypothesis to explain the limited growth of the fungus involves the absence of nuclear DNA replication during spore germination (Burggraaf and Beringer, 1989). A static cytofluorimetric analysis has confirmed that nuclei extracted from spores of *G. margarita* and *Glomus versiforme* are rather homogeneous with regard to DNA quantity (0.75 and 0.25 pg of DNA respectively). Peaks with a double quantity of DNA were not found, suggesting that duplication events, if any, are very rare; nuclei in the spore may be blocked at the G_1 phase (Bianciotto and Bonfante, 1992). On the contrary, in the intraradical mycelium a small population of nuclei with a double quantity of DNA exists even if the quantity of DNA per nucleus does not change. This result is particularly evident in the nuclei of *Glomus mosseae*.

A combination of morphological observations and cell biology techniques may lead to a better understanding of the cell cycle of the fungus.

The contact phase

The in vitro system for culturing VAM fungi has demonstrated that the root is required for fungal growth since it produces not only the signal molecules which may trigger the infection process, but provides the right surface on which the fungus can develop (Bonfante and Perotto, 1992). The fungal germ tubes or extraradical mycelium make contact with the root, branching and growing on the root surface. Appressoria, lens shaped structures 20–40 µm long, are formed usually between adjacent epidermal cells as a result of the contact (Garriock et al., 1989). *G. mosseae* requires living root surfaces to produce appressoria in about 42 hours, while they are not formed either when roots are killed or a non-host plant is used (Giovannetti et al., 1991). In addition, in pea mutants which do not form either nodules or mycorrhizas (nod$^-$ and myc$^-$ phenotype), normal appressoria were not observed (Gianinazzi-Pearson et al., 1991) and wall thickenings, similar to the papillae formed in response to pathogens, were present at the point where the VAM fungus attempted penetration. In seedlings of non-nodulating alfalfa which exhibited resistance to VAM colonization, appressoria varying in size and shape, but typically larger than those on nod$^+$fix$^+$ genotypes, developed on the root surface but failed to form internal structures (Bradbury et al., 1991). The formation of appressoria and penetrating hyphae are the first morphological

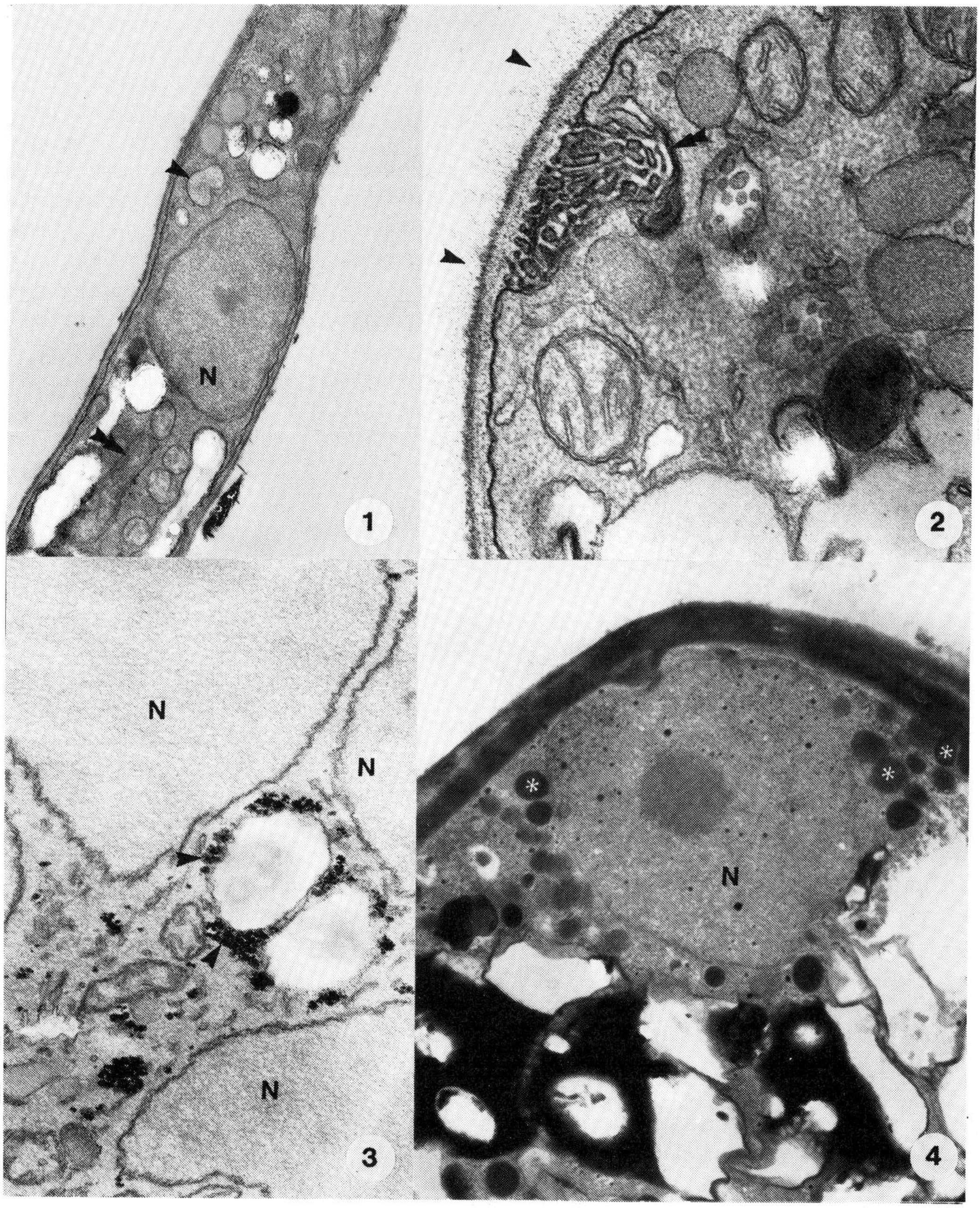

Plate 1. Figs. 1–4. Ultrastructural features of vesicular-arbuscular fungi in the presymbiotic phase.
Fig. 1. Presymbiotic hyphae of *Glomus mosseae*, showing a nucleus (N) with diffuse chromatin, abundant mitochondria (double arrowhead), a small bacterium-like organelle (BLO) (arrowhead) and a thin cell wall. ×11,000
Fig. 2. Enlargement of a presymbiotic hypha, showing details of the thin wall, which is covered by fine radiating fibrils (arrowheads). Vesicles (double arrowheads), rich in electron dense material, are associated with the plasma membrane. ×36,000
Fig. 3. Detail of a young auxiliary cell from a germinating spore of *Gigaspora margarita*. Nuclei (N) with diffuse chromatin and glycogen granules (arrowheads) which surround small lipid globules are evident. ×30,000
Fig. 4. Detail of a mature auxiliary cell. The nucleus (N) occupies a peripheral position and is surrounded by a highly vacuolated cytoplasm. Abundant electron dense granules (probably pigment) (*) are found near the cell wall. ×30,000

signs of the compatibility between plant and fungus.

The penetration events are influenced by root characteristics since the outer layers of the root often possess thickened walls containing abundant cellulose, phenols, and UV autofluorescent components (Bonfante and Perotto, 1992). These components may act as structural barriers to the penetration of the fungus, which may penetrate through junctions between the epidermal cells, directly through tangential walls, or root hairs only before the deposition of such substances. The deposition of suberin on the radial cell walls of the hypodermal layer may channel fungal growth and development inside the root. Characteristics of the root surface together with root anatomical features (Brundrett and Kendrick, 1990) may, therefore, influence the first steps of the plant–fungal interactions. It would be interesting to determine whether the nod^{-}/myc^{-} phenotype found in peas and alfalfa is the result of changes in the regulatory control of cell wall synthesis, leading to an abnormal deposition of wall components which blocks fungal penetration.

The colonization process

Once the fungus overcomes the barrier represented by the outer root layers it mostly develops intercellular hyphae and highly branched structures called arbuscules. Each arbuscule begins as a trunk with few branching hyphae and with subsequent repeated branching completes its development in 4–5 days (Brundrett et al., 1985). Arbuscules rapidly collapse forming clumps which are initially limited to the apical part of the arbuscular branches, but quickly extend to the whole complex.

Ultrastructural observations have revealed important modifications of both partners during the symbiotic phase. Intraradical hyphae are rich in nuclei, mitochondria, secretion granules, glycogen particles and lipid globules. Cell wall structure has been studied in detail since there is a dramatic reduction in thickness (from about 12 µm in the spore wall to 20–30 nm in the arbuscular walls) and changes in chitin organization. Nuclei of extraradical or intercellular hyphae show a chromocentric organization with dispersed chromatin and a well developed nucleolus. On aging and particularly during arbuscule collapse, the chromatin condenses and the nucleus becomes pycnotic. DNA localization by using a monoclonal antibody, which binds to single and double DNA molecules, shows a strong decrease in the number of gold particles during this stage (Balestrini et al., 1992a). Arbuscules are not only key sites for nutrient exchange, but also represent a dead end for VAM fungi. Their ephemeral life, branching pattern and nuclear degeneration pattern suggest that arbuscules are the first step in a programmed death. Intercellular hyphae might represent the surviving structures in roots.

Morphological analysis shows that concomitant with arbuscule development, there is also a dramatic modification of the host cell architecture: invagination of the plant plasmalemma, fragmentation of the vacuole and increase in the number of organelles, such as Golgi bodies. The effects of the fungus over the plant nucleus are of particular interest. The nucleus increases in size without changing its ploidy (Blair et al., 1988; Berta et al., 1990), and shows an unfolding of its chromatin. In addition, there is a positional effect caused by the fungus: the nucleus occupies a median position in the infected cells, differently from the peripheral position shown in the non-infected cells (Balestrini et al., 1992b). This suggests that the plant cytoskeleton may be involved in the response of the plant to fungal penetration and in exchange between the partners.

The interface as a new apoplastic compartment

When the fungus penetrates root cells, either to form coils or arbuscules, the host plasmalemma invaginates and proliferates around the fungus. This creates a new compartment, an interface where interfacial material is laid down. Affinity probes, such as enzymes, lectins and antibodies, have been used to characterize this compartment as a space of high molecular complexity (Bonfante et al., 1992). These methods revealed the presence of structural and enzymatic molecules such as ß(1–4) glucans, non-esterified polygalacturonans and HRGP molecules which are common to the host primary wall. On the contrary, callose or chitin were not found.

The apoplastic space, which is a structural expression of the symbiotic status, prevents direct contact between the plant and fungus, but does not represent a drawback for nutrient exchange between the partners (Smith et al., Chapter 10).

Structural features of ectomycorrhizas

Initial colonization of roots

The earliest stages of root colonization by ectomycorrhizal fungi have been studied infrequently partly because many of the synthesis methods used are not conducive to the recognition of the first contact of hyphae with the root. This problem has been largely overcome by using methods such as growth pouches (Fortin et al., 1980), the paper-sandwich technique (Chilvers et al., 1986) and simple in vitro methods (Malajczuk et al., 1990). Piché and Peterson (1988) timed colonization of *Pinus strobus* by *Pisolithus tinctorius* using the growth pouch system, and found fungal attachment to roots 1–2 days after inoculation, mantle and Hartig net establishment 2–4 days post inoculation, and root dichotomy after 5 days. Alterations in ultrastructure and response to the PATAG reaction for polysaccharides as hyphae contacted the root surface were described earlier (Piché et al., 1983a, b). Similar results have been obtained with *Eucalyptus urophylla* roots colonized by *P. tinctorius* using a simple in vitro method (Lei et al., 1990). Also Lei et al. 1991 have characterized the surface mucilage as containing a-D-glucose and a-D-mannose residues using the lectin concanavalin A.

Dramatic alterations in branching and diameter of hyphae occur in the earliest stages of colonization of *Eucalyptus pilularis* roots by *P. tinctorius* (Jacobs et al., 1989) and *Picea abies* roots colonized by *Hebeloma crustuliniforme* (Brunner and Scheidegger, 1992); it was suggested by Jacobs et al. (1989) that these early morphogenetic changes might be a reliable indicator of a compatible interaction between symbionts.

Recognition phenomena, the initial site of root contact by hyphae, and the subsequent establishment of the mantle and Hartig net are still in question. Horan et al. (1988) suggested that initial colonization of *Eucalyptus globulus* roots occurs at the root cap and that substances emanating from the root apex are responsible for a chemotropic response of the hyphae (Horan and Chilvers, 1990). This is an important observation since the dogma in the ectomycorrhiza literature is that there is a 'mycorrhiza infection zone' located proximal to the root apical meristem. There is evidence that colonization does appear to begin in such a region in some ectomycorrhizas (Melville et al., 1987).

Establishment and structure of the mantle

Mantle morphology, which is the result of the interaction between mycobiont and phycobiont genomes (Massicotte et al., 1987), plays a major role in identifying the mycobiont species of field-collected ectomycorrhizas. Recently, Wong et al. (1990) have shown that the extent of mantle development by closely related genotypes of *Laccaria bicolor* on *Pinus banksiana* roots depends on whether sib-monokaryotic or dikaryotic mycelium is used as the inoculum. This study emphasized that there is considerable intraspecific genetic variability in mantle formation. The control by various abiotic factors on mantle formation has received little attention, but it is reasonable to suggest that it is the interaction between the environment, the fungal genome and the host genome that controls mantle morphology.

The organization of hyphae within the mantle and the nature of the hyphal walls obviously play a major role in determining the mechanism of transport of water and nutrients from the soil solution to root tissue. Ashford et al. (1988, 1989) showed that the mantle of *Pisonia* and *Eucalyptus pilularis* ectomycorrhizas is impermeable to cellufluor, a fluorochrome used to test for apoplastic permeability. The presence of a barrier to apoplastic transport in the mantle would necessitate the uptake of substances from the soil solution into the fungal symplast and set up a shared enclosed apoplastic nutrient exchange compartment (Ashford et al., 1989). Recently, however, Behrmann and Heyser (1991) have shown that the mantle of *Pinus sylvestris–Suillus bovinus* ectomycorrhizas is permeable to the apoplastic tracers, sulphorhodamine G and lanthanum nitrate, suggesting that mantle permeability may be molecule or species specific.

Hyphae within the mantle are capable of synthesizing storage reserves thought to include glycogen and polyphosphate. The presence of the metabolic machinery for nitrogen assimilation, the rapid accumulation of free amino acids in ectomycorrhizas (Chalot et al., 1991), and evidence that ectomycorrhizal fungi have a role in the uptake of nitrogen compounds, suggest that the mantle could serve as a depository for storage protein.

Further research is needed to determine the role and variability of these storage reserves and mantle permeability in the overall functioning of mycorrhizas.

Establishment and structure of the Hartig net

The usual criteria used in defining an ectomycorrhiza are the presence of a mantle and an Hartig net, although Warcup (1980) suggests that only a mantle is required. The majority of ectomycorrhizas do, however, develop an intercellular network of hyphae, the Hartig net, which is confined to the epidermis in most angiosperm species (Fig. 10), but pervades the cortex in most conifer species (Figs. 6, 7). A number of papers have provided detailed accounts of Hartig net structure (see Kottke and Oberwinkler, 1986). Hyphae in the Hartig net show labyrinthic branching which increases the apoplastic and symplastic exchange interface between the symbionts. In non-senescent regions of an ectomycorrhiza the cytoplasm of these highly branched hyphal tips often contains large numbers of mitochondria and rough endoplasmic reticulum cisternae (Massicotte et al., 1990; Kottke and Oberwinkler, 1986), suggesting an enhancement of an energy-requiring step in the uptake or release of nutrients in this interface. Recently, Lei and Dexheimer (1988) have shown the localization of ATPase activity along the Hartig net hyphal plasma membranes and plasma membranes of contiguous living cortical cells and suggest that this provides evidence for the existence of a mechanism for bidirectional transport between the symbionts.

Since there is no direct symplastic continuity between root cells and Hartig net hyphae, materials must pass from the symplast to the apoplast and vice versa, facilitated by permeability of the walls of either or both symbionts in the Hartig net region. There is some evidence that wall changes do occur in this interface (Duddridge, 1986; Paris et al., 1991), but whether these are related to transport phenomena has not been determined.

As the Hartig net is established it is reasonable to suppose that the symplastic continuity between root cells via plasmodesmata may be altered. Nylund (1980) describes regions of plasmodesmata between cortical cells in the Hartig net region of Norway spruce ectomycorrhizas. Fungal hyphae were apparently unable to penetrate between these regions so that they were maintained until cortical cells senesced. The author was uncertain as to whether the Hartig net hyphae eventually separated cell walls in these regions. In a recent study of *Dryas integrifolia–Hebeloma cylindrosporum* ectomycorrhizas, Melville et al. (1988) described primary pitfields with electron-lucent thickenings and plasmodesmata which impeded Hartig net hyphal growth initially (see Fig. 8); these were eventually breached to form a continuous Hartig net. The question still remains as to whether in the functional portion of an ectomycorrhiza, epidermal and cortical cells retain their symplastic continuity. Experiments using fluorochromes of known size could be used to test for symplastic continuity.

Recent results using microautoradiography (Warner and Heyser, 1991) have shown that the Hartig net is a major sink for ^{14}C labelled assimilates but the transport route across the cortex was not shown.

Changes in root morphology and structure

The most obvious morphological changes that

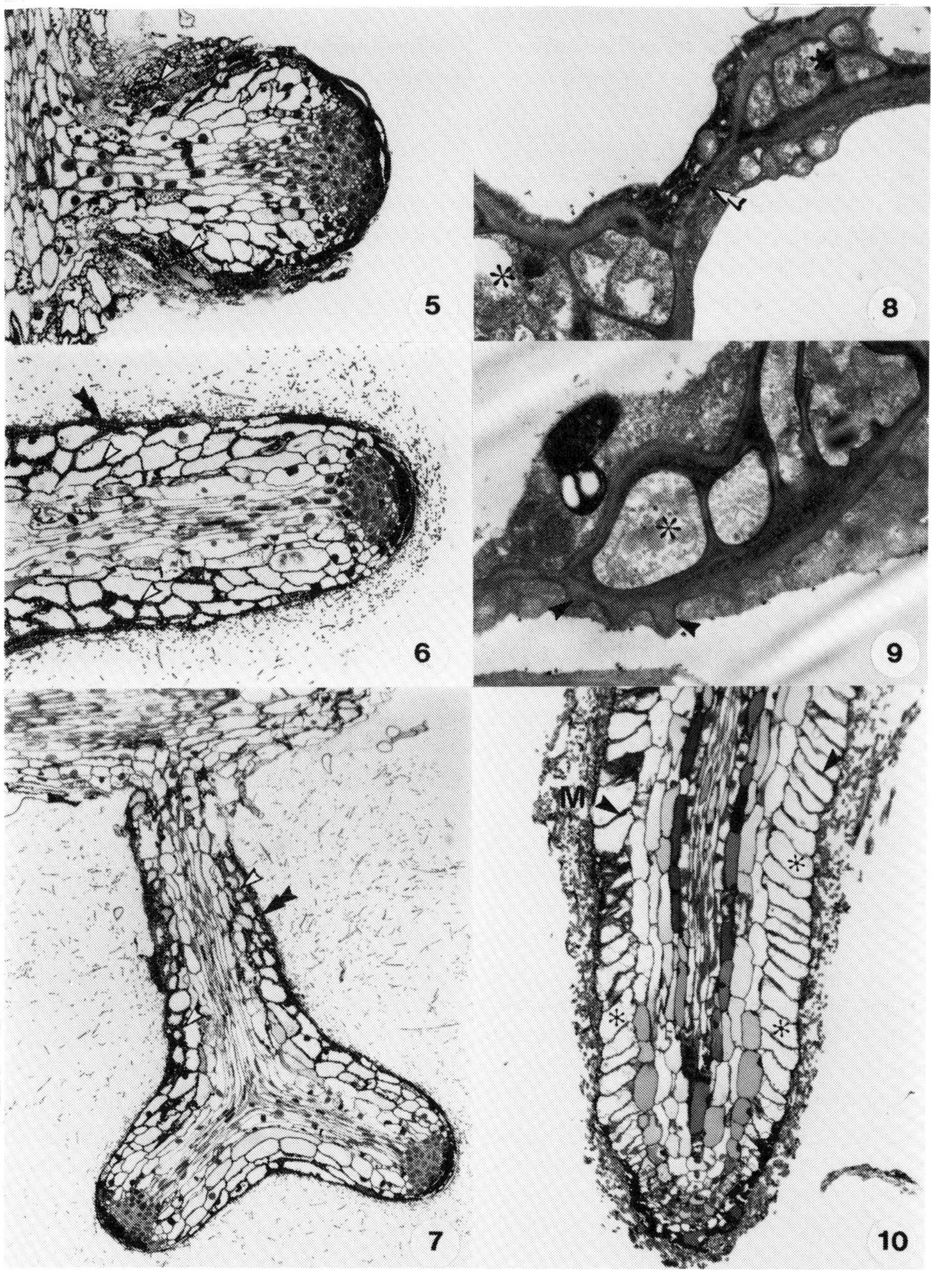

Plate 2. Figs. 5–7. First-order lateral roots of *Pinus resinosa* colonized by *Piloderma bicolor* showing change from monopodial to dichotomous branching. Hartig net hyphae (arrowheads) and a thin mantle (double arrowheads) are evident.
Fig 5. ×340. Fig 6. ×340. Fig 7. ×200. Figs. 8–9. *Dryas integrifolia–Hebeloma cylindrosporum* ectomycorrhizae.
Fig. 8. Hartig net hyphae (asterisks) between cortical cells. A primary pit field with plasmodesmata (double arrowhead) has not been breached by hyphae. ×9,000. Fig. 9. Wall ingrowths (arrowheads) in cortical cell adjacent to Hartig net hyphae (asterisk). ×11,000. Fig. 10. Non-median L.S. *Betula alleghaniensis* root colonized by *Piloderma bicolor*. Marked radial elongation of epidermal cells (*) is apparent. Arrowheads indicate Hartig net hyphae. Mantle–M. ×300.

occur in roots as a result of colonization by ectomycorrhizal fungi are induced dichotomy of the apical meristem in the genus *Pinus* (Figs. 5–7) and the precocious formation of lateral roots in many species (Massicotte et al., 1987). Other structural changes are more subtle. Root epidermal cells of most angiosperm species show a dramatic reorientation of growth subsequent to the initiation of a Hartig net (Fig. 10). These cells elongate radially instead of axially (Clowes, 1951; Massicotte et al., 1990) but apparently, at least in *Fagus,* without any overall increase in cell volume as compared to epidermal cells of control roots (Clowes, 1951). The effect of ectomycorrhizal fungi on reorientation of cortical microtubules and cellulose microfibrils in epidermal cells is currently being explored; methods to study the cytoskeleton of ectomycorrhizal partners have been developed (Timonin et al., 1991). Allaway et al. (1985) calculated from published micrographs that the Hartig net caused the epidermal cell surface in contact with the fungus to be greater than the area of the outer face of the epidermal cells by about 13-fold in *Fagus sylvatica* and about 10-fold in *Eucalyptus fastigata.*

Wall ingrowths may occur in epidermal or cortical cells (Fig. 9) adjacent to Hartig net hyphae (Ashford and Allaway, 1985; Massicotte et al., 1986). These modifications of wall and plasma membrane increase the surface area for nutrient exchange between the symbionts (Allaway et al., 1985). With the development of freeze substitution techniques accompanied by microautoradiography for mycorrhizal systems (Paris et al., 1991), it should be possible to explore how nutrients are exchanged at the cellular level in the interface between symbionts.

Conclusions

Notwithstanding many morphological differences, vesicular-arbuscular mycorrhizas and ectomycorrhizas share some substantive structural features which allow them to become functioning symbioses. In this brief review, the following points have been recognized as common and relevant to function: the extraradical, contact, and the colonization phase. The first, which is essential for the uptake and transport of water and nutrients, consists of morphologically heterogenous hyphae in VAMs and in well-organized hyphae and rhizomorphs in ectomycorrhizas.

The contact phase plays an important role for recognition and subsequent acceptance or rejection between symbionts in both mycorrhizal types. Acceptance is manifested by the formation of appressoria and infection hyphae in VAMs and by the modification of hyphal growth and mantle establishment in ectomycorrhizas.

The colonization phase, during which nutrient exchange occurs, involves the formation of intercellular hyphae, intercellular hyphae and arbuscules in VAMs and Hartig net hyphae in ectomycorrhizas. Arbuscules and Hartig net hyphae provide a substantial increase in the contact surface area between fungus and host. Events leading to the establishment of these interfaces are closely regulated and involve modifications to the morphogenesis of both symbionts. The interfacial zone, particularly in VAMs, has well defined molecular characteristics.

In the complex interaction between fungi and plant roots which forms mycorrhizas, modifications in the morphogenesis and structure of each symbiont may provide clues as to how the system functions physiologically. Understanding the role of the presymbiotic phase and the changes which occur in both symbionts during the earliest stages of hyphal contact with the root surface still requires considerable work. Plant mutants which are being used to explore these early events in VAMs represent an important approach in studying the signals involved and the mechanisms regulating this symbiotic association.

Acknowledgments

Support for some of the research presented was provided by the Natural Sciences and Engineering Research Council of Canada and by the National Council of Italy, Special project RAISA, Subproject No. 2, paper N. 773. We thank Andrew Moore, Lewis Melville, Carol Schlaht and Laurie Winn for help in assembling the manuscript. Hugues Massicotte provided some of the micrographs on ectomycorrhizas.

References

Allaway W G, Carpenter J L and Ashford A E 1985 Amplification of inter-symbiont surface by root epidermal transfer cells in the *Pisonia* mycorrhiza. Protoplasma 128, 227–231.

Ashford A E and Allaway W E 1985 Transfer cells and Hartig net in the root epidermis of the sheathing mycorrhiza of *Pisonia grandis* R.Br. from Seychelles. New Phytol. 100, 595–612.

Ashford A E, Peterson C A, Carpenter J L, Cairney J W G and Allaway W G 1988 Structure and permeability of the fungal sheath in the *Pisonia* mycorrhiza. Protoplasma 147, 149–161.

Ashford A E, Allaway W G, Peterson C A and Cairney J W G 1989 Nutrient transfer and the fungus–root interface. Aust. J. Plant Physiol. 16, 85–97.

Balestrini R, Bianciotto V and Bonfante-Fasolo P 1992a Nuclear architecture and DNA location in two VAM fungi. Mycorrhiza 1, 105–112.

Balestrini R, Berta G and Bonfante P 1992b The plant nucelus in mycorrhizal roots: positional and structural modifications. Biol. Cell. 75, 235–243.

Bécard G and Piché Y 1989 New aspects on the acquisition of biotrophic status by a vesicular arbuscular mycorrhizal fungus *Gigaspora margarita*. New Phytol. 112, 77–83.

Bécard G, Douds D D and Pfeffer P E 1992 Extensive in vitro growth of vesicular-arbuscular mycorrhizal fungi in the presence of $C0_2$ and flavonols. Appl. Environ. Microbiol. 58, 821–825.

Behrmann P and Heyser W 1991 Apoplastic transport through the fungal sheath of *Pinus sylvestris / Suillus bovinus* mycorrhizae. *In* Mycorrhizas in Ecosystems—Structure and Function, Abstracts 3rd ESM, Sheffield.

Berta G, Sgorbati S, Soler V, Fusconi A, Trotta A, Citterio A, Bottone MG, Sparvoli E and Scannerini S 1990 Variations in chromatin structure in host nuclei of a vesicular arbuscular mycorrhiza. New Phytol. 114, 199–205.

Bianciotto V and Bonfante P 1992 Quantification of the nuclear DNA content of two VAM fungi. Mycol. Res. 96, 1071–1076.

Blair D A, Peterson R L and Bowley S R 1988 Nuclear DNA content in root cells of *Lotus* and *Trifolium* colonized by the VAM fungus, *Glomus versiforme*. New Phytol. 109, 167–170.

Bonfante P 1984 Anatomy and morphology. *In* V.A. Mycorrhizas. Eds. C L Powell and D J Bagyaraj. pp 5–33. CRC Press, Boca Raton, USA.

Bonfante P and Perotto S 1992 Plants and endomycorrhizal fungi: the cellular and molecular basis of their interaction. *In* Molecular signals in Plant–Microbe communication. Ed. D P Verma. pp 445–470. CRC Press, Boca Raton, USA.

Bonfante-Fasolo P and Scannerini S 1992 The cellular basis of plant–fungus interchanges in mycorrhizal associations. *In* Mycorrhizal Functioning. Ed. M Allen. pp 65–101. Chapman and Hall, New York.

Bonfante P, Perotto R and Perotto S 1992 Cell surface interactions in endomycorrhizal symbiosis. *In* Perspectives in Plant Cell recognition. Eds. J A Callow and J R Green. Cambridge University Press (*in press*).

Bradbury S M, Peterson R L and Bowley S R 1991 Interactions between three alfalfa nodulation genotypes and two *Glomus* species. New Phytol. 119, 115–120.

Brundrett, M C and Kendrick B 1990 The roots and mycorrhizas of herbaceous woodland plants. II. Structural aspects of morphology. New Phytol. 114, 469–479.

Brundrett M C, Piché Y and Peterson R L 1985 A developmental study of the early stages in vesicular-arbuscular mycorrhiza formation. Can. J. Bot. 63, 194–194.

Brunner I and Scheidegger C 1992 Ontogeny of synthesized *Picea abies* (L.) Karst.–*Hebeloma crustuliniforme* (Bull. ex St Amans) Quél. ectomycorrhizas. New Phytol. 120, 359–369.

Burggraaf A J P and Beringer J E 1989 Absence of nuclear DNA synthesis in vesicular-arbuscular mycorrhizal fungi during in vitro development. New Phytol. 111, 25–33.

Chalot M, Stewart G R, Brun A, Martin F and Botton B 1991 Ammonium assimilation by Spruce–*Hebeloma* sp. ectomycorrhizas. New Phytol. 119, 541–550.

Chilvers G A, Douglass P A and Lapeyrie F F 1986 A paper-sandwich technique for rapid synthesis of ectomycorrhizas. New Phytol. 103, 397–402.

Clowes F A L 1951 The structure of mycorrhizal roots of *Fagus sylvatica*. New Phytol. 50, 1–16.

Duddridge J A 1986 Specificity and recognition in mycorrhizal associations. *In* Physiological and Essential Aspects of Mycorrhizae. Eds. V Gianinazzi-Pearson and S Gianinazzi. pp 45–48. Institut National de la Recherche Agronomique, Paris.

Fortin J A, Piché Y and Lalonde M 1980 Technique for the observation of early morphological changes during ectomycorrhiza formation. Can. J. Bot. 58, 361–365.

Garriock M L, Peterson R L and Ackerley C A 1989 Early stages in colonization of *Allium porrum* (leek) roots by the VAM fungus, *Glomus versiforme*. New Phytol. 112, 85–92.

Gianinazzi-Pearson V, Gianinazzi S, Guillemin J P, Trouvelot A and Duc G 1991 Genetic and cellular analysis of resistance to vesicular-arbuscular (VA) mycorrhizal fungi in pea mutants, *In* Advances in Molecular Genetics of Plant–Microbe Interactions. Eds. H Hennecke and P S Verma. pp 36–342. Kluwer Academic Publishers, The Netherlands.

Giovannetti M, Sbraba C, Avio L and Citernesi S 1991 Appressorium formation in VAM fungi in presence of host and non-host plants. Abstracts, Fungal Cell Biology: Cytology and Ultrastructure, April 1991, Portsmouth, U.K.

Hepper M C 1984 Isolation and culture of VA mycorrhizal (VAM) fungi. *In* VA mycorrhizas. Eds. C L Powell and D J Bagyaraj. pp 95–112. CRC Press, Boca Raton.

Horan D P and Chilvers G A 1990 Chemotropism—the key to ectomycorrhizal formation? New Phytol. 116, 297–301.

Horan D P, Chilvers G A and Lapeyrie F F 1988 Time sequence of the infection process in eucalypt ectomycorrhizas. New Phytol. 109, 451–458.

Jacobs P F, Peterson R L and Massicotte H B 1989 Altered fungal morphogenesis during early stages of ectomycorrhiza formation in *Eucalyptus pilularis*. Scanning Microsc. 3, 249–255.

Kottke I and Oberwinkler F 1986 Mycorrhiza of forest trees—structure and function. Trees 1, 1–24.

Lei J and Dexheimer J 1988 Ultrastructural localization of ATPase activity in the *Pinus sylvestris/Laccaria laccata* ectomycorrhizal association. New Phytol. 108, 329–334.

Lei J, Lapeyrie F, Malajczuk N and Dexheimer J 1990 Infectivity of pine and eucalypt isolates of *Pisolithus tinctorius* (Pers.) Coker and Couch on roots of *Eucalyptus urophylla* S.T. Blake in vitro. II Ultrastructural and biochemical changes at the early stage of mycorrhiza formation. New Phytol. 116, 115–122.

Lei J, Wong K K Y and Piché Y 1991 Extracellular concanavalin A-binding sites during early interactions between *Pinus banksiana* and two closely related genotypes of the ectomycorrhizal basidiomycete *Laccaria bicolor*. Mycol. Res. 952, 357–363.

Malajczuk N, Lapeyrie F and Garbaye J 1990 Infectivity of pine and eucalypt isolates of *Pisolithus tinctorius* on roots of *Eucalyptus urophylla* in vitro. 1. Mycorrhiza formation in model systems. New Phytol. 114, 627–631.

Massicotte H B, Peterson R L, Ackerley C A and Piché Y 1986 Structure and ontogeny of *Alnus crispa–Alpova diplophloeus* ectomycorrhizae. Can. J. Bot. 64, 177–192.

Massicotte H B, Melville L H and Peterson R L 1987 Scanning electron microscopy of ectomycorrhizae potential and limitations. Scanning Microsc. 1, 1439–1454.

Massicotte H B, Peterson R L, Ackerley C A and Melville L H 1990 Structure and ontogeny of *Betula alleghaniensis–Pisolithus tinctorius* ectomycorrhizae. Can. J. Bot. 68, 579–583.

Melville L H, Massicotte H B and Peterson R L 1987 Ontogeny of ectomycorrhiza synthesized between *Dryas integrifolia* and *Hebeloma cylindrosporum*. Bot. Gaz. 148, 332–341.

Melville L H, Massicotte H B, Ackerley C A and Peterson R L 1988 An ultrastructural study of modifications in *Dryas integrifolia* and *Hebeloma cylindrosporum* during ectomycorrhiza formation. Bot. Gaz. 149, 408–418.

Nylund J E 1980 Symplastic continuity during Hartig net formation in Norway spruce ectomycorrhizae. New Phytol. 86, 373–378.

Paris F, Dexheimer J and Lapeyrie F 1991 Cytochemical evidence of a fungal cell wall modification during infection of *Eucalyptus* roots by the ectomycorrhizal fungus *Cenococcum geophilum*. *In* Mycorrhizas in Ecosystems—Structure and Function, Abstracts 3rd ESM, Sheffield, UK.

Piché Y and Peterson R L 1988 Mycorrhiza initiation: an example of plant–microbial interactions. *In* Forest and Crop Biotechnology Progress and Prospects. Ed. F A Valentine. pp 298–313. Springer-Verlag, New York.

Piché Y, Peterson R L and Ackerley C A 1983a Early development of ectomycorrhizal short roots of pine. Scanning Electron Microsc. III, 1467–1474.

Piché Y, Peterson R L, Howarth M J and Fortin J A 1983b A structural study of the interaction between the ectomycorrhizal fungus *Pisolithus tinctorius* and *Pinus strobus* roots. Can. J. Bot. 61, 1185–1193.

Smith S E and Smith F A 1990 Structure and function of the interfaces in biotrophic symbioses as they relate to nutrient transport. New Phytol. 114, 1–38.

Timonin S, Finlay R D, Söderström B and Raudaskoski M 1991 Identification of cytoskeletal proteins in pine ectomycorrhizas. *In* Mycorrhizas in Ecosystems—Structure and Function, Abstracts, 3rd ESM, Sheffield, UK.

Warcup J H 1980 Ectomycorrhizal associations of Australian indigenous plants. New Phytol. 85, 531–535.

Warner J and Heyser W 1991 Solute transfer in *Pinus sylvestris / Suillus bovinus* ectomycorrhizae. I. Microautoradiography of ^{14}C-labelled photoassimilate. *In* Mycorrhizas in Ecosystems—Structure and Function, Abstracts, 3rd ESM, Sheffield, UK.

Wong K K, Piché Y and Fortin J A 1990 Differential development of root colonization among four closely related genotypes of ectomycorrhizal *Laccaria laccata*. Mycol. Res. 94, 876–884.

Reprinted from *Plant and Soil*, Volume 159, No. 1 (1994), pp. 79–88.

Nutrient uptake in mycorrhizal symbiosis

H. MARSCHNER[1] and B. DELL[2]
[1]*Institut fur Pflanzenernahrung, Universitat Hohenheim, 7000 Stuttgart, Germany*
[2]*School of Biological Sciences, Murdoch University, Perth, WA 6150, Australia*

Key words: copper, ectomycorrhizas, hyphal uptake, phosphorus, vesicular-arbuscular mycorrhizas, zinc

Abstract

The role of mycorrhizal fungi in acquisition of mineral nutrients by host plants is examined for three groups of mycorrhizas. These are; the ectomycorrhizas (ECM), the ericoid mycorrhizas (EM), and the vesicular-arbuscular mycorrhizas (VAM). Mycorrhizal infection may affect the mineral nutrition of the host plant directly by enhancing plant growth through nutrient acquisition by the fungus, or indirectly by modifying transpiration rates and the composition of rhizosphere microflora.

A capacity for the external hyphae to take up and deliver nutrients to the plant has been demonstrated for the following nutrients and mycorrhizas; P (VAM, EM, ECM), NH_4^+ (VAM, EM, ECM), $N0_3^-$ (ECM), K (VAM, ECM), Ca (VAM, EM), $S0_4^{2-}$ (VAM), Cu (VAM), Zn (VAM) and Fe (EM). In experimental chambers, the external hyphae of VAM can deliver up to 80% of plant P, 25% of plant N, 10% of plant K, 25% of plant Zn and 60% of plant Cu. Knowledge of the role of mycorrhiza in the uptake of nutrients other than P and N is limited because definitive studies are few, especially for the ECM. Although further quantification is required, it is feasible that the external hyphae may provide a significant delivery system for N, K, Cu and Zn in addition to P in many soils. Proposals that ECM and VAM fungi contribute substantially to the Mg, B and Fe nutrition of the host plant have not been substantiated.

ECM and EM fungi produce ectoenzymes which provide host plants with the potential to access organic N and P forms that are normally unavailable to VAM fungi or to non mycorrhizal roots. The relative contribution of these nutrient sources requires quantification in the field.

Further basic research, including the quantification of nutrient uptake and transport by fungal hyphae in soil and regulation at the fungal–plant interface, is essential to support the selection and utilization of mycorrhizal fungi on a commercial scale.

Introduction

As a rule, mycorrhizal infection enhances plant growth by increasing nutrient uptake via increases in the absorbing surface area, by mobilizing sparingly available nutrient sources, or by excretion of chelating compounds or ectoenzymes. Mycorrhizal infection may also protect roots from soil pathogens (Perrin, 1990) and thereby increase root growth and nutrient acquisition of the host root. A higher nutrient uptake in mycorrhizal plants can therefore not exclusively be attributed to uptake by the fungal hyphae.

For these beneficial effects, mycorrhizal plants have to pay a price—between 10 and 20 percent of the net photosynthates are required for formation, maintenance and function of mycorrhizal structures (Jakobsen and Rosendahl, 1990). Competition between the fungus and the roots for photosynthates is the main factor responsible for the typical larger shoot:root dry weight ratio in favour of the shoots in mycorrhizal plants (Berta et al., 1990). The

A.D. Robson, L.K. Abbott & N. Malajczuk (eds.), Management of mycorrhizas in agriculture, horticulture and forestry, 89–102.

impact of this shift for host plant growth depends on various soil and environmental factors and in particular on the capacity of the fungus to fully compensate for or even surpass the root functions in nutrient (and water) uptake. In this respect, important differences exist between ectomycorrhiza (ECM), ericoid mycorrhiza (EM) and vesicular-arbuscular mycorrhiza (VAM). Some of the main differences between ECM and VAM are indicated in Fig. 1.

In ECM of plants such as Norway spruce, more than 90 percent of the root apical zones are enclosed by a fungal sheath. In some broad leafed species, such as eucalyptus, this proportion may not exceed 40–50 percent. Depending on the tree species, a varied proportion of the root supply of mineral nutrients from the soil may occur via the fungal hyphae. In contrast, in VAM the apical root zones are free of fungal structures and may only be indirectly affected in their uptake capacity, for example by competition from the fungus for carbohydrates or feedback regulation according to the nutritional status of the plants.

As VAM fungi are obligatorily dependent on host plant carbon for their source of energy, the capacity and mechanism of VAM fungi in nutrient uptake can only be evaluated in association with the host plant. This is an important limitation for gaining a better understanding of the role of VAM and in quantification of the contribution of VAM in nutrient uptake. Furthermore, in VAM symbiosis, growth of both partners is influenced by the host plant genotype, by the VAM species and ecotype, and by the soil, but is not necessarily related to soil fertility (Bethlenfalvay et al., 1985) or the percentage of root infection (Vierheilig and Ocampo, 1991). At a given infection level, nutrient uptake efficiency can be

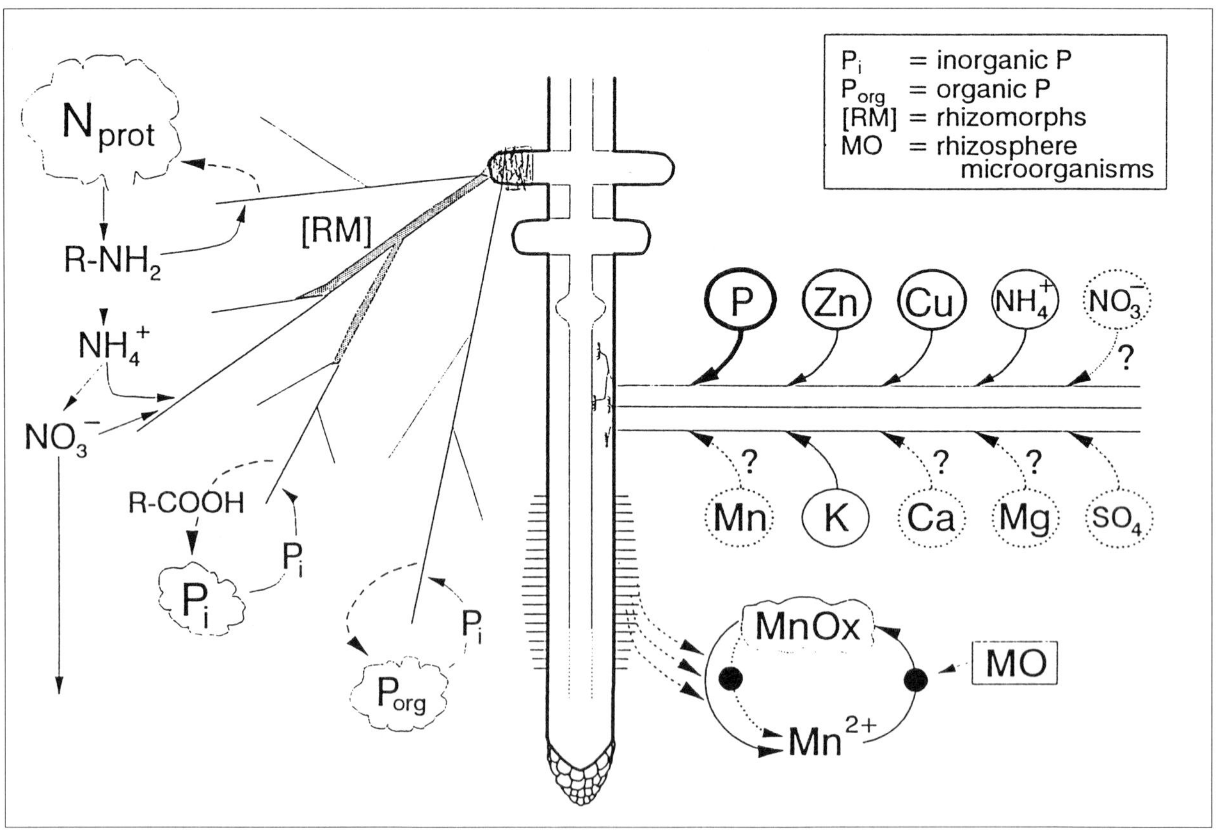

Fig. 1. Schematic presentation of components of the nutrient dynamics in, and acquisition from, the 'hyphosphere' of endo- (VA) mycorrhizal roots and of additional components found in ectomycorrhizal roots.

affected by the parameters of nutrient exchange at the fungus–root cell interface (arbuscules), and the extension, viability and transport capacity of the external hyphae. Whereas a close correlation exists between root colonization and active hyphae within the roots or those attached to the root surface, the external hyphal length differs much between VAM species and is often not well correlated with the infected root length (Abbott and Robson, 1985; Jakobsen et al., 1992). Furthermore, the activity of the external hyphae may vary considerably with distance from the root surface and time after infection (Sylvia, 1988). In the majority of studies on nutrient uptake of non-mycorrhizal and mycorrhizal plants, the infection level is the only root parameter determined and higher nutrient uptake in mycorrhizal plants is attributed to contribution of the external hyphae. Because measurements are not taken of nutrient fluxes through the hyphal delivery system, this 'difference method' is as inadequate as in the legume–*Rhizobium* symbiosis when the contribution of N_2 fixation is calculated from the differences in N uptake of nodulated plants supplied with or without mineral N fertilizer.

The fungal mantle and Hartig net are features of all ECM (Peterson, Chapter 8). There is, however, wide variation in the amount of extramatrical mycelium, mantle thickness and degree of Hartig net formation which ranges from the shallow Hartig net in the superficial sheathing ECM in some eucalypt associations to the extensive deep Hartig net in most pine ECM. Thus the Hartig net provides a large but variable surface of contact between the fungus and the root through which nutrient exchange can occur. Except where the mantle is diffuse, most nutrients delivered to the Hartig net are likely to be transported through the living external hyphae. Some ECM have mycelial cords which transport water and nutrients over long distances and, in some cases, via specialized non-living hyphae (Read, 1984). Given this variation in the morphology of ECM, it is surprising that so few studies have examined the extent and nature of hyphal development in soil, and the degree of Hartig net development in relation to nutrient uptake and transfer capabilities of ECM. Data from severed root experiments in this regard are of little value as the external hyphae connections are broken, hence both the main uptake and transport facilities are missing.

Despite these difficulties and limitations for some mineral nutrients and plant species, the important role of mycorrhizal fungi for nutrient uptake and growth of the host plant is well defined and the degree of growth enhancement effects by mycorrhiza inoculation predictable. Fig. 1 highlights the components of nutrient acquisition by VAM and ECM where experimental data is available. In the following, some examples are given with an emphasis on the VAM, because the nutrient dynamics of the VAM have been investigated with a wide range of nutrients.

Phosphorus

In general, most large growth enhancement effects of root infection with mycorrhizal fungi are caused by increases in P absorption, particularly from sparingly soluble P sources (Bolan et al., 1987). When root exploration is restricted, up to 80 percent of the plant P can be delivered by the external VAM hyphae to the host plant over a distance of more than 10 cm from the root surface (Li et al., 1991a). Comparable data on ECM and EM are lacking although translocation rates have been calculated for three ECM rhizomorphs (Finlay and Read, 1986; Kammerbauer et al., 1989). In general, ECM hyphae may extend for much greater distances into soil than VAM hyphae and mycelial strands have been shown to transport P over several metres (Finlay and Read, 1986). Besides this increase in spatial availability of P, effective P acquisition by the external hyphae is related to (a) formation of polyphosphates in the hyphae and, thus, maintaining low internal phosphate (Pi) concentrations, (b) the small hyphal diameter leading to a relatively larger soil volume delivering P per unit surface area compared to the root surface area (Jungk and Claassen, 1989) and a correspondingly 2–6 times higher P influx rate per unit length of hyphae (Jakobsen et al., 1992), and (c) production of extracellular acid phosphatases which catalyse the release of P from organic complexes in the soil (Fig. 1). Production of acid phosphatases

Table 1. Growth and P uptake and root length—shoot dry weight ratio in non-mycorrhizal and mycorrhizal (*G. fasciculatum*) sorghum plants grown at different P levels (Raju et al., 1990)

Treatment P (mg kg^{-1})	+/- VAM	Shoot dry wt (g plant^{-1})	Shoot P content (g plant^{-1})	Root length (m plan^{-1})	Root length / shoot dry wt ratio
0	-	0.43	0.29	1010	2.34
	+	2.59	4.66	3000	1.16
12.5	-	2.77	5.09	3910	1.41
	+	4.33	10.28	5300	1.22
37.5	-	3.21	7.61	4360	1.36
	+	4.17	12.39	3730	0.89

and phytase as ectoenzymes in external hyphae is well established in ECM fungi (Ho, 1989; Hilger and Krause, 1989) and EM fungi (Straker and Mitchell, 1986). In contrast, in VAM fungi evidence is lacking for extracellular phytase activity, and evidence is still poor for extracellular acid phosphatase. Slight increases in acid phosphatase activity in the rhizosphere of VA mycorrhizal roots may derive from the mycorrhizal root surface rather than the external hyphae (Dodd et al., 1987). However, using vessels with separate hyphal compartments, a distinct increase in extracellular acid phosphatase activity in the hyphal compartment has recently been demonstrated (Tarafdar et al., pers. comm.). Thus, the external hyphae of VAM may also contribute to the better utilization of certain organic P sources by mycorrhizal plants.

There had been many efforts to study whether VAM plants obtain P from sources not available to non-mycorrhizal plants (Bolan et al., 1983). However, as has been recently summarized by Bolan (1991), convincing experimental evidence is still lacking for such an assumption. The situation is different in ECM where, for example, some fungal species release large quantities of oxalic acid (Cromack et al., 1979; Lapeyrie et al., 1990) which, at least in calcareous soils, may mobilize P from sparingly soluble Ca phosphates and simultaneously prevent excessive Ca uptake by formation of Ca oxalate at the fungal–soil interface (Lapeyrie, 1990; Lapeyrie et al., 1990). It has not yet been shown whether the release of oxalic acid occurs in acid soils where it could facilitate P acquisition by hyphae from sparingly soluble Fe and Al phosphates.

With respect to P nutrition, the growth response to VAM infection depends on soil and plant factors which determine the P acquisition of the host plant (e.g. potential growth rate), extent of P deficiency-induced root responses (acid phosphatase activity, organic acid excretion) and the P status of the soil. An example for the P status of the soil is given in Table 1.

With increasing soil P, the growth enhancement effect of VAM declines and may either be abolished or lead to growth depressions (shift from mutualism to parasitism) if, for example, the carbon costs of the symbiosis are not compensated by other beneficial effects. The shift in root length : shoot dry weight ratio (Table 1) is a typical response to improved P nutrition in both non-mycorrhizal and mycorrhizal plants. Typically, in mycorrhizal plants, the P concentrations per unit dry weight

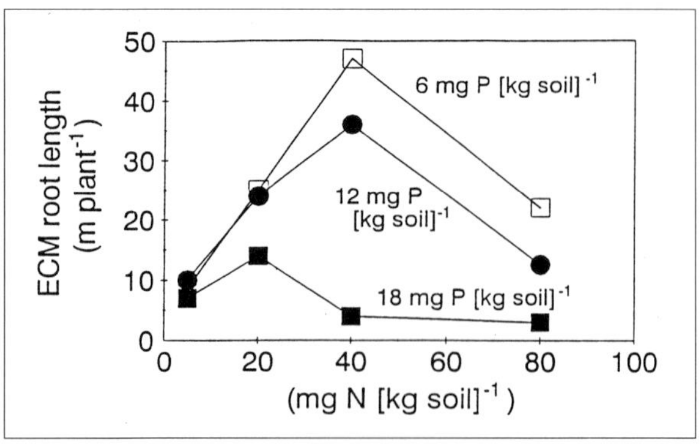

Fig. 2. Effect of application of P and N as $(NH_4)_2SO_4$ on mycorrhizal root length in *Eucalyptus diversicolor* inoculated with *Pisolithus tinctorius* (Grove et al., unpublished results).

Table 2. Growth and P uptake and ectomycorrhizal root length in *Eucalyptus diversicolor* seedlings inoculated with *Laccaria laccata*. (Bougher et al., 1990)

Treatment P (mg kg^{-1})	+/- ECM	Plant dry wt (g $plant^{-1}$)	Plant P content (mg $plant^{-1}$)	P uptake (mg P g^{-1} fine root)	ECM root length (m $plant^{-1}$)
0	-	0.09	0.02	0.38	-
	+	0.16	0.07	0.74	0.22
4	-	0.09	0.03	0.45	-
	+	1.90	1.62	1.49	4.33
8	-	0.32	1.73	0.58	-
	+	2.22	2.41	2.17	3.07
12	-	1.04	0.95	1.19	-
	+	2.49	2.68	2.19	4.29
16	-	2.46	2.03	1.42	-
	+	3.46	4.26	2.14	4.94
20	-	7.15	5.14	1.93	-
	+	4.86	5.14	2.05	7.07
36	-	8.58	10.56	3.75	-
	+	8.69	11.57	3.59	0.81

are higher and, thus, the P utilization efficiency lower than in non-mycorrhizal plants (Koide, 1991). This could indicate that other growth factors (carbohydrates, mineral nutrients) become limiting, or that in mycorrhizal root systems the feed-back regulation between nutrient uptake rates and shoot demand are less well-regulated than in non-mycorrhizal root systems (Douds et al., 1988). Similar effects of soil P on the growth enhancement of ECM have also been observed (Table 2). Given the wide variation in ECM and hyphal morphological traits between species, studies on the type and amount of external hyphae in relation to P acquisition are warranted to extend the observation of Jones et al. (1990).

Depression of VAM and ECM infection levels by high soil P is often observed. The degree of depression not only depends on the fungal species but also on the level of N supply (Fig. 2), NH_4-N in particular (Baath and Spokes, 1989), probably related to both lower carbohydrate levels in the roots and lower total root length per plant. Commonly, mycorrhizas give little benefit to plants which have other adaptations for getting P from soil. These include thin roots, long root hairs and cluster roots. Thus, there is a tendency, with respect to P uptake, for mycorrhizal dependency of plants to be negatively related to root parameters such as root length and surface area. This holds true both for genotypes within a species as well as between species including wild and cultivated ones (Bryla and Koide, 1990; Brundrett and Abbott, 1991). Speculations upon co-evolution of certain host plant–VAM combinations (Grey, 1991) are attractive but require experimental support.

Nitrogen

In VAM research on N nutrition, the most attention has been paid to legumes. When VAM improve the P nutrition of the host plant there may be a corresponding increase in nodulation, N_2 fixation and growth (Robson et al., 1981). In view of the high P requirement for nodulation, many legume species growing on low P soils are highly dependent on VAM infection. However, the symbioses impose a strong competition for photosynthates, usually at expense of root growth (Piccini et al., 1988). Accordingly, the beneficial effects on N_2 fixation are either confined to, or at least most distinct, in low P soils (Table 3).

The existence of extra-radical hyphal bridges between individual plants permits transfer of nutrients such as N. Although this is also true for legumes in mixed stands, such transfer, for

Table 3. Plant dry weight and P content, number and specific mass of nodules and nitrogenase activity (ARA) in nodules of soybeans grown at low and high P (Brown et al., 1988)

	Low P	High P	Low P + VAM
Shoot dry wt (g)	2.8	3.8	5.6
Root dry wt (g)	1.7	1.9	2.0
P content (mg $plant^{-1}$)	2.9	6.0	5.8
Nodules (N_0 $plant^{-1}$)	33	30	97
ARA (μmol C_2H_4 $plant^{-1}$ h^{-1})	4.6	22.8	9.0

example from soybean to maize, takes place in substantial quantities only when the legume is supplied with mineral N (NH_4NO_3) but not when relying on N_2 fixation (Bethlenfalvay et al., 1991). In view of the high carbon costs for N_2 fixation, it is not surprising that legumes have mechanisms to prevent the drain of fixed N, via VAM hyphae, to the non-legumes.

For non-legumes, reports of a direct role of VAM in N uptake from the soil are scarce. This is so even though natural and agricultural ecosystems are often limited by N not P. The most relevant work is that of Ames et al. (1983) who used compartmented vessels and ^{15}N to measure directly hyphal uptake and transport of ^{15}N to the host plant (celery). After 30 days, 25% and 3.5% of the total N in the mycorrhizal and non-mycorrhizal plants respectively, was derived from the ^{15}N in the hyphal compartment. The calculated flux rate of N in the hyphae was 7.4×10^{-8} moles N $cm^{-2}s^{-1}$ which is in a similar order as those reported for P (Sanders and Tinker, 1973). Interestingly, neither total N nor shoot dry weight differed between mycorrhizal and non-mycorrhizal plants, indicating a correspondingly lower N uptake from the root compartment of the mycorrhizal plants which also had a lower root dry weight.

With a similar experimental approach with coach grass (*Agropyron repens*) growing in pots with or without hyphal compartments, we also showed a considerable capacity of the external VAM hyphae for uptake and transport not only of P but also N supplied as NH_4NO_3 to the hyphal compartment (Table 4). About 53% of the total P uptake and 24% of the total N uptake in mycorrhizal plants could be attributed to uptake and delivery by the external hyphae. Changes in the N concentrations in the hyphal compartment indicated preferential uptake of NH_4-N. In agreement with this, in mycorrhizal, non-nodulated white clover supplied with NH_4-N, in the hyphal compartment the soil pH at the hyphae–soil interface ('hyphosphere') decreased similar to that at the root–soil interface of non-mycorrhizal plants (Li et al., 1991b). Thus, in mycorrhizal plants preferential NH_4-N (cation) uptake and corresponding substrate acidification is not confined to the root–soil interface but also takes place at the hyphae–soil interface and may contribute to the P acquisition of the external hyphae, especially in calcareous soils. Enhancement effects of VAM on the N content in nodulated legumes are not necessarily derived only from better N_2 fixation but in part stem from higher uptake of soil N (Azeon and Barea, 1992). Up to now, information is lacking on the capacity of external hyphae for uptake and transport of NO_3-N or NO_3-reduction. In view of the high mobility of NO_3^- but generally low mobility of NH_4^+ in soils, an important contribution of the external VAM hyphae to the N nutrition of the host plant can be expected from NH_4-N rather than NO_3-N uptake. Also hyphal transport of N between individual plants, for example of *Plantago lanceolata*, can be substantial (Newman et al., 1992) and may play a role in competition between plants as well as N cycling between species sharing the same type of VAM fungi. Hyphal transport of N may also be involved in the 'nursery effect' of ECM Douglas fir overstory trees on the establishment of Douglas fir seedlings in the understorey (Griffith et al., 1991).

In contrast to VAM, many studies have shown that ECM and EM fungi can utilize organic N sources through the production of extracellular acid proteinases (Leake and Read, 1990; Zhu et al., 1990; Maijala et al., 1991; Fig. 1). Thus, mycorrhizal infection by these fungi can provide

Table 4. Nutrient contents of mycorrhizal (*G. mosseae*) and non-mycorrhizal *Agropyron repens* plants grown in pots without and with hyphal compartments (= soil only accessible to the VAM hyphae, not the roots) (George et al., 1992)

Treatment	+/- VAM	Nutrient content (mg pot^{-1})		
		P	N	K
Without hyphal compartment	-	4.9	41.6	41.7
	+	5.0	40.0	40.2
With hyphal compartment	-	4.6	47.3	41.9
	+	9.8	61.9	45.9
Nutrient supply in root compartment (mg pot^{-1})		7.6	48.0	50.0

host plants with access to N sources which are normally unavailable to non-mycorrhizal roots (Bajwa and Read, 1985; Abuzinadah and Read, 1989; Finlay et al., 1992). This is likely to be of particular significance in soils where most of the N is in organic forms such as in many eucalypt and conifer forests and peat bogs. In agreement with this, in general, VAM dominate in P limited forests and ECM and EM dominate in N limited forests (Vogt et al., 1991). However, there is scant data on either the relative importance of the various organic N sources (amino acids, peptides, proteins) to mycorrhizal plants under these conditions or on the contribution of organic N to the total N intake of the plant. Further, the proteolytic capacity may vary greatly between fungal isolates (Finlay et al., 1992). This has yet to be exploited in the selection of commercial inocula for any host species.

Perhaps all ECM fungi in pure culture can assimilate ammonia (Jongbloed et al., 1991) by a combination of the glutamate dehydrogenase and glutamine synthetase pathways (Martin et al., 1988; Ahmad et al., 1990). In contrast, only a smaller number of species can efficiently reduce nitrate (Scheromm et al., 1990). The nitrate reducing fungi may prove to be especially suited for the afforestation of agricultural soils in which nitrate dominates during the early years of tree establishment. This could be particularly important for tree species which are better able to metabolize ammonium than nitrate. However, genetic variation of the fungal nitrate reductase activity has not been correlated with the effect of the fungus on nitrate assimilation by the host tree. Extensive areas of non-leguminous trees, such as eucalypts in China, are being planted on highly eroded acid soils and it is here that hyphal acquisition of ammonia is likely to be of benefit to the host plant. Mycorrhizas can enhance the inflow of N into root tips (Hogberg, 1989) although this has not been universally observed for ECM because of restraints in experimental design. Ammonia is rapidly assimilated in the extramatrical hyphae (Finlay et al., 1988, 1989), and N is transferred to the host primarily as glutamine. The extent to which other nitrogenous compounds are transferred across the mantle is unclear and may depend on the relative activities of glutamate dehydrogenase/glutamine synthetase (Chalot et al., 1991) and the extent of interaction between the symbionts on gene expression in the extramatrical hyphae. ECM symbiosis alters metabolic pathways of N assimilation (Martin et al., 1986; Dell et al., 1989; Chalot et al., 1990) in the fungal symbiont, hence, assimilation pathways differ between mycorrhizal roots of spruce and beech. It is not yet clear though whether the changes in enzyme expression extend for long distances from the ECM to the hyphal front where N uptake is expected to occur.

Other macronutrients

Not very much is known on the role of mycorrhiza in uptake of K, Ca, Mg and S. Although for VAM there are many results on effects of infection on concentrations and amounts of K in shoots, these results are inconsistent and difficult to interpret (Sieverding and Toro, 1988). Most of these experiments were carried out in P deficient soils to demonstrate the role of VAM in P acquisition.

There were therefore correspondingly large differences between mycorrhizal and non-mycorrhizal plants in growth and shoot:root dry weight ratio. Plants of equivalent growth are necessary to examine effects of infection on nutrient uptake.

The capacity of external hyphae for uptake and transport of K can also be demonstrated in compartmented pots (Table 4). About 10% of the total K uptake in mycorrhizal coach grass is attributable to hyphal uptake and transport. This is a small contribution but does not necessarily reflect the potential transport capacity for K of the external hyphae. Remarkable differences in growth response of soybean to VAM inoculation with different geographic isolates of *Glomus mosseae* seemed to be more related to improved K rather than P nutrition of the host plant (Bethlenfalvay et al., 1989). In conifers, higher uptake rates of K (^{86}Rb) in ECM compared to non-mycorrhizal seedlings were attributed to lower efflux and higher influx rates as well as higher vacuolar pool size (Rygiewicz and Bledsoe, 1984).

By supplying radioisotopes (^{45}Ca, $^{35}SO_4$) a certain capacity of the external hyphae of VAM for uptake and transport to the host root has been shown for Ca (Rhodes and Gerdemann, 1978) and SO_4-S (Cooper and Tinker, 1978). Compared to P, the uptake and transport rates particularly of Ca were very low. It has been suggested that polyphosphate granules could serve as the means for hyphal transport of Ca (White and Brown, 1979). However, due to the necessity to maintain very low cytosolic free Ca^{2+} concentrations, delivery of Ca-loaded polyphosphates to the arbuscules would be harmful to their functioning. In view of the high mobility of Ca^{2+} and SO_4^{2-} in soil, a substantial contribution of hyphal delivery to the host plant is not likely under most circumstances. The situation might be different for ericaceous plants growing on acid peat soils which are typically very low in Ca. As shown by Leake and Read (1989), EM can considerably increase acquisition and concentrations of Ca in the host plants. For Mg, direct experimental evidence for uptake and transport in mycorrhizal hyphae is either lacking or, in case of VAM, inconclusive (Kothari et al., 1990a).

Micronutrients

There are numerous reports on the enhancement of Zn and Cu uptake by VAM infection. At least in part, this enhancement can be attributed to uptake and transport in external hyphae to the host plant. For example, in maize grown in calcareous soil, the hyphal contribution (*Glomus mosseae*) to the total uptake ranged between 16 and 25% for Zn and 13 to 20% for P (Kothari et al., 1991a). In the same soil, in white clover with restricted rooting space, the delivery of Cu from the hyphal compartment ranged from 52% to 62% of the total Cu uptake (Li et al., 1991c).

Increasing P supply to the hyphal compartment enhanced hyphal delivery of P and slightly depressed that of Cu with corresponding increase in the P:Cu molar ratio from 37 to 912, a factor of about 25 (Table 5). Thus, hyphal uptake and transport of P and Cu appear to be rather independent. In contrast, partitioning of Cu between roots and shoots was strongly affected by P. Phosphorus enhanced not only the content but also the concentration of Cu in the shoot dry matter indicating that the enhancement effect of P on Cu translocation was not exclusively regulated by the shoot demand. The results also demonstrate that, particularly with Cu, the role of VAM in uptake cannot properly

Table 5. Hyphal delivery and amount of P and Cu in roots and shoots of mycorrhizal white clover (*T. repens*) plants, supplied with various P levels in the hyphal compartment (Li et al., 1991c)

P supply (mg kg^{-1} soil)		Hyphal delivery (μmol pot^{-1})		Amount of P and Cu (μmol plant^{-1})			
Roots	Hyphae	P	Cu	Root P	Root Cu	Shoot P	Shoot Cu
50	0	27	0.73	29	97	52	0.23
50	20	226	0.57	102	61	184	0.43
50	50	465	0.51	152	45	382	0.52

Table 6. Effect of increasing P fertilizer levels on shoot growth, P and micronutrient concentrations in non-mycorrhizal soybean (Lambert and Weidensaul, 1991)

P supply (mg kg^{-1} soil)	Shoot dry wt (g)		Concentrations per g shoot dry matter							
			P (mg)		Cu (μg)		Zn (μg)		Mn (μg)	
	NM	M	NM	M	NM	M	NM	M	NM	M
0	1.25	2.80	0.7	4.3	3.3	10.3	21	44	366	111
60	1.61	3.21	1.1	6.5	3.7	7.9	27	35	415	109
150	1.85	3.42	1.5	6.6	2.9	6.3	30	36	412	115
270	2.78	3.83	3.7	6.8	3.5	4.6	29	33	556	123

be evaluated from only the shoot content or concentration.

In agreement with the importance of the external hyphae for uptake and transport of Zn and Cu, in mycorrhizal plants the shoot concentrations of Zn and Cu are usually higher than in non-mycorrhizal plants (Table 6). Decrease in mycorrhizal infection with increasing P supply is usually compensated by higher root uptake of P. This is not necessarily so for Zn and Cu in soils low in Zn and Cu. Accordingly, depressing effects of P fertilization on concentrations and amounts of Cu and Zn in plants often indicate the importance of VAM in acquisition of these nutrients from the soil under the particular conditions.

In contrast, Mn uptake and concentrations in plants are sometimes not affected but more often are lower in VAM plants (Table 6). This mycorrhizal effect can contribute to higher Mn tolerance in plants, for example in soybean grown in soils high in Mn (Bethlenfalvay and Franson, 1989). The decrease in Mn concentrations is most likely an indirect effect caused by VAM-induced changes in rhizosphere microorganisms in general and decrease in population of Mn reducers in particular (Kothari et al., 1991b). Changes in number and composition of the rhizosphere microflora due to VAM infection are well documented (Secilia and Bagyaraj, 1987; Paulitz and Linderman, 1989) and are another example of the various direct and indirect effects VAM infection may have in the rhizosphere ('mycorrhizosphere'; Linderman, 1988) and on uptake of nutrients.

The majority of studies on ECM and micronutrient uptake are focussed on the protection from excessive uptake of Cu and Zn on soils high in heavy metals. As a rule, heavy metal tolerance of the host plant is increased by preferential binding of these metals in fungal structures (Denny and Wilkins, 1987; Wilkins, 1991). In contrast, only a few reports are available on the role of ECM and EM on micronutrient uptake under conditions when supply from the soil is low or growth limiting. Reduction of sparingly soluble Mn oxides and substrate acidification is well documented in pure cultures of *Pisolithus tinctorius* and *Paxillus involutus* (Cairney and Ashford, 1991) and has also been observed in ECM root zones (Cairney and Ashford, 1989), but effects on the Mn nutrition on the host plant are not known.

Production of siderophores is widespread among ECM (Powell et al., 1982) and EM (Schuler and Haselwandter, 1988) fungi; for VAM fungi the evidence is less clear (Cress et al., 1986). Conclusions upon the importance of siderophore production of ECM fungi for the Fe nutrition of the host plant and its adaptation to calcareous soils (Lapeyrie, 1990) are merely based on observations rather than measurements. In EM, however, experimental evidence has been presented for higher affinity of EM roots for Fe and higher uptake rates at low external concentrations (Shaw et al., 1990).

Boron (B) is not essential for the growth of fungi and experimental evidence of a role of mycorrhiza in B nutrition of the host plant is either lacking or the data are inconclusive. ECM may increase (Mitchell et al., 1990) and VAM may decrease (Table 7) B concentrations in the host plants. In view of the widespread occurrence of B deficiency in exotic plantations

Table 7. Nutrient concentrations in non-mycorrhizal and mycorrhizal (*G. mosseae*) maize (*Zea mays*) grown in calcareous soil with root and hyphal compartments (Data from Kothari et al., 1990a, 1991b)

+/-VAM	Concentrations in shoot dry matter									Mn reducer (no. g^{-1} soil) ¥ 10^5
	mg g^{-1}				$\mu g\ g^{-1}$					
	K	P	Mg	Ca	Zn	Cu	Mn	Fe	B	
-VAM	17	2.1	4.0	9.0	10	5.6	139	88	46	44.1
+VAM	12	3.7	4.1	5.3	36	7.1	95	58	35	1.7

of eucalypts and pines (Stone, 1990) and the higher responsiveness of ECM trees compared to non-ECM trees to B fertilization (Mitchell et al., 1990), a substantial contribution of mycorrhiza to the B nutrition of the host plants is questionable.

Direct and indirect effects on nutrient acquisition

Mycorrhizas may affect mineral nutrition of the host plant either directly or indirectly. Causal interpretations of the mechanisms involved require distinction between direct and indirect effects as summarized in an example for VAM in maize (Table 8). Shoot and root dry weight were similar in non-mycorrhizal and mycorrhizal plants but, in mycorrhizal plants, root length and number and length of root hairs (i.e. root surface area) were much lower. As mycorrhizal plants had higher transpiration rates, the water uptake per unit root length, and thus the mass flow of soil solution to the root surface, was about twice as high as in non-mycorrhizal plants. The lower K concentrations in mycorrhizal plants are in accordance with the reduction in root surface area and indicate a relatively low hyphal capacity for K delivery. As is to be expected, in mycorrhizal plants the P concentrations are much higher. Magnesium concentrations are not affected. This may indicate either substantial contribution of the external hyphae or higher Mg supply via mass flow to the mycorrhizal roots. In view of the higher mass flow of soil solution to the roots, the distinctly lower Ca concentrations in mycorrhizal plants were unexpected, but in accord with the much coarser root system and lower proportions of apical root zones—the main sites of Ca uptake. In mycorrhizal plants, the concentrations of Cu and particularly Zn were higher and of Mn lower, and so was the number of Mn reducers in the rhizosphere soil. Also the shoot concentrations of Fe and B were lower in the mycorrhizal plants. In principle, the impact of mycorrhizal infection on concentrations of macro- and micro-nutrients were similar in roots and shoots (Kothari et al., 1990a) with the exception of much higher concentrations of Cu, Ca and Si in the roots. The higher Cu concentrations in roots are in agreement with Table 5, and the higher concentrations of Ca and Si in agreement with the higher mass flow of soil solution to mycorrhizal roots.

In VAM plants, the soil structure may be altered in the rhizosphere (Kothari et al., 1991b). The higher uptake rate of water per unit root

Table 8. Growth and water relations in non-mycorrhizal and mycorrhizal (*G. mosseae*) maize (*Zea mays*) grown in calcareous soil with root and hyphal compartments (Data from Kothari et al., 1990b)

+/-VAM	Dry weight (g $plant^{-1}$)		Root length (m $plant^{-1}$)	Root hairs		Transpiration (l $plant^{-1}$)	Water uptake (ml cm $root^{-1}$ s^{-1}) ¥ 10^7
	Shoot	Root		no. mm^{-1}	Length (μm)		
-VAM	20.0	4.8	619	35	347	3.40	0.61
+VAM	22.8	4.6	367	25	235	4.08	1.34

length in mycorrhizal plants may deplete soil water more rapidly in the rhizosphere. This will impair nutrient supply via diffusion, unless hydraulic conductivity of the soil is increased by external hyphae (Davies et al., 1992) or water transport within the hyphae is important. Arguments and evidence against hyphal water transport are published elsewhere (George et al., 1992).

Outlook

The role of mycorrhiza, in general, and VAM in particular, in P acquisition of plants has been well documented for more than two decades. Despite worldwide research interest, however, progress is slow in understanding the mechanisms involved, the differences between mycorrhizal fungi in their capacity to deliver P to the host plant, and in quantification of the benefit mycorrhizal plants have under field and in natural ecosystems. In future, more attention should be paid to the role of mycorrhiza, especially ECM, in the acquisition of mineral nutrients other than P, and on other mycorrhizal effects such as changes in rhizosphere microflora and in plant water balance. In some instances (especially tree nurseries) mycorrhizas are already utilized, and there is considerable scope for increasing the use especially of ECM fungi in areas where exotic families of trees are being planted because of the generally low soil fertility and low populations of compatible fungal symbionts. However, better matching of ECM fungi with host and site conditions is required for the full potential of large scale inoculation programmes to be realised. Utilization of mycorrhizal fungi on a large scale strongly depends on progress in basic research in (a) development of new methods for identification of mycorrhizal fungi (biochemical and molecular markers), (b) quantification of the external mycelium and its capacity to supply not only P but also other nutrients and perhaps also water to the host root through the application of techniques allowing the separation of root and hyphal compartments, (c) characterizing the mechanisms regulating the exchange of nutrients at the fungus–plant–root interface, and (d) studies on carbon demand of mycorrhizal structures and on side effects of mycorrhizal infections on morphology and physiology of the host root and its rhizoplane and rhizosphere microflora.

References

Abbott L K and Robson A D 1985 Formation of external hyphae in soil by four species of vesicular-arbuscular mycorrhizal fungi. New Phytol. 99, 245–255.

Abuzinadah R A and Read D J 1989 The role of proteins in the nitrogen nutrition of ectomycorrhizal plants IV. The utilization of peptides by birch (*Betula pendula* L) infected with different mycorrhizal fungi. New Phytol. 112, 55–60.

Ahmad I, Carleton T J, Malloch D W and Hellebust J.A. 1990 Nitrogen metabolism in the ectomycorrhizal fungus *Laccaria bicolor* (R. Mre.) Orton. New Phytol. 116, 431–441.

Ames R N, Reid C P P, Porter L K and Cambardella C 1983 Hyphal uptake and transport of nitrogen from two ^{15}N labelled sources by *Glomus mosseae*, a vesicular-arbuscular mycorrhizal fungus. New Phytol. 95, 381–396.

Azeon R and Barea J M 1992 Nodulation, N_2 fixation (^{15}N) and N nutrition relationships in mycorrhizal or phosphate-amended alfalfa plants. Symbiosis 12, 33–41.

Baath E and Spokes J 1989 The effect of added nitrogen and phosphorus on mycorrhizal growth response and infection in *Allium schoenoprasum*. Can. J. Bot. 67, 3227–3232.

Bajwa R and Read D J 1985 The biology of mycorrhiza in the Ericaceae. IX. Peptides as nitrogen sources for the erecoid endophyte and for mycorrhizal and non-mycorrhizal plants. New Phytol. 101, 459–467.

Berta G, Fusconi A, Trotta A and Scannerini S 1990 Morphogenetic modifications induced by the mycorrhizal fungus *Glomus* strain E3 in the root system of *Allium porrum* L. New Phytol. 114, 207–215.

Bethlenfalvay G J and Franson R L 1989 Manganese toxicity alleviated by mycorrhizae in soybean. J. Plant Nutr. 12, 953–970.

Bethlenfalvay G J, Ulrich J M and Brown M S 1985 Plant response to mycorrhizal fungi: host, endophyte, and soil effects. Soil Sci. Soc. Am. J. 49, 1164–1168.

Bethlenfalvay G J, Franson R L, Brown M S and Mihara K L 1989 The *Glycine–Glomus–Bradyrhizobium* symbiosis. IX. Nutritional morphological and physiological responses of nodulated soybean to geographic isolates of the mycorrhizal fungus *Glomus mosseae*. Physiol. Plant 76, 226–232.

Bethlenfalvay G J, Reyes-Solis M G, Camel S B and Ferrera-Cerrato R 1991 Nutrient transfer between the root zones of soybean and maize plants connected by a common mycorrhizal mycelium. Physiol. Plant. 82, 423–432.

Bolan N S 1991 A critical review of the role of mycorrhizal fungi in the uptake of phosphorus by plants. Plant Soil 134, 189–207.

Bolan N S, Robson A D, Barrow N J and Aylmore L A G 1983 Specific activity of phosphorus in mycorrhizal and non-mycorrhizal plants in relation to the availability of phosphorus to plants. Soil Biol. Biochem. 16, 299–304.

Bolan N S, Robson A D and Barrow N J 1987 Effects of vesicular-arbuscular mycorrhiza on the availability of iron phosphates to plants. Plant Soil 22, 401–410.

Bougher N L, Grove T S and Malajczuk N 1990 Growth and phosphorus acquisition of karri (*Eucalyptus diversicolor* F. Muell.) seedlings inoculated with ectomycorrhizal fungi in relation to phosphorus supply. New Phytol. 114, 77–85.

Brown M S, Thamsurakul S and Bethlenfalvay G J 1988 The *Glycine–Glomus–Bradyrhizobium* symbiosis. IX. Phosphorus—use efficiency of $C0_2$ and N_2 fixation in mycorrhizal soybean. Physiol. Plant. 74, 159–163.

Brundrett M C and Abbott L K 1991 Roots of jarrah forest plants. I. Mycorrhizal associations of shrubs and herbaceous plants. Aust. J. Bot. 39, 445–457.

Bryla D R and Koide R T 1990 Role of mycorrhizal infection in the growth and reproduction of wild vs. cultivated plants. II. Eight wild accessions and two cultivars of *Lycopersicon esculentum* Mill. Oecologia 84, 82–92.

Cairney J WG and Ashford A E 1989 Reducing activity at the root surface in *Eucalyptus pilularis–Pisolithus tinctorius* ectomycorrhizas. Aust. J. Plant Physiol. 16, 99–105.

Cairney J W G and Ashford A E 1991 Release of a reducing substance by the ectomycorrhizal fungi *Pisolithus tinctorius* and *Paxillus involutus*. Plant Soil 135, 147–150.

Chalot M, Brun A, Khalid A, Dell B, Roh R and Botton B 1990 Occurrence and distribution of aspartate aminotransferases in spruce and beech ectomycorrhizas. Can. J. Bot. 68, 1756–1762.

Chalot M, Stewart G R, Brun A, Martin F and Botton B 1991 Ammonium assimilation by spruce–*Hebeloma* sp. ectomycorrhizas. New Phytol. 119, 541–550.

Cooper K M and Tinker P B 1978 Translocation and transfer of nutrients in vesicular-arbuscular mycorrhizas. II. Uptake and translocation of phosphorus, zinc and sulphur. New Phytol. 81, 43–52.

Cress W A, Johnson G V and Barton L L 1986 The role of endomycorrhizal fungi in iron uptake by *Hilaria jamesii*. J. Plant Nutr. 2, 547–556.

Cromack K, Sollins P, Cranstein W C, Speidel K, Todd A W, Spycher G, Ching Y-Li and Todd R L 1979 Calcium oxalate accumulation and soil weathering in mats of the hypogeous fungus *Hysterangium crassum*. Soil Biol. Biochem 11, 463–468.

Davies F T Jr, Potter J R and Linderman R G 1992 Mycorrhiza and repeated drought exposure affect drought resistance and extraradical hyphae development of pepper plants independent of plant size and nutrient content. J. Plant Physiol. 132, 289–294.

Dell B, Botton B, Martin F and Le Tacon F 1989 Glutamate dehydrogenases in ectomycorrhizas of spruce (*Picea excelsa* L) and beech (*Fagus sylvatica* L). New Phytol. 111, 683–692.

Denny H J and Wilkins D A 1987 Zinc tolerance in *Betula* spp. IV. The mechanism of ectomycorrhizal amelioration of zinc toxicity. New Phytol. 106, 545–553.

Dodd J C, Burton C C, Burns R G and Jeffries P 1987 Phosphatase activity associated with the roots and the rhizosphere of plants infected with vesicular-arbuscular mycorrhizal fungi. New Phytol. 107, 163–171.

Douds D D Jr, Johnson C R and Koch K E 1988 Carbon cost of the fungal symbiont relative to net leaf P accumulation in a split-root VA mycorrhizal symbiosis. Plant Physiol. 86, 491–496.

Finlay R D and Read D J 1986 The structure and function of the vegetative mycelium of ectomycorrhizal plants. II. The up-take and distribution of phosphorus by mycelial strands in the connecting host plants. New Phytol. 103, 157–165 .

Finlay R D, Ek H, Odham G and Söderström B 1988 Mycelial uptake translocation and assimilation of nitrogen from ^{15}N-labelled ammonium by *Pinus sylvestris* plants infected with four different ectomycorrhizal fungi. New Phytol. 110, 59–66.

Finlay R D, Ek H, Odham G and Söderström B 1989 Uptake, translocation and assimilation of nitrogen from ^{15}N-labelled ammonium nitrate sources by intact ectomycorrhizal systems of *Fagus sylvatica* infected with *Paxillus involutus*. New Phytol. 113, 47–55.

Finlay R D, Frostegard A and Sonnerfeldt A M 1992 Utilization of organic and inorganic nitrogen sources by ectomycorrhizal fungi in pure culture and in symbiosis with *Pinus contorta* Dougl. ex. Loud. New Phytol. 120, 105–115.

George E, Haussler K U, Vetterlein D, Gorgus E and Marschner H 1992 Water and nutrient translocation by hyphae of *Glomus mosseae*. Can. J. Bot. 70, 2130–2137.

Grey W E 1991 Influence of temperature on colonization of spring barleys by vesicular arbuscular mycorrhizal fungi. Plant Soil 137, 181–190.

Griffith R P, Castellano M A and Caldwell B A 1991 Hyphal mats formed by two ectomycorrhizal fungi and their association with Douglas-fir seedlings. A case study. Plant Soil 134, 255–259.

Hilger A B and Krause H H 1989 Growth characteristics of *Laccaria laccata* and *Paxillus involutus* in liquid culture media with inorganic and organic phosphorus sources. Can. J. Bot. 67, 1782–1789.

Ho I 1989 Acid phosphatase, alkaline phosphatase, and nitrate reductase activity of selected ectomycorrhizal fungi. Can. J. Bot. 67, 750–753.

Hogberg P 1989 Growth and nitrogen inflow rates in mycorrhizal and non-mycorrhizal seedlings of *Pinus sylvestris*. Forest Ecol. Manag. 28, 7–17.

Jakobsen I and Rosendahl L 1990 Carbon flow into soil and external hyphae from roots of mycorrhizal cucumber plants. New Phytol. 115, 77–83.

Jakobsen I, Abbott L K and Robson A D 1992 External hyphae of vesicular-arbuscular mycorrhizal fungi associated with *Trifolium subterraneum* L I. Spread of hyphae and phosphorus inflow into roots. New Phytol. 120, 371–380.

Jones M D, Durall D M and Tinker P B 1990 Phosphorus relationships and production of extramatrical hyphae by two types of willow ectomycorrhizas at different soil phosphorus levels. New Phytol. 115, 259–267.

Jongbloed R H, Clement J M A M and Borst-Pauwels G W F H 1991 Kinetics of NH_4^+ and K^+ uptake by ectomycorrhizal fungi. Effect of NH_4^+ on K^+ uptake. Physiol. Plant 83, 437–432.

Jungk A and Claassen N 1989 Availability in soil and acquisition by plants as the basis for phosphorus and potassium supply to plants. Z Pflanzenernaehr. Bodenkd.

152, 151–157.

Kammerbauer H, Agerer R and Sandermann H Jr 1989 Studies on ectomycorrhiza XXII. Mycorrhizal rhizomorphs of *Telephora terrestris* and *Pisolithus tinctorius* in association with Norway spruce (*Picea abies*): formation in vitro and translocation of phosphate. Trees 3, 78–84.

Koide R 1991 Nutrient supply, nutrient demand and plant response to mycorrhizal infection. New Phytol. 117, 365–386.

Kothari S K, Marschner H and Römheld V 1990a Direct and indirect effects of VA mycorrhiza and rhizosphere microorganisms on mineral nutrient acquisition by maize (*Zea mays* L) in a calcareous soil. New Phytol. 116, 637–645.

Kothari S K, Marschner H and George E 1990b Effect of VA mycorrhizal fungi and rhizosphere microorganisms on root and shoot morphology, growth and water relations in maize. New Phytol. 116, 303–311.

Kothari S K, Marschner H and Römheld V 1991a Contribution of VA mycorrhizal hyphae in acquisition of phosphorus and zinc by maize grown in a calcareous soil. Plant Soil 131, 177–185.

Kothari S K, Marschner H and Römheld V 1991b Effect of a vesicular-arbuscular mycorrhizal fungus and rhizosphere microorganisms on manganese reduction in the rhizosphere and manganese concentrations in maize (*Zea mays* L). New Phytol. 117, 649–655.

Lambert D H and Weidensaul T C 1991 Element uptake by mycorrhizal soybean from sewage-sludge-treated soil. Soil Sci. Am. J. 55, 393–398.

Lapeyrie F 1990 The role of ectomycorrhizal fungi in calcareous soil tolerance by 'symbiocalcicole' woody plants. Ann. Sci. For. 21, 579–589.

Lapeyrie F, Picatto C, Gerard J and Dexheimer J 1990 T.E.M. study of intracellular and extracellular calcium oxalate accumulation by ectomycorrhizal fungi in pure culture or in association with *Eucalyptus* seedlings. Symbiosis 2, 163–166.

Leake J R and Read D J 1989 The biology of mycorrhiza in the Ericaceae XV. The effect of mycorrhizal infection on calcium uptake by *Calluna vulgris* (L) Hull. New Phytol. 113, 535–544.

Leake J R and Read D J 1990 Proteinase activity in mycorrhizal fungi I. The effect of extracellular pH on the production and activity of proteinase by ericoid endophytes from soils of contrasted pH. New Phytol. 115, 243–250.

Li X-L, George E and Marschner H 1991a Extension of the phosphorus depletion zone in VA-mycorrhizal white clover in a calcareous soil. Plant Soil 136, 41–48.

Li X-L, George E and Marschner H 1991b Phosphorus depletion and pH decrease at the root–soil and hyphae–soil interfaces of VA mycorrhizal white clover fertilized with ammonium. New Phytol. 119, 397–404.

Li X-L, Marschner H and George E 1991c Acquisition of phosphorus and copper by VA-mycorrhizal hyphae and root-to-shoot transport in white clover. Plant Soil 136, 49–57.

Linderman R G 1988 Mycorrhizal interactions with the rhizosphere microflora: the mycorrhizosphere effect. Phytopathology 78, 366–371.

Maijala P, Fagerstedt K F and Raudaskoski M 1991 Detection of extracellular cellulolytic and proteolytic activity in ectomycorrhizal fungi and *Heterobasidion annosum* (Fr.) Bref. New Phytol. 117, 643–648.

Martin F, Stewart G, Genetet I and Le Tacon F 1986 Assimilation of $^{15}NH_4^+$ by beech (*Fagus sylvatica* L) ectomycorrhizas. New Phytol. 102, 85–94.

Martin F, Stewart G R, Genetet I and Mourot B 1988 The involvement of glutamate dehydrogenase and glutamine synthetase in ammonia assimilation by the rapidly growing ectomycorrhizal ascomycete, *Conococcum geophilum* Fr. New Phytol. 110, 541–550.

Mitchell R J, H E Garrett, Cox G S and Atalay A 1990 Boron and ectomycorrhizal influences on mineral nutrition of container-grown *Pinus ehinata* Mill. J. Plant Nutr. 13, 1555–1574.

Newman E I, Eason W R, Eissenstat D M and Ramos M I R F 1992 Interactions between plants: the role of mycorrhizae. Mycorrhiza 1, 47–53.

Perrin R 1990 Interactions between mycorrhizae and diseases caused by soil-borne fungi. Soil Use Manag. 6, 189–195.

Paulitz T C and Linderman R G 1989 Interactions between fluorescent pseudomonads and VA mycorrhizal fungi. New Phytol. 113, 37–45.

Piccini D, Ocampo J A and Bedmar E J 1988 Possible influence of *Rhizobium* on VA mycorrhiza metabolic activity in double symbiosis of alfalfa plants (*Medicago sativa* L) grown in a pot experiment. Biol. Fert. Soils 6, 65–67.

Powell P E, Szaniszlo P J, Cline G R and Reid C P P 1982 Hydroxamate siderophores in the iron nutrition of plants. J. Plant Nutr. 5, 653–573.

Raju P S, Clark R B, Ellis J R and Maranville J W 1990 Mineral uptake and growth of sorghum colonized with VA mycorrhiza at varied soil phosphorus levels. J. Plant Nutr. 13, 843–859.

Read D J 1984 The structure and function of the vegetative mycelium of mycorrhizal roots. In The Ecology and Physiology of the Fungal Mycelium. Ed. D H Jennings and A D M Rayner. pp 215–240. Cambridge University Press. Cambridge.

Rhodes L H and Gerdemann J W 1978 Translocation of calcium and phosphate by external hyphae of vesicular-arbuscular mycorrhizae. Soil Sci. 126, 125–126.

Robson A D, O'Hara G W and Abbott L K 1981 Involvement of phosphorus in nitrogen fixation by subterranean clover (*Trifolium subterraneum* L). Aust. J. Plant Physiol. 8, 427–436.

Rygiewicz P T and Bledsoe C S 1984 Mycorrhizal effects on potassium fluxes by northwest coniferous seedlings. Plant Physiol. 76, 918–923.

Sanders F E and Tinker P B 1973 Phosphate inflow into mycorrhizal roots. Pestic. Sci. 4, 385–395.

Scheromm P, Plassard C and Salsac L 1990 Nitrate nutrition of maritime pine (*Pinus pinaster* Soland *in* Ait.) ectomycorrhizal with *Hebeloma cylindrosporum* Romagn. New Phytol. 114, 93–89.

Schuler R and Haselwandter K 1988 Hydroxamate siderophore production by ericoid mycorrhizal fungi. J. Plant Nutr. 11, 907–913.

Secilia J and Bagyaraj DJ 1987 Bacteria and actinomycetes associated with pot cultures of vesicular-arbuscular mycorrhizas. Can. J. Bot. 33, 1069–1073.

Shaw G, Leake J R, Baker A J M and Read D J 1990 The biology of mycorrhiza in the Ericaceae. XVII. The role of mycorrhizal infection in the regulation of iron uptake by ericaceous plants. New Phytol. 115, 251–258.

Sieverding E and Toro S 1988 Influence of soil water regime on VA mycorrhiza. V. Performance of different VAM fungal species with cassava. J. Agron. Crop Sci. 161, 322–332.

Stone E L 1990 Boron deficiency and excess in forest trees: A review. For. Ecol. Manag. 37, 49–75.

Straker C J and Mitchell D T 1986 The activity and characterization of acid phosphatases in endomycorrhizal fungi of the Ericaceae. New Phytol. 104, 243–256.

Sylvia D 1988 Activity of external hyphae of vesicular-arbuscular mycorrhizal fungi. Soil Biol. Biochem. 20, 39–43.

Vierheilig H and Ocampo J A 1991 Receptivity of various wheat cultivars to infection by VA-mycorrhizal fungus influenced by inoculum potential and the relation of VAM-effectiveness to succinic dehydrogenase activity of the mycelium in the root. Plant Soil 133, 291–296.

Vogt K A, Publicover D A and Vogt D J 1991 A critique of the role of ectomycorrhizas in forest ecology. Agric. Ecos. Environ. 35, 171–190.

White J A and Brown M F 1979 Ultrastructural and X-ray analysis of phosphorus granules in a vesicular-arbuscular mycorrhizal fungus. Can. J. Bot. 57, 2812–2818.

Wilkins D A 1991 The influence of sheathing (ecto-) mycorrhizas of trees on the uptake and toxicity of metals. Agric. Ecos. Environ. 35, 145–260.

Zhu H, Guo D and Dancik B P 1990 Purification and characterization of an extracellular acid proteinase from the ectomycorrhizal fungus *Hebeloma crustuliniforme*. Appl.Environ. Microbiol. 56, 837–843.

Reprinted from *Plant and Soil*, Volume 159, No. 1 (1994), pp. 89–102.

Nutrient transport in mycorrhizas: structure, physiology and consequences for efficiency of the symbiosis

S.E. SMITH[1], V. GIANINAZZI-PEARSON[2], R. KOIDE[3] and J.W.G. CAIRNEY[1]
[1]*Department of Soil Science, Waite Institute, The University of Adelaide, Adelaide, SA 5064, Australia*
[2]*Station d'Amelioration des Plantes, INRA, BV 1540, Dijon 21034, France*
[3]*Department of Biology, Pennsylvania State University, University Park, PA 16802, USA*

Key words: interface, mycorrhizas, nutrient acquisition, nutrient transport, nutrient use

Abstract

Nutrient transport in mycorrhizas occurs across specialized interfaces which are the result of coordinated development of the organisms. The structural modifications give rise to large areas of either inter- or intra-cellular interface in which wall synthesis is frequently modified and in which altered distribution of membrane bound ATPases is important, particularly with respect to mechanisms that may be involved in bidirectional transfer of nutrients. Except in orchid mycorrhizas, net movement of organic carbon from plant to fungus occurs, complemented by mineral nutrient movement in the opposite direction. The general consensus is that sustained transfer at rates that will maintain the growth and development of the organisms requires increases in the rates at which nutrients are lost from the organisms; possible mechanisms for this are discussed. The transfer processes are essential in determining both plant and fungal productivity and an approach to calculating the efficiency of the symbiosis in terms of the expenditure of carbon (or of phosphorus) is discussed.

Introduction

Bidirectional transfer of nutrients between plant and fungus is typical of ecto-, vesicular-arbuscular (VA) and ericoid mycorrhizal associations and is the basis for the prolonged compatible interactions typical of these symbioses. The transfer occurs across interfaces which develop in different ways in the different mycorrhizas, although the end result in terms of the spatial arrangement of membrane and wall components varies only in detail. The effectiveness of the symbiosis with respect to plant growth and yield is influenced by the magnitude of the flux of phosphate (or other nutrients) to the plant and of carbohydrate to the fungus.

In orchid mycorrhizas the relationship between the symbionts is different; nutrient transfer is unidirectional, in favour of the plant and involves both carbohydrate and mineral nutrients. Nevertheless, this symbiosis is important in nature and may have horticultural potential. It will therefore be considered briefly.

In this review we consider the structure and development of the interfaces as the sites at which transfer occurs, the mechanisms and rates of transfer and the implications of this for plant productivity.

Plant/fungus interfaces: development and structure in relation to transport

At the cellular level interfaces in all types of mycorrhiza are composed of the membranes of both partners, separated by an apoplastic region. The various types differ in whether the interfaces are inter- or intra-cellular (or both), their developmental and structural complexity and

A.D. Robson, L.K. Abbott & N. Malajczuk (eds.), Management of mycorrhizas in agriculture, horticulture and forestry, 103–113.

the degree to which wall and membrane modifications contribute to specialization and to large areas of contact between the symbionts (Smith and Smith, 1990). The spectrum ranges from simple, intercellular, wall–wall contact in ectomycorrhiza to specialized fungal wall–plant membrane structure in VA mycorrhiza, but in every case the fungal partner remains in the apoplast outside the plant protoplast. Interfaces also vary ontogenetically and the developmental changes may be of importance for nutrient transport (Duddridge, 1986; Gianinazzi-Pearson et al., 1991; Scannerini and Bonfante-Fasolo, 1983).

Among endomycorrhizal associations, orchid and ericoid mycorrhizal fungi have little intercellular growth and hyphal coils or peletons are formed within the plant cells (Gianinazzi-Pearson, 1984; Harley and Smith, 1983). Intracellular development affects fungal wall structure, metabolism and physiology, including the repression of surface and extracellular enzymes (Gianinazzi-Pearson et al., 1986; Smith, 1974) potentially detrimental to the plant cell.

The morphogenesis of VA mycorrhizal fungi within host tissues is more complex (Bonfante-Fasolo, 1984; Gianinazzi-Pearson, 1984). Colonization of the outermost root cells involves the sparse formation of simple intracellular coils, while in the cortical parenchyma infection is much more intense. Numerous highly branched intracellular arbuscules arise from longitudinally proliferating intercellular hyphae, so that two structurally different types of interface are present (Gianinazzi-Pearson et al., 1991; Smith and Dickson, 1991) which may have different functions (see below). Within the root, hyphae are thinner compared with external hyphae and, particularly in arbuscules, do not develop the multi-layered, crystalline chitin structure (Bonfante-Fasolo and Grippiolo, 1982; Gianinazzi-Pearson, 1986).

Host reaction to penetration and proliferation of the fungus appears similar in different endomycorrhizas (see for example Bonfante-Fasolo, 1984; Bonfante-Fasolo and Gianinazzi-Pearson, 1982; Gianinazzi-Pearson, 1984; Hadley, 1975; Serrigny, 1982) and it is in the presence of the intracellular haustoria that the host cell shows the most specific modifications. In all cases, the interface formed between plant cell and fungus involves a newly formed plant membrane extending from the peripheral plasmamembrane around the fungal haustorium as the latter grows intra cellularly. It has been estimated that this can cause up to a 20-fold increase in the surface area-to-volume ratio of the host in VA mycorrhiza (e.g. Alexander et al., 1988) and although increases have not been calculated for coiled haustoria they must be considerable. Invading hyphae of all types are surrounded by plant-derived material in the apoplast (Bonfante-Fasolo et al., 1991; Dexheimer et al., 1979). In orchids, cell wall material persists around intracellular stages, but in ericoid and VA mycorrhizas it decreases and can be reduced to scattered protein and polysaccharide fibrils (Bonfante-Fasolo et al., 1991; Dexheimer et al., 1979). The presence of pectins, cellulose and glycocalyx constituents in the arbuscular interface (Gianinazzi-Pearson et al., 1990; Bonfante-Fasolo et al., 1990), together with neutral phosphatase activity (Jeanmaire et al., 1985) indicates that the plant protoplast retains wall synthesizing activity. Production of wall precursors, together with small amounts of pectinolytic and cellulolytic enzymes (Garcia Romera et al., 1991; Pearson and Read, 1975) suggests that the material may represent a carbon source for the fungi (see Harley and Smith, 1983). Carbohydrate transfer in orchid mycorrhizas is from fungus to plant and synthesis of plant wall components will not contribute to fungal nutrition in this case, but may be of significance in 'controlling' fungal invasion.

A very important modification in terms of nutrient transport is the ATPase activity in the host membrane formed around coils or arbuscules (Marx et al., 1982; Gianinazzi-Pearson et al., 1991). In VA mycorrhizas, part of this activity is attributable to an H-ion ATPase present in the periarbuscular membrane, but cytochemically undetectable along other plant membranes adjacent to hyphae (intercellular hyphae in the cortex, hyphal coils etc.) or the plant plasmamembranes adjacent to the walls of either infected or uninfected parenchyma cells (Gianinazzi-Pearson et al., 1991). The distribution of activity around living arbuscules suggests that the infected root cells have increased ability to absorb nutrients, via the symbiotic fungus, which in non-mycorrhizal roots is restricted to the outer cell layers.

With respect to the VA mycorrhizal fungus, H-ion ATPase activity is observed on the plasmamembrane of intercellular hyphae and in coils, but is variable on the arbuscule branches (Gianinazzi-Pearson et al., 1991).

In ectomycorrhizal associations, fungal colonization of host root tissues is entirely extracellular. A pseudoparenchymatous sheath envelops the absorbing rootlets and is continuous with intercellular hyphae growing between epidermal and cortical cells (Hartig net) and the extra-radical network of mycelium. Host/fungus contact results in changes in hyphal morphology. In the Hartig net region hyphae branch profusely and septa formation becomes irregular or incomplete to give a characteristic labyrinthine system (see Harley and Smith, 1983). Surface fibrils and acid phosphatase activity, present in the mantle, disappear as hyphae become tightly pressed against host cell walls (Dexheimer et al., 1986; Duddridge, 1986; Lei et al., 1990). Adjacent fungal and host walls become indistinguishable from each other, forming a homogeneous interfacial matrix (Duddridge, 1986; Duddridge and Read, 1984; Scannerini and Bonfante-Fasolo, 1983). Structural changes in the host tissues are small, but in some cases wall ingrowths appear to be induced by fungal infection (Ashford and Allaway, 1982). These resemble those of transfer cells, increasing the area available for transport between the symbionts. ATPase activity is associated with the plasmamembranes of both symbionts when these are active and closely associated in the Hartig net region, but it disappears as plant cells senesce (Lei and Dexheimer, 1988).

Ultrastructural information for all types of mycorrhiza indicates that deposition of wall material in the interface is reduced with possible consequences in terms of the permeability of the apoplast. However, extracellular material is also deposited around hyphae of endomycorrhizas at points of entry to host cells (Scannerini and Bonfante-Fasolo, 1983) and accumulates in intercellular spaces of the fungal sheath in ectomycorrhizas of *Eucalyptus* and *Pisonia* (Ashford et al., 1988, 1989). Such material is impermeable to fluorescent tracer dyes and, if also impermeable to solutes, would restrict the passage of nutrients to the cortex except via the fungal symplast, prevent loss of solutes from the apoplast to the soil and, along with the exodermis (hypodermis), may permit control of the apoplastic nutrient concentrations (Ashford et al., 1989; Smith et al., 1989; Smith and Smith, 1990).

Compounds transferred and mechanisms of transfer

Evidence (or lack of it) for the nature of the compounds transferred between the symbionts has been reviewed previously (Harley and Smith, 1983; Martin et al., 1987; Smith and Smith, 1990). Sugars are important in carbohydrate transfer, with hydrolysis of sucrose (or trehalose in orchid mycorrhizas), together with synthesis of characteristic 'non-recyclable' carbohydrates (e.g. mannitol in ectomycorrhizas), as important steps in polarizing transport in favour of one symbiont (see Smith et al., 1969). Hydrolysis could potentially occur at the host surface, in the apoplast and/or at the fungal wall, but only immunocytochemical methods will confirm the precise location. Other potential sources of carbon include wall precursors, which might also require hydrolysis (see above), and organic anions or organic N compounds, but evidence for their involvement is lacking.

The soil-derived nutrients which are transferred from fungus to plant vary in importance in the different types of association. Ecto- and ericoid mycorrhizas have a major influence on nitrogen (N) and a lesser effect on phosphorus (P) nutrition. In VA mycorrhizas this situation is reversed. N transfer from fungus to plant has been demonstrated in ecto- and ericoid mycorrhizas and is also likely in VA associations. Organic N (as amino acids or amides), rather than inorganic NH_4^+, is probably involved (see Martin et al., 1987). Such transfer will result in movement of organic carbon against the general flow of carbohydrate and will affect the net flux of reduced carbon, providing a possible route for carbon movement during inter-plant transfer (see Smith and Smith, 1990).

Transfer of P from fungus to plant is well documented in ecto-, VA, ericoid, monotropoid and orchid mycorrhizas. In ectomycorrhizas there is evidence that inorganic orthophosphate is the major form in which phosphorus is transferred (Harley and Loughman, 1963) but direct evidence for endomycorrhizas is lacking, largely because of technical difficulties. Given that the pH of the interfacial apoplast is probably acid due to the polarized activity of H-ion ATPases, $H_2PO_4^-$ are likely to be the predominant ionic species involved. Other nutrients (S, Zn, Cu, Ca and Na) have been shown by experiments with radio-tracers to be transferred between symbionts in various types of mycorrhizas and it is likely that these cross the interface as free ions in solution.

Transfer of nutrients must be bidirectional in the symbiosis as a whole. Carbohydrate movement from plant to fungus and mineral nutrient movement from fungus to plant certainly occurs at the whole plant level. However, whether this transfer occurs bidirectionally *at the same interface* (as has been generally assumed) is open to question (see Smith and Smith, 1990). At a single interface, transfer between symbionts in either direction depends on export through the membrane of one, movement through the interfacial apoplast and import by the other. Distribution of ATPase activity has given some new insights with respect to the processes involved, but we still have very limited information on the magnitude of the fluxes of nutrients in either direction or on the conditions that may affect these fluxes.

Assuming that bidirectional nutrient transfer occurs at a single interface, both symbionts must possess functional plasmamembranes capable of uptake of nutrients from the apoplast. In ectomycorrhizas the coexistence of ATPase activity on plant and fungal plasmamembranes at the Hartig net interface suggests that the two systems work cooperatively in bidirectional nutrient exchange. This is the scenario that has been assumed for all types of mycorrhiza, but we cannot entirely discount the possibility that two spatially separate interfaces such as arbuscule and intercellular hypha in VA mycorrhizas (see diagram in Smith and Smith, 1990) may be involved and this is explored below using P as an example.

Models of P transfer across a single interface generally invoke passive efflux of P_i from the fungus and active absorption across the plant plasmamembrane (Smith and Smith, 1986, 1989, 1990; Woolhouse, 1975). Net loss of P_i from free living fungi is normally regarded as slight, with membrane transport processes favouring absorption (Beever and Burns, 1980), so that conditions promoting efflux are likely to be important at the fungus/root interface. Perhaps the simplest mechanism for this is where loss of P_i across the fungal plasmamembrane to the interfacial apoplast is unmodified, but reabsorption from the apoplast by the fungus is reduced. This would result in an increase in net efflux from the fungus. There is good evidence that intracellular P_i concentrations in both fungi (Beever and Burns, 1980; Clipson et al., 1987) and plants (Clarkson and Scattergood, 1982; Lee, 1982) affect P_i transport. High intracellular concentrations result in reduced P uptake relative to mycelium with a lower P_i status in ecto- (Cairney and Smith, 1992a), ericoid (Straker and Mitchell, 1987) and VA (Thomson et al., 1990) mycorrhizal fungi. High P concentrations in hyphae at the interface (Clarkson, 1985) and a low interfacial P_i concentration would, along with reduced reabsorption, maximize loss from the fungus. Lower P_i status of the root relative to the fungus (maintained by metabolism and/or translocation within the plant) and the concomitant increase in P_i transporting activity at the plant plasmamembrane, would permit net movement of P_i into the plant. P_i transfer to host tissue in excised beech ectomycorrhizas was around 10–20% of P_i absorbed by the sheath hyphae (Harley and Brierley, 1955; see Strullu et al., 1986) and in mycelium of *Pisolithus tinctorius* net efflux of P_i over an 8 h period was also about 10% of P_i absorbed during the preceding 16 h (Cairney and Smith, 1992b; unpublished results). This implies that efflux across the unmodified fungal membrane might be sufficient to account for transfer in the ectomycorrhiza. Preliminary estimates in the range $1.4–4.6 \times 10^{-8}$ mol.m^{-2}.s^{-1} can be calculated from data for efflux from *P. tinctorius* (see

Cairney and Smith, 1993) which suggest relatively high efflux from the cultured fungal mycelium. These data require confirmation but it is certainly an approach that merits further investigation.

As with the fungus, membrane transport processes in roots normally promote net absorption from the apoplast rather than efflux (see Patrick, 1989). While coupled sugar-phosphate transporters have been proposed for root cortical cell membranes at the interface (Schwab et al., 1991; Woolhouse, 1975), there remains no evidence for their existence. Indeed the lack of demonstrable stoichiometry between sugar and P transfer and the need for transfer of other nutrients and regulation of pH and charge balance across the interface suggests independent transfer of the two (see Smith and Smith, 1990). Furthermore, the general assumption that bidirectional nutrient transfer occurs at the same interface in all mycorrhizas is now being questioned. It is conceivable, however, that loss of carbohydrate from the root, like loss of phosphate from the fungus could be explained simply in terms of decreased rates of reabsorption into the plant cells. This might be brought about by a reduction in activity of the plant hexose/proton cotransporter, resulting from fungal infection. Hydrolysis of sucrose to hexoses by invertase in the apoplast (Patrick, 1989) would allow a net sugar transport to the fungus. This would be supported by rapid conversion of hexoses to fungal metabolites (trehalose and mannitol in ectomycorrhizas) which, not being readily utilized by the plant, would maintain the flux of carbohydrate in favour of the fungus (Smith et al., 1969). The fact that individual ecto- (Bevege et al., 1975; Cairney et al., 1989) and VA (Bevege et al., 1975) mycorrhizal roots act as greater sinks for photosynthetically-fixed carbon than non-mycorrhizal roots strongly suggests greater concentration at the interface and the probability of increased loss from the root cells.

It is important to gain more evidence on the overall fluxes of nutrients (whether plant- or fungus-derived) across the interface as a whole in order to judge whether 'normal' efflux, associated with reduced reabsorption, would be sufficient to account for the flux across the interface. There is evidence that this may not be so in VA mycorrhizas, where overall fluxes are of similar order to measured rates of P uptake by plant cells (Cox and Tinker, 1976; Morris, Dickson and Smith, unpublished results), making it likely that some condition enhancing efflux must exist at the interface.

Efflux could be promoted in a number of ways, including reversal of H-ion co-transporters for phosphate and sugar, increases in frequency or rates of operation of transmembrane carriers or by opening of channels (Patrick, 1989; Smith and Smith, 1986, 1989, 1990; Tester et al., 1992). Perhaps the most direct mechanism would involve localized accumulation of particular ions (e.g. H^+, K^+, Ca^{2+}) in the interfacial apoplast, which could certainly occur, possibly as a result of the presence of apoplastic barriers (see above). High extracellular concentrations of monovalent cations, in particular K^+, have recently been shown to stimulate efflux of P from mycelium of the ectomycorrhizal fungus *Pisolithus tinctorius* (Cairney and Smith, 1992b; unpublished results). Efflux was stimulated at concentrations of 10–20 m*M*, known to cause depolarization of transmembrane electric potentials in fungal hyphae (Slayman, 1965) and algal cells (Keifer and Lucas, 1982). This could stimulate opening of a number of ion channels (Tester, 1990), raising the possibility that the monovalent cation-induced efflux of P from *P. tinctorius* may be mediated by opening of a channel. Work of this nature is still at an early stage. We do not know how general the effect is and it must be stressed that the involvement of ion channels in such cation-induced efflux is speculative. Release of phytohormones at the interface might also influence nutrient transport (see Schwab et al., 1991; Smith and Smith, 1990). Auxin (for example) can increase ATPase activity in plant tissue and may have a role in the mycorrhizal situation (see Gianinazzi-Pearson et al., 1991) and the cytokinin N^{6-2}-isopentyl-adenosine (2iPA) has been shown to influence both absorption and loss of ions from mycelium of the ectomycorrhizal fungus *Suillus variegatus* (Pohleven, 1989). Alternatively, changes in osmotic potential in the apoplast or increased cytoplasmic sugar concentrations in infected host cells may influence transmembrane transport either directly, by influencing ATPase activity (Wyse et al., 1986), or indirectly via stretching of the plasmamembrane. Stretch-activated channels specific for both cations and anions are known to exist on higher plant plasmamembranes (Cosgrove and Hedrich, 1991; Hedrich and Scroeder, 1989) and could be potentially important in controlling fluxes from both symbionts at the mycorrhizal interface.

While nutrient exchange is generally interpreted as occurring simultaneously in both directions across a single interface, it is theoretically possible that exchange occurs across interfaces which are separated temporally or spatially. Temporal separation would involve net movement of nutrients in different directions at different developmental stages. In VA mycorrhizas for example, absorption of sugar into young arbuscules and loss of nutrients such as P from older arbuscules cannot totally be ruled out at present, although it seems unlikely (Smith and Smith, 1990). In ectomycorrhizas, where turnover of the exchange interface is much less rapid, it becomes even more difficult to envisage. Functional ectomycorrhizas persist for months, during which time they act as sinks for current photosynthate and effect P transfer to the plant (Cairney and Alexander, 1992a, b; Downes et al., 1992). This, together with the distribution of ATPase activity on the membranes (see above), strongly implies simultaneous transfer in active ectomycorrhizas.

Separation of the components of bidirectional nutrient transfer in space seems more likely than separation in time. Based on differences in H-ion ATPase distribution between arbuscules and intercellular hyphae, Gianinazzi-Pearson et al. (1991) have proposed that spatial separation might occur in VA mycorrhizas. The result would be net loss of nutrients such as P from arbuscule branches associated with uptake by the plant via the highly energized periarbuscular membrane. Carbohydrate transfer would occur via loss from cortical cells to intercellular spaces and uptake by intercellular hyphae, which have consistent H-ion ATPase activity. As stressed by the authors this interpretation is speculative, but it challenges the generally held view of a single transport interface in these mycorrhizas. We clearly need more information on ATPase distribution (using immunological as well as cytochemical techniques), apoplastic pH and solute concentrations as well as data for fluxes of different nutrients across the interface.

The importance of nutrient transfer to plant productivity

From an agricultural or ecological standpoint, and therefore in the context of this symposium, the interest in nutrient transfer between mycorrhizal symbionts is based on the potential effects of infection on plant yield, survival or species composition of plant communities. When plant growth and reproduction are limited by the rate of nutrient acquisition and when mycorrhizal infection increases this (by uptake and transfer processes), it is clear that infection is of importance and that the net result of bidirectional transfer is a gain in plant biomass or other measurable parameter. However, mycorrhizal infection does not always increase plant growth or reproduction (Fitter, 1985; McGonigle, 1988). Possible reasons include high efficiency of P acquisition or low P requirement (Koide, 1991; Smith et al., 1992) of the plants. P acquisition can be reduced when external hyphae are destroyed by grazing soil animals (Finlay, 1985; McGonigle and Fitter, 1988; Rabatin and Stinner, 1988), soil disturbance (see below) or fungicides (Hale and Sanders, 1982) or where incomplete physiological integration of the symbionts results in low rates of transfer (see below). In these and other situations it is important to be able to analyse the effectiveness of a symbiosis in terms which can be related to the quantitative aspects of the nutrient transfer processes, so that predictions can be made of the importance of mycorrhizal associations in particular plant species or in particular agricultural or natural systems. The outcome of the increased understanding will be potential reductions in application of chemical fertilizers and improvement of the efficiency with which they are used. The reasons are both economic and concerned with environmental quality.

Measurement of the efficiency of nutrient acquisition

Efficient plants might be defined as those which, either with or without mycorrhizal infection, thrive in nutrient deficient soils. They can be detected by trial and error in natural environments and in screening programs, but a more mechanistic approach is needed in plant breeding to produce 'efficient genotypes'. This requires a knowledge of what makes a particular genotype successful and it may therefore make sense to speak of a physiological efficiency with which resources (such as P) are either acquired or utilized. These can be defined in terms of the rate or amount of nutrient absorbed per unit of root and in the amount of dry matter produced per unit of nutrient absorbed, respectively (Baon et al., 1992; Smith et al., 1992).

Koide and Elliott (1989) defined both a P utilization efficiency ($\Delta C^w/\Delta P^w$) and a P acquisition efficiency ($\Delta P^w/\Delta C^b$), where ΔC^w is the total carbon accumulated in the whole plant during unit time, ΔP^w is the P accumulated in the plant during the same period, and ΔC^b is the carbon expended below-ground for the acquisition of that P. Calculating the efficiency of P acquisition as the ratio of P acquired to carbon expended is a logical approach since carbon can be regarded as the energy currency of the plant (Mooney, 1972). However, carbon is not the only possible currency with which to express the cost of P acquisition. For example, if P were the limiting resource, P would also be an appropriate currency. P acquisition efficiency would then be: ($\Delta P^w/\Delta P^b$), where ΔP^b is that increment of P expended below-ground in the acquisition of ΔP^w. P^b includes two components: that which is lost from the root system through leakage, death, exudation, etc., and that which is retained in the root and hyphal tissues of the below-ground system. One might argue that P and carbon are so closely linked physiologically that the P for P approach contributes nothing to our understanding beyond that contributed by the P for carbon approach. Indeed, when P deficiency limits plant growth it could be argued that it does so by limiting carbon acquisition (Heldt et al., 1977; see Smith and Gianinazzi-Pearson, 1988). Moreover, in the construction of roots, in exudation and tissue death, etc., P as well as carbon are 'expended'. However, many studies show that P-deficient plants accumulate starch in their leaves (e.g. Heldt et al., 1977), suggesting that P limits growth more directly than carbon when plants are P-deficient. Either the P for P or the P for carbon approach may prove useful, depending on the objectives of the research, but clearly this type of mechanistic approach is required if we are to be able to understand the basis for variation among plants in terms of efficiency.

Natural vs agricultural systems

Although most research on mycorrhiza has been carried out on cultivated plants, there is much to be learned from natural ecosystems and the approach just outlined may be useful in making comparisons. Most research has been concerned with short-lived annuals and the relations hip between these and mycorrhizal fungi may differ from that in longer-lived perennials. Annuals may be able to defer some of their payments (in carbon) to mycorrhizal fungi until after reproduction has occurred and a large expenditure on fungal sporulation may occur during senescence without negative consequences in terms of harvestable yield. Perennials cannot do this and so may face additional carbon costs. Moreover, the relationships between root and shoot growth in terms of temporal changes, allometry and plasticity of root/shoot ratio, together with the exploratory size of the root system in relation to P sources (Campbell et al., 1991; Koide, 1991) and turnover of roots may be quite different in short and long-lived species. These differences are likely to have a strong influence on the contribution of mycorrhizas to nutrient acquisition, as well as to the 'costs' of maintaining the symbiosis in terms of the proportion of total C (or P) used in the process.

Another important difference between agricultural and natural systems is the degree of disturbance of the soil. This has a negative effect on the rate and extent of mycorrhizal infection, and hence on nutrient acquisition and growth because of the disturbance of the mycelial network (Evans and Miller, 1988; Fairchild and Miller, 1988; Jasper et al., 1989a, b). Even no-till agricultural systems may have far less diverse assemblages of mycorrhizal species than natural ecosystems, brought about by the degree of ecological specificity between host plants and mycorrhizal fungi (e.g. McGonigle and Fitter, 1990). The consequences of this for efficiency of nutrient uptake and use are presently unknown, but its relevance is now being appreciated and work is in progress to determine what attributes of the symbionts (such as length of external hyphae or nutrient fluxes across the interface) are important determinants of fungal 'effectiveness' (Jakobsen and Pearson, unpublished results; Morris, Dickson and Smith, unpublished results).

In mixed cropping systems (economically significant in some countries) mycorrhizas may have important effects on yield (Pugna and Janos, unpublished results), just as they have on the community structure and species composition in natural communities (Allen and Allen, 1984; Eissenstat and Newman, 1990; Fitter, 1977; Gange et al., 1990; Grime et al., 1987; Koide and Mooney, 1987). The basis for this may be found in terms of interplant transport of both carbon and mineral nutrients and it will be necessary to determine the community efficiency in a similar way as that outlined above for individual plants.

Conclusions

Identification of efficient genotypes of both fungus and host and understanding the way their function is integrated depends, as we have said before (Smith and Gianinazzi-Pearson, 1988), on the identification and quantification of the key processes involved in nutrient uptake and use. Much of the emphasis of mycorrhizal research in an agricultural context has been on the way in which mycorrhizas increase nutrient uptake via the mycelial network in soil. This is a crucial process in exploitation of soil resources, but is useless unless we also have a quantitative understanding of the complementary processes of nutrient transfer to root cells and the way the symbiosis is maintained with respect to allocation of carbon and other nutrients between the partners.

References

Alexander T, Meier R, Toth R and Weber H C, 1988 Dynamics of arbuscule development and degeneration in mycorrhizas of *Triticum aestivum* L. and *Avena sativa* L. with reference to *Zea mays*. New Phytol. 110, 363–370.

Allen E B and Allen M F 1984 Competition between plants of different successional stages: mycorrhizas as regulators. Can. J. Bot. 62, 2625–2629.

Ashford A E and Allaway W G 1982 A sheathing mycorrhiza on *Pisonia grandis* R. BR. (Nyctaginaceae) with development of transfer cells rather than a Hartig net. New Phytol. 90, 511–517.

Ashford A E, Petersen C A, Carpenter J L, Cairney J W G and Allaway W G 1988 Structure and permeability of the fungal sheath in the *Pisonia* mycorrhiza. Protoplasma 147, 149–161.

Ashford A E, Allaway W G, Peterson C A and Cairney J W G 1989 Nutrient transfer and the fungus-root interface. Aust. J. Plant Physiol. 16, 85–97.

Baon J B, Smith S E, Alston A M and Wheeler R D 1982 Phosphorus efficiency of three cereals as related to indigenous mycorrhizal infection. Aust. J. Agric. Res. 43, 479–491.

Beever R E and Burns D J W 1980 Phosphorus uptake, storage and utilization by fungi. Adv. Bot. Res. 8, 127–129.

Bevege D I, Bowen G D and Skinner M F 1975 Comparative carbohydrate physiology of ecto- and endo-mycorrhizas. *In* Endomycorrhizas. Eds. F E Sanders, B Mosse and P B Tinker. pp 149–174. Academic Press, London.

Bonfante-Fasolo P 1984 Anatomy and morphology. *In* VA Mycorrhizas. Eds. C L Powell and D J Bagyaraj . pp 5–33. CRC Press, Boca Raton, FL.

Bonfante-Fasolo P and Gianinazzi-Pearson V 1982 Ultrastructural aspects of endomycorrhiza in the Ericaceae. III. Morphology of the dissociated symbionts and modifications occurring during their reassociation in axenic culture. New Phytol. 91, 691–704.

Bonfante-Fasolo P and Grippiolo R 1982 Ultrastructural and cytochemical changes in the wall of a vesicular-arbuscular mycorrhizal fungus during symbiosis. Can. J. Bot. 60, 2303–2312.

Bonfante-Fasolo P, Vian B, Perotto S, Faccio A and Knox J P 1990 Cellulose and pectin localization in roots of mycorrhizal *Allium porrum* : labelling continuity between host cell wall and interfacial material. Planta 180, 537–547.

Bonfante-Fasolo P, Tamagnone L, Peretto R, Esquerré-Tugayé D, Mazau D, Mosiniak M and Vian B 1991 Immunocytochemical location of hydroxyproline rich glycoproteins at the interface between a mycorrhizal fungus and its host plants. Protoplasma 165, 127–138.

Cairney J W G and Alexander I J 1992a A study of ageing of spruce (*Picea sitchensis* Bong. Carr.) ectomycorrhizas. II. Carbohydrate allocation in ageing *Picea sitchensis/Tylospora fibrillosa* (Burt.) Donk ectomycorrhizas. New Phytol. 122, 153–158.

Cairney J W G and Alexander I J 1992b A study of ageing of spruce (*Picea sitchensis* Bong. Carr.) ectomycorrhizas. III. Phosphate absorption and transfer in ageing *Picea sitchensis/Tylospora fibrillosa* (Burt.) Donk ectomycorrhizas. New Phytol. 122, 159–164.

Cairney J W G and Smith S E 1992a Influence of intracellular phosphorus concentration on phosphate absorption by the ectomycorrhizal basidiomycete *Pisolithus tinctorius*. Mycol. Res. 96. 673–676.

Cairney J W G and Smith S E 1992b Effect of monovalent cations on efflux of phosphate from the ectomycorrhizal fungus *Pisolithus tinctorius*. *In* Mycorrhiza in Ecosystems. Eds. D J Read, D H Lewis, A H Fitter and I J Alexander. pp 352–355. CAB International, Oxon, UK.

Cairney, J.W.G. and Smith, S.E. 1993 Efflux of phosphate from the ectomycorrhizal basidiomycete *Pisolithus tinctorius*: general characteristics and the influence of intracellular phosphorus concentration. Mycol. Res. (*in press*).

Cairney J W G, Ashford A E and Allaway W G 1989 Distribution of photosynthetically fixed carbon within root systems of *Eucalyptus pilularis* ectomycorrhizal with *Pisolithus tinctorius*. New Phytol. 112, 495–500.

Campbell B D, Grime J P and Mackey J M L 1991 A trade-off between scale and precision in resource foraging. Oecologia 87, 532–538.

Clarkson D T 1985 Factors affecting mineral acquisition by plants. Ann. Rev. Plant Physiol. 36, 77–115.

Clarkson D T and Scattergood C B 1982 Growth and phosphate transport in barley and tomato plants during the development of, and recovery from, phosphate stress. J. Exp. Bot. 33, 865–875.

Clipson N J W, Cairney J W G and Jennings D H 1987 The physiology of basidiomycete linear organs. I. Phosphate uptake by cords and mycelium in the laboratory and in the field. New Phytol. 105, 449–457.

Cosgrove D J and Hedrich R 1991 Stretch-activated chloride, potassium, and calcium channels coexisting in plasmamembranes of guard cells of *Vicia faba* L. Planta 186, 143–153.

Cox G and Tinker P B 1976 Translocation and transfer of nutrients in vesicular-arbuscular mycorrhizas. I. The arbuscule and phosphorus transfer: a quantitative ultrastructural study. New Phytol. 77, 371–378.

Dexheimer J, Gianinazzi S and Gianinazzi-Pearson V 1979 Ultrastructural cytochemistry of the host-fungus interfaces in the endomycorrhizal association *Glomus mosseae/Allium cepa*. Z. Pflanzenphysiol. 92, 191–206.

Dexheimer J, Aubert-Dufresne M-P, Gérard J, Le Tacon F and Mousain D 1986 Étude de la localisation ultrastructurale des activités phosphatasiques acides dans deux types d'ectomycorrhizes. Bull. Soc. Bot. Fr. 133, 343–352.

Downes G M, Alexander I J and Cairney J W G 1992 A study of ageing of spruce (*Picea sitchensis* Bong. Carr.) ectomycorrhizas. I. Morphological and cellular changes in mycorrhizas formed by *Tylospora fibrillosa* (Burt.) Donk and *Paxillus involutus* (Batsch.) Fr. New Phytol. 122, 141–152.

Duddridge J A 1986 Specificity and recognition in mycorrhizal associations. *In* Physiological and Genetical Aspects of Mycorrhizae. Eds. V Gianinazzi-Pearson and S Gianinazzi. pp 45–58. INRA, Paris.

Duddridge J A and Read D J 1984 Modification of the host fungus interface in mycorrhizas synthesized between *Suillus bovinus* (FR.) O. Kuntz and *Pinus sylvestris* L. New Phytol. 96, 583–588.

Eissenstat D M and Newman E I 1990 Seedling establishment near large plants: effects of vesicular-arbuscular mycorrhizas on the intensity of plant competition. Functional Ecol. 4, 95–99.

Evans D G and Miller M H 1988 Vesicular-arbuscular mycorrhizas and the soil-disturbance-induced reduction of nutrient absorption in maize. I. Causal relations. New Phytol. 110, 67–74.

Fairchild G L and Miller M H 1988 Vesicular-arbuscular mycorrhizas and the soil-disturbance-induced reduction of nutrient absorption in maize. II. Development of the effect. New Phytol. 110, 75–84.

Finlay R D 1985 Interactions between soil micro-arthropods and endomycorrhizal associations of higher plants. *In* Ecological Interactions in Soil. Eds. A H Fitter, D Atkinson, D J Read and MB Usher. pp 319–331. Blackwell Scientific Publications, Oxford.

Fitter A H 1985 Functioning of vesicular-arbuscular mycorrhizas under field conditions. New Phytol. 99, 257–265.

Fitter A H 1977 Influence of mycorrhizal infection on competition for phosphorus and potassium by two grasses. New Phytol. 79, 119–125.

Gange A C, Brown V K and Farmer L M 1990 A test for mycorrhizal benefit in an early successional plant community. New Phytol. 115, 85–91.

Garcia Romera I, Garcia Garrido J M, Martinez-Molina E and Ocampo J A 1991 Production of pectolytic enzymes in lettuce root colonized by *Glomus mosse*ae. Soil Biol. Biochem. 23, 597–601.

Gianinazzi-Pearson V 1984 Host-fungus specificity, recognition and compatibility in mycorrhizae. *In* Genes Involved in Microbe Plant Interactions. Eds. ES Dennis, B Hohn, Th Hohn, P King, I Schell and DPS Verma. pp 225–253. Springer-Verlag, Vienna.

Gianinazzi-Pearson V 1986 Cellular modifications during host-fungus interactions in endomycorrhizae. NATO-ASI, Series H, Vol. 1, 29–37.

Gianinazzi-Pearson V, Bonfante-Fasolo P and Dexheimer J 1986 Ultrastructural studies of surface interactions during adhesion and infection by ericoid endomycorrhizal fungi. NATO-ASI, Series H, Vol. 4, 273–282.

Gianinazzi-Pearson V, Gianinazzi S and Brewin N J 1990 Immunocytochemical localization of antigenic sites in the perisymbiotic membrane of endomycorrhiza using monoclonal antibodies reacting against the peribacteroid membranes of nodules. Endocytobiology IV, pp 127–131. INRA, Paris .

Gianinazzi-Pearson V, Smith SE, Gianinazzi S and Smith FA 1991 Enzymatic studies on the metabolism of vesicular-arbuscular mycorrhizas. V. Is H^+-ATPase a component of ATP-hydrolysing enzyme activities in plant–fungus interface? New Phytol. 117, 61–76.

Grime J P, Mackey J M L, Hillier S H and Read D J 1987 Floristic diversity in a model system using experimental microcosms. Nature 328, 420–422.

Hadley G 1975 Organization and fine structure of orchid mycorrhiza. *In* Endomycorrhizas. Eds. F E Sanders, B Mosse and P B Tinker. pp 335–351. Academic Press, London.

Hale K A and Sanders F E 1982 Effects of benomyl on

vesicular-arbuscular mycorrhizal infection of red clover (*Trifolium pratense* L.) and consequences for phosphorus inflow. J. Plant Nutr. 5, 1355–1367.

Harley J L and Smith S E 1983 Mycorrhizal Symbiosis. Academic Press, London. 483 p.

Harley J L and Brierley J K 1955 The uptake of phosphate by excised mycorrhizal roots of the beech. VII. Active transport of ^{32}P from fungus to host during uptake of phosphate from solution. New Phytol. 54, 296–301.

Harley J L and Loughman B C 1963 The uptake of phosphate by excised mycorrhizal roots of the beech. IX. The nature of the phosphate compounds passing into the host. New Phytol. 62, 350–359.

Hedrich R and Schroeder J I 1989 The physiology of ion channels and electrogenic pumps in higher plants. Ann. Rev. Plant Physiol. and Plant Mol. Biol. 40, 539–569.

Heldt H H, Chon C J, Maronde D, Herold A, Stankovic Z S and Walker D A 1977 Role of orthophosphate and other factors in the regulation of starch formation in leaves and isolated chloroplasts. Plant Physiol. 59, 1146–1155.

Jasper D A, Abbott L K and Robson A D 1989 Soil disturbance reduces the infectivity of external hyphae of vesicular-arbuscular mycorrhizal fungi. New Phytol. 112, 93–99.

Jasper D A, Abbott L K and Robson A D 1989 Hyphae of a vesicular-arbuscular mycorrhizal fungus maintain infectivity in dry soil, except when the soil is disturbed. New Phytol. 112, 101–107.

Jeanmaire C, Dexheimer J, Marx C, Gianinazzi S and Gianinazzi-Pearson V 1985 Effect of vesicular-arbuscular mycorrhizal infection on the distribution of neutral phosphatase activities in root cortical cells. J. Plant Physiol. 119, 285–293.

Keifer D W and Lucas W J 1982 Potassium channels in *Chara corallina*. Control and interaction with the electrogenic H^+ pump. Plant Physiol. 86, 841–847.

Koide R T 1991 Nutrient supply, nutrient demand and plant response to mycorrhizal infection. New Phytol. 117, 365–386.

Koide R T and Elliott G 1989 Cost, benefit and efficiency of the vesicular-arbuscular mycorrhizal symbiosis. Funct. Ecol. 3, 252–255.

Koide R T and Mooney H A 1987 Spatial variation in inoculum potential of vesicular-arbuscular mycorrhizal fungi caused by formation of gopher mounds. New Phytol. 107, 173–182.

Lee R B 1982 Selectivity and kinetics of ion uptake by barley plants following nutrient deficiency. Ann. Bot. 50, 429–449.

Lei J and Dexheimer J 1988 Ultrastructural localization of ATPase activity in the *Pinus sylvestris/Laccaria laccata* ectomycorrhizal association. New Phytol. 108, 329–334.

Lei J, Lapeyrie F, Malajczuk N and Dexheimer J, 1990 Infectivity of pine and eucalypt isolates of *Pisolithus tinctorius* (Pers.) Coker and Couch on roots of *Eucalyptus urophylla* S.T. Blake in vitro. II. Ultrastructural and biochemical changes at the early stage of mycorrhiza formation. New Phytol. 116, 115–122.

Martin F, Ramstedt M and Söderhäll K 1987 Carbon and nitrogen metabolism in ectomycorrhizal fungi and ectomycorrhizas. Biochimie 69, 569–581.

Marx C, Dexheimer J, Gianinazzi-Pearson V and Gianinazzi S 1982 Enzymatic studies on the metabolism of vesicular-arbuscular mycorrhizas. IV. Ultrastructural evidence (ATPase) for active transfer processes in the host–arbuscule interface. New Phytol. 90, 37–43.

McGonigle T P 1988 A numerical analysis of published field trials with vesicular-arbuscular mycorrhizal fungi. Functional Ecol. 2, 473–478.

McGonigle T P and Fitter A H 1988 Ecological consequences of arthropod grazing on VA mycorrhizal fungi. Proc. R. Soc. Edinburgh 94B, 25–32.

McGonigle T P and Fitter A H 1990 Ecological specificity of vesicular-arbuscular mycorrhizal associations. Mycol. Res. 94, 120–122.

Mooney H A 1972 The carbon balance of plants. Annu. Rev. Ecol. Syst. 3, 315–346.

Patrick J W 1989 Solute efflux from the host at plant–microorganism interfaces. Aust. J. Plant Physiol. 16, 53–67.

Pearson V and Read D J 1975 The physiology of the mycorrhizal endophyte of *Calluna vulgaris* L. Hull. Trans. Br. Mycol. Soc. 64, 1–7.

Pohlaven F 1989 The influence of cytokinin 2iPA on growth, ion transport and membrane fluidity in mycelia of the mycorrhizal fungus *Suillus variegatus*. Agric. Ecos. Environ. 28, 399–402.

Rabatin S C and Stinner B R 1988 Indirect effects of interactions between VAM fungi and soil-inhabiting invertebrates on plant processes. Agric. Ecos. Environ. 24, 135–146.

Scannerini S and Bonfante-Fasolo P 1983 Comparative ultrastructural analysis of mycorrhizal associations. Can. J. Bot. 61, 917–943.

Schwab S M Menge J A and Tinker P B 1991 Regulation of nutrient transfer between host and fungus in vesicular-arbuscular mycorrhizas. New Phytol. 117, 387–398.

Serrigny J 1982 Etudes cytologique, cytochimique, et cytoenzymologique d'une orchidée tropicale : *Epidendrum* sp. Diplôme d'Etude Approfondie, Université of Nancy. 32 p.

Slayman C L 1965 Electrical properties of *Neurospora crassa*. Effects of external cations on the intracellular potential. J. Gen. Physiol. 49, 69–92.

Smith D C, Muscatine L and Lewis D 1969 Carbohydrate movement from autotrophs to heterotrophs in parasitic and mutualistic symbiosis. Biol. Rev. 44, 17–90.

Smith F A and Smith S E 1986 Movement across membranes: physiology and biochemistry. *In* Physiological and General Aspects of Mycorrhizae. Eds. V Gianinazzi-Pearson and S Gianinazzi. pp 75–84. INRA, Paris.

Smith F A and Smith S E 1989. Membrane transport at the biotrophic interface: an overview. Aust. J. Plant Physiol. 16, 33–43.

Smith S E 1974. Mycorrhizal fungi. CRC Critical Reviews in Microbiology, 275–313.

Smith S E and Dickson S 1991 Quantification of active vesicular-arbuscular mycorrhizal infection using image analysis and other techniques. Aust. J. Plant Physiol. 18, 637–648.

Smith S E and Gianinazzi-Pearson V 1988 Physiological interactions between symbionts in vesicular-arbuscular mycorrhizal plants. Ann. Rev. Plant Physiol. Mol. Biol. 39, 211–244.

Smith S E and Smith F A 1990 Structure and function of the interfaces in biotrophic symbioses as they relate to nutrient transport. New Phytol. 114, 1–38.

Smith S E, Long C M and Smith F A 1989 Infection of roots with a dimorphic hypodermis: possible effects on solute uptake. Agric. Ecosy. Environ. 29, 403–407.

Smith S E, Robson A D and Abbott L K 1992 The involvement of mycorrhizas in the assessment of genetically dependent efficiency of nutrient uptake and use. Plant Soil 146, 169–179.

Straker C J and Mitchell D T 1987 Kinetic characterization of a dual phosphate uptake system in the endomycorrhizal fungus of *Erica hispidula* L. New Phytol. 106, 129–137.

Strullu D G, Grellier B, Garrec J P, McCready c c and Harley J L 1986 Effects of monovalent and divalent cations on phosphate absorption by beech mycorrhizas. New Phytol. 103, 403–416.

Tester M 1990 Plant ion channels: whole-cell and single-channel studies. New Phytol. 114, 305–340.

Tester M A, Smith F A and Smith S E 1992 The role of ion channels in controlling solute exchange in mycorrhizal associations. *In* Mycorrhiza in Ecosystems. Eds. DJ Read, DH Lewis, AH Fitter and IJ Alexander. pp 348–351. CAB International, Oxon, UK.

Thomson B D, Clarkson D T and Brain P 1990 Kinetics of phosphorus uptake by the germ-tubes of the vesicular-arbuscular mycorrhizal fungus, *Gigaspora margarita*. New Phytol. 116, 647–653.

Woolhouse H W 1975 Membrane structure and transport problems considered in relation to phosphorus and carbohydrate movements and the regulation of endotrophic mycorrhizal associations. *In* Endomycorrhizas. Eds. F E Sanders, B Mosse and P B Tinker. pp 209–239. Academic Press, London.

Wyse R E, Zamski E and Tomos, A D 1986 Effect of turgor on kinetics of sucrose uptake. Plant Physiol. 81, 478–481.

Reprinted from *Plant and Soil*, Volume 159, No. 1 (1994), pp. 103–113.

Possible role of soil microorganisms in aggregation in soils

J.M. TISDALL
Department of Food and Agriculture, Victoria, Institute for Sustainable Agriculture, Tatura, Vic. 3616, Australia

Key words: mycorrhizal fungi, roots, soil bacteria, soil macroaggregates, soil microaggregates

Abstract

In many soils, roots and fungal hyphae, especially those of vesicular arbuscular mycorrhizal (VAM) fungi, stabilize macroaggregates (> 250 μm diameter); organic residues, bacteria, polysaccharides and inorganic materials stabilize microaggregates (< 250 μm). This review discusses the factors (including other organisms) which affect VAM hyphae and their extracellular polysaccharides in soil, and the subsequent effect on stability of aggregates. The review also discusses the possible role of other organisms, including ectomycorrhizal fungi, in the stability of soil, and suggests future research.

Introduction

Soils consist of particles of sand, silt and clay, held together into aggregates of various sizes by organic and inorganic materials. In the field, the stability to water of these aggregates and the pores between them affect infiltration, drainage and storage of water, the activity of biota in the soil, erosion of the topsoil and the growth of crops. This review discusses mainly the role of vesicular arbuscular mycorrhizal (VAM) fungi, in the stabilization of aggregates, but also examines these organisms as part of the total microbial population. Where mechanisms are not understood, future research is suggested.

Aggregation

Edwards and Bremner (1967) proposed that soils consist of microaggregates (< 250 μm diameter) bound into macroaggregates (> 250 μm), and that bonds within microaggregates are stronger than those between microaggregates. Several other alfisols and mollisols have also been shown to contain macroaggregates and microaggregates of these sizes (Oades and Waters, 1991). Tisdall and Oades (1982) proposed that, in an alfisol, microaggregates themselves are also built up in stages with different bonds at each stage. Oades and Waters (1991) modified the model further alfisols and mollisols, whose aggregates are mainly stabilized by organic materials. In this model, the stages of aggregation are ; < 20 μm —> 20–90 μm —> 90–250 μm —> > 250 μm. Not all soils fit this model. Unstable macroaggregates of a vertisol broke down in water directly to stable microaggregates 20–35 μm diameter (Collis-George and Lal, 1970). An oxisol, stabilized mainly by oxides, contained macroaggregates which were stable to water but broke down directly to microaggregates < 20 μm diameter when subjected in water to prolonged end-over-end shaking or ultrasound (Oades and Waters, 1991). In the model of Oades and Waters (1991), roots (up to about 1 metre length per gram of soil) and fungal hyphae (up to about 20 metre length per gram) fit in the pores between microaggregates and so stabilize macroaggregates (Oades and Waters, 1991). Because they are smaller (up to a few micrometres diameter), plant and microbial debris and bacteria fit in pores within microaggregates, and with their extracellular

A.D. Robson, L.K. Abbott & N. Malajczuk (eds.), Management of mycorrhizas in agriculture, horticulture and forestry, 115–121.

polysaccharides, contribute to the stability of the microaggregates. Inorganic materials also stabilize microaggregates.

Macroaggregation

In soils stabilized by organic matter, the stability depends on the amount of organic material or carbohydrate added to soil and to the microbial biomass and length of fungal hyphae in the soil (Lynch, 1984; Tisdall, 1991). Therefore the stability of macroaggregates is greater in soil under pasture than under crops, in direct-drilled soil than in untilled soil and in soil cropped every year than in soil fallowed every second or so years. For example, alfisols under pasture, wheat every year or a wheat-fallow rotation (each for over 30 years) contained 2.3%, 1.7% and 1.1% organic carbon and consisted of 33%, 17% and 5% water-stable aggregates > 2000 μm diameter respectively (Tisdall and Oades, 1980). After three years with cereals under systems of direct-drill, reduced tillage and traditional tillage, alfisols contained, 46% and 42% and 38% water-stable macroaggregates respectively (Burch et al., 1986). Theoretically, aggregates > 250 μm are separated by macropores > 25 μm diameter, which water, air and roots can follow (Oades, 1984). When unstable macroaggregates break up into microaggregates, macropores are theoretically replaced with micropores 0.2–2.5 μm. These micropores store water which theoretically plants could extract from soil, but micropores restrict aeration, hydraulic conductivity and the growth of roots. Agricultural systems affect the macroporosity not only through their effect on the stability of macroaggregates, but also through their effect on faunal activity in the soil and on compaction from above the ground (Mullins et al., 1990).

In soils under plants, macroaggregates are stabilized mainly by roots and VAM hyphae (Tisdall and Oades, 1982). The relative effect of roots and (VAM hyphae plus roots) on the stabilization of macroaggregates could be determined in soil divided by a screen which allowed hyphae but not roots to grow through (Camel et al., 1991). Saprophytic fungi also stabilize macroaggregates (Lynch and Bragg, 1985), and Emerson et al. (1986) suggested that ectomycorrhizal hyphae stabilized macroaggregates in several forest soils.

Roots and VAM hyphae form an extensive network in soil and are covered with extracellular polysaccharides to which microaggregates are firmly held. The network of encrusted roots and hyphae hold the macroaggregates intact, so that they do not collapse in water. The clay on the surface protects the roots and hyphae from microbial decomposition, but once the roots and hyphae die the network is broken by fauna or tillage, and the macroaggregates are disrupted in water; the encrusted fragments remain as microaggregates (Oades and Waters, 1991). Macroaggregates are formed quickly by growing roots and hyphae but are easily destabilized by tillage. For example, ryegrass grown in the field for six months increased the % of stable macroaggregates of an alfisol from 39% to 78%; several similar soils under tilled tomatoes for one year after old pasture, contained 58% stable macroaggregates compared with 90% in several soils under old pasture (Adem and Tisdall, 1984).

Reid et al. (1982) showed that maize grown in a silt loam destabilized soil aggregates. They suggested that organic acids released into the rhizosphere destabilized the macroaggregates by breaking the polycation bridges between organic matter and clay by reducing the pH and/or chelating the cations (Yeoh and Oades, 1981). This hypothesis is not supported by the results of Lynch and Bragg (1985) or Pojasak and Kay (1990). Pojasak and Kay (1990) showed that root exudates of maize and bromegrass incubated with macroaggregates of an alfisol increased the stability of the macroaggregates. They concluded that the exudates released unchelated polycations into the soil solution which strengthened bonds between organic matter and clay and so increased the stability. This conclusion is similar to the hypothesis of Tisdall (1991) for the formation of bonds between fungal hyphae and clay. She suggested that hyphal exudates dissolve polyvalent cations from the surface of hyphae or from mineral soil. In this model, yet to be tested, the polyvalent cations then coagulate the clay and form cation bridges between the hyphal polysaccharides and the clay, forming stable aggregates.

Microaggregation

In alfisols and mollisols, microaggregates are very stable to water, but are not produced quickly. Even after a wheat-fallow rotation for 45 years, 70% of an alfisol consisted of water-stable microaggregates 20–250 μm diameter (Oades and Waters, 1991), as it had 12 years earlier (Tisdall and Oades, 1980). However, high energy, such as prolonged end-over-end shaking or ultrasound in the laboratory or heavy rain or sodicity in the field, disrupts microaggregates.

In electron micrographs, microaggregates about 100–200 μm diameter appeared to be fragments of plants encrusted with particles of soil, whereas microaggregates < 90 μm contained highly degraded plant material or were empty (Oades and Waters, 1991). Aggregates < 20 μm diameter are also highly stable in water, and appear to be clusters of clay stabilized by inorganic and organic materials (Tisdall and Oades, 1982). The organic materials in microaggregates < 20 μm diameter include polysaccharides, hyphal fragments, and bacterial cells or colonies which are encrusted with particles of clay. Particles of clay attached to extracellular polysaccharides around microbial cells also formed microaggregates 5–20 μm in pure kaolinite or montmorillonite (Dorioz et al., 1993). Foster (1988) suggested that some empty microaggregates < 20 μm diameter seen in micrographs were once bacteria encrusted with particles of clay.

VA mycorrhizal fungi

Vesicular arbuscular (VA) mycorrhizas form on most species of plants and produce spores or sporocarps in most soils (Harley and Smith, 1983). Individual plants may form VA mycorrhizas with several species of fungi, with more diverse fungal species being found in native grasslands than in agricultural soils (Schenck et al., 1989). Most VAM fungi produce simple branched hyphae or hyphal strands in the soil and extend the root systems of plants. The stability of macroaggregates of several soils was related to the length of these hyphae in soil (Tisdall and Oades, 1979, 1980; Elliott and Coleman, 1988; Miller and Jastrow, 1990). Within each macroaggregate, fungal hyphae form a network with up to 50 metre of hyphae per gram of stable aggregate, or up to 14 metre of external hyphae per centimetre of root (Barea, 1991; Tisdall, 1991). These hyphae produce extracellular polysaccharides to which microaggregates are attached and bound into stable macroaggregates by the network of hyphae. Thomas et al. (1986) suggested that onion roots rather than their VAM hyphae stabilized macroaggregates of a silty clay loam and that the hyphae stabilized the aggregates indirectly by stimulating the growth of the roots. However, the stability of macroaggregates of an alfisol under grassland was directly related to the length of external VAM hyphae and to the length of fine roots (0.2–1 mm diameter); the stability was not directly related to the length of very fine roots (< 0.2 mm diameter), but was indirectly related through the external hyphae supported by the very fine roots (Miller and Jastrow, 1990). The different effects of fine and very fine roots on stability were probably because plants with mainly very fine roots < 0.1 mm diameter form fewer mycorrhizas than those with mainly coarse roots (> 0.5 mm diameter) (Barea, 1991).

Little is known about the growth of external hyphae of VA mycorrhizas, which fungal species are the most effective stabilizers, whether some species are more effective in some soils than others, and the mechanisms by which they form and stabilize macroaggregates. Research is needed to determine whether the most effective stabilizers; (a) produce more mucilage, (b) produce more persistent or stickier mucilage, (c) are bonded by hydrophobic bonds or bridges of polyvalent cations or have greater electrostatic attraction, to clay platelets, (d) persist longer in the soil, (e) interact better with some species of plants, microorganisms or animals than others, (f) orientate particles of clay so that they lie closer together and are strongly bound together, (g) invade the soil more readily, or (h) produce more external hyphae or more of one type of hypha than another (Tisdall, 1991).

External hyphae of some VAM fungi can grow at least 90 mm from the surface of the root, but it is not known whether all species can (Camel et al., 1991). As with other types of fungi, research is needed to determine whether all the hyphae within one species are completely covered with polysaccharide with particles of clay attached. Within any one species of VAM fungus, the morphology of the external hyphae varies considerably (Harley and Smith, 1983), yet it is not known which type of hypha, if any, is the most effective stabilizer. Although some VAM fungi can adapt to physical and chemical changes in soil, the amount, and possibly type, of external hyphae of some species of VAM fungi is affected by other microorganisms, root exudates, pests, clay content, soil pH, organic matter, fungicides, pesticides and the phosphorus content of the plant, and possibly the charges on the surfaces of the hyphae and of the clay (Gianinazzi-Pearson, 1989; Barea, 1991; Tisdall, 1991). Research is needed to determine whether these factors also affect the stability of macroaggregates through their effect on the fungi.

Tillage, long fallow, biocides and compacted or waterlogged soil can each significantly decrease the growth and colonization of roots by mycorrhizal fungi. Research is needed to determined whether these factors slow the rate at which external hyphae can subsequently stabilize aggregates (Tisdall, 1991). Agricultural soils contain fewer species of VAM fungi than native grasslands do (Skinner and Bowen, 1974; Schenck et al., 1989), yet it is not known whether those species that survive in agriculture are the most effective stabilizers, or whether combination of species of fungi are more effective than a single species. Roots of the same or different species of plant can be connected through hyphae of the same fungus; more research is needed to determine whether mixed or relay crops, or agroforests are more effective than sole crops (Tisdall, 1991). Research is also needed to determine whether crop rotation affects the stability of macroaggregates through its effect on VAM fungi. For example, the roots of a direct-drilled crop may become connected to the hyphal network of the previous crop; the effect on the stability of macroaggregates is unknown. Also, when crops which rarely form VA mycorrhizas, e.g. lupins or Cruciferae, are grown, the species of VAM fungi may change restricting infection of subsequent crops and growth of external hyphae (Ocampo and Hayman, 1981; Harinkumar and Bagyaraj, 1988). Research is needed to determine whether these non-mycorrhizal crops affect the stability of macroaggregates, and whether crop rotation is important.

Mycorrhizal roots translocate carbon to the external hyphae which in turn produce external polysaccharides. Research is needed to determine whether factors which affect photosynthesis and distribution of photosynthate within the plant, affect stabilization of aggregates by these hyphae and roots. These factors include shaded, diseased, grazed or cut shoots, nutrient status or age of the plant. Also, research is needed to determine whether some combinations of plant, fungus and other organisms lead to more external hyphae or more polysaccharide than others. For example, *Pseudomonas* species have been shown to stimulate, and actinomycetes to depress, the formation of VA mycorrhizas (Krishna et al., 1982), and possibly affect the stability of macroaggregates. Also *Lolium perenne* and C4 grasses supported more external hyphae and stabilized soil more effectively than *Trifolium repens* and C3 grasses respectively did (Tisdall and Oades, 1979; Miller and Jastrow, 1990).

Other fungi

In forests, the lateral roots of many trees are covered by a sheath of ectomycorrhizal hyphae, usually basidiomycetes (Harley and Smith, 1983). At the same time these plants may also be infected with VAM fungi. The ectomycorrhizal hyphae sometimes extend at least 120 mm into the soil, and are covered with extracellular polysaccharides; the hyphae often have particles of soil enmeshed between them (Skinner and Bowen, 1974; Foster, 1981). Emerson et al. (1986) suggested that ectomycorrhizal hyphae helped to explain the more stable aggregates from several soils under forests than from under pasture in several arable soils. Research is needed to determine whether ectomycorrhizal hyphae stabilize forest soils, and produce stronger bonds with clays than other fungi do, as suggested by Emerson et al. (1986). If so, as with VAM fungi, research is needed to determine whether; (a) ectomycorrhizal fungi compete successfully for nutrients retarding decomposition of organic bonds by other microorganisms, (b) ectomycorrhizal fungi produce less labile polysaccharides or organic bonds than those produced in soils under herbaceous plants, (c) ectomycorrhizal roots have more tips and potentially release more organic materials than non-mycorrhizal roots do, or (d) polyphenols which are known to exist in ectomycorrhizas interact with polysaccharides protecting them chemically from degradation, or e) bacteria which are known to colonize dead or dying cells in the outer sheath contribute to stable aggregation. However, trees add much organic material each year from above and below the ground potentially increasing the production of bonds in soil; the organic material is often rich in polyphenols and resins which may chemically protect them from degradation. Forest soils are untilled and the polysaccharides may physically be more protected from microbial attack than polysaccharides in tilled soils are. As in other soils, the relative abundance of different fungal species may affect stabilization of aggregates (Foster and Marks, 1967; Harley and Smith, 1983; Gosz and Fischer, 1984; Emerson et al., 1986; Foster, 1988; Schenck et al., 1989). The relative effects of these factors need to be determined. As with other fungi, we need to determine whether hyphal growth is affected by physical, chemical and biological properties of the soil. The hyphal growth of one ectomycorrhizal fungus was 80% less in compacted soil than in uncompacted soil (bulk density 1.6 and 1.2 Mg m^{-3} respectively, and with high than low levels of added phosphate (250 and 20 parts 10^{-6} respectively); a *Pseudomonas* species stimulated the hyphal growth (Skinner and Bowen, 1974). The effects on stability of macroaggregates are unknown.

Many papers have reported that saprophytic fungi can also stabilize macroaggregates with some species being more efficient stabilizers than others. Experiments have shown such stabilization with incubated soil or clay; (a) with or without added organic residues or simple sugars, (b) inoculated with pure cultures of microorganisms, and (c) with or without chemical inhibition of selected bacteria or fungi (Harris et al., 1966; Molope et al., 1982; Lynch and Bragg, 1985; Kinsbursky et al., 1989). The stability did not appear to persist once the organic substrate had disappeared. In an irrigated peach orchard in south-eastern Australia, 11 t ha^{-1} of straw had to be added to the untilled alfisol each year to maintain stable aggregates (Cockroft and Tisdall, 1978). Research is needed to determine whether stabilization by saprophytic fungi is affected by other organisms in the soil, and whether the mechanisms of stabilization are similar to those of mycorrhizal fungi. The effect on the stability of macroaggregates is unknown.

General microbiology

Living roots release many types of organic materials into the rhizosphere stimulating microorganisms, especially Gram negative bacteria, within 50 μm of the root surface. Research is needed to determine which bacterial species, if any, are the most effective stabilizers. On agar plates or in electron micrographs, many bacteria (especially of Gram negative bacteria) are surrounded by a distinct capsule made of polysaccharides (Hepper, 1975; Foster, 1983). Many of these bacteria also produce extracellular polysaccharides which form a mucoid layer around the organisms (Hepper, 1975). As seen in electron micrographs, bacterial cells or colonies are surrounded by a layer of clay platelets which are tangential to the surface of the colony and physically protect from microbial attack the extracellular polysaccharides which hold the orientated clay together forming stable microaggregates < 20 μm diameter (Foster et al., 1983). However, some of the polysaccharide interacts with cations or phenolic polymers in soil and is chemically protected from microbial degradation (Martin, 1971; Verma and Martin, 1976).

Research is needed to determine whether mycorrhizal roots stimulate or suppress specific groups of bacteria to produce capsules or extracellular polysaccharides in soil and hence promote or suppress bacterial stabilization of aggregates. Some mycorrhizal roots produce exudates which differ from those of non-mycorrhizal roots (Gosz and Fisher, 1984), and some mycorrhizal roots may stimulate specific groups of bacteria, including those which are known to produce extracellular polysaccharide in soil (Meyer and Linderman, 1986; Bagyaraj and Menge, 1978).

Much of the mucilage from roots is eventually transformed into microbial polysaccharide, forming the mucigel which coats particles of soil and organic fragments (Emerson et al., 1986). A synthetic polymer which strengthened the coarse pores (mainly 15–50 μm diameter) in an old arable soil prevented the aggregates from collapsing in water (Quirk and Williams, 1974), suggesting that roots and microorganisms stabilize aggregates by producing polysaccharides in accessible coarse pores Oades (1984).

Conclusions

Soils are made up of aggregates of different sizes with different bonds for each size. Roots and hyphae, mainly mycorrhizal hyphae, stabilize macroaggregates (> 250 μm diameter), and microbial and plant residues, polysaccharides, bacteria and inorganic materials stabilize microaggregates (< 250 μm diameter). Roots release many types of organic material into the soil and support active microorganisms in the rhizosphere. The roots, bacteria, mycorrhizal fungi and other organisms compete for space, carbon and nutrients. However, we understand little about the role of each type of organism and how they all interact in soil to stabilize aggregates. For each soil, we need to determine the best combinations and management of plant and organisms which will best stabilize aggregates.

Acknowledgment

I thank Dr H J Downing for helpful criticism of the manuscript, and Grains Research and Development Corporation for financial support.

References

Adem H H and Tisdall J M 1984 Management of tillage and crop residues for double-cropping in fragile soils of south eastern Australia. Soil Tillage Res. 4, 577–589.

Bagyaraj D J and Menge J A 1978 Interactions between a VA mycorrhiza and *Azotobacter* and their effects on rhizosphere microflora and plant growth. New Phytol. 80, 567–573.

Barea J M 1991 Vesicular-arbuscular mycorrhizae as modifiers of soil fertility. Adv. Soil Sci. 15, 1–40.

Burch G J, Mason I B, Fischer R A and Moore I D 1986 Tillage effects on soils: Physical and hydraulic responses to direct drilling at Lockhart, N.S.W. Aust. J. Soil Res. 24, 377–391.

Camel S B, Reyes-Solis M G, Ferrera-Cerrato R, Franson R L, Brown M S and Bethlenfalvay G J 1991 Growth of vesicular-arbuscular mycorrhizal mycelium through bulk soil. Soil Sci. Soc. Am. J. 55, 389–393.

Cockroft B and Tisdall J M 1978 Soil management, soil structure and root activity. *In* Modification of Soil Structure. Eds. W W Emerson R D Bond and A R Dexter. pp 387–391. Wiley, London.

Collis-George N and Lal R 1970 Infiltration into columns of swelling soil as studied by high speed photography. Aust. J. Soil Res. 8, 195–207.

Dorioz J M, Robert M and Chenu C 1993 The role of roots, fungi, and bacteria on clay particle organization. An experimental approach. Geoderma 54 (in press).

Edwards A P and Bremner J M 1967 Microaggregates in soils. J. Soil Sci. 18, 64–73.

Elliott E T and Coleman D C 1988 Let the soil work for us. Ecol. Bull. 39, 23–32.

Emerson W W, Foster R C and Oades J M 1986 Organo–mineral complexes in relation to soil aggregation and structure. *In* Interactions of Soil Minerals with Natural Organics and Microbes, Soil Science Society of Australia, Spec. Publ. No. 17. pp 521–548.

Foster R C 1981 Polysaccharides in soil fabrics. Science 214, 665–667.

Foster R C 1983 The plant root environment. *In* Soils: An Australian Viewpoint. pp 673–684. CSIRO, Melbourne/Academic Press, London.

Foster R C 1988 Microenvironments of soil microorganisms. Biol. Fert. Soils. 6, 189–203.

Foster R C and Marks G C 1967 Observations on the mycorrhizas of forest trees. Aust. J. Biol. Sci. 20, 915–926.

Foster R C, Rovira A D and Cock T W 1983 Ultrastructure of the root-soil interface. Am. Phytopathol. Soc., St. Paul, Minnesota. 157 p.

Gianinazzi-Pearson V 1989 Vesicular arbuscular endomycorrhizae as determinants for plant growth and survival. Dev. Soil Sci. 18, 69–76.

Gosz J R and Fisher F M 1984 Influence of clear-cutting on selected microbial processes in forest soils. *In* Current Perspectives in Microbial Ecology. Eds. M J Klug and C A Reddy pp 523–530. Amer. Soc. Microbiol., Washington.

Harinkumar K M and Bagyaraj D J 1988 Effect of crop rotation on native vesicular arbuscular mycorrhizal propagules. Plant Soil 110, 77–80.

Harley J L and Smith S E 1983 Mycorrhizal Symbiosis. Academic Press, London. 483 p.

Harris R F, Chesters G and Allen O N 1966 Dynamics of soil aggregation. Adv. Agron. 18, 107–169.

Hepper C M 1975 Extracellular polysaccharides of soil bacteria. *In* Soil Microbiology. Ed. N Walker. pp 93–110.

Butterworth, London.
Kinsbursky R S, Levanon D and Yaron B 1989 Role of fungi in stabilizing aggregates of sewage sludge amended soils. Soil Sci. Soc. Am. J. 53, 1086–1091.
Krishna K R, Balakrishna A N and Bagyaraj D J 1982 Interaction between a vesicular-arbuscular mycorrhizal fungus and *Streptomyctes cinnamomeous* and their effects on finger millet. New Phytol. 92, 401–405.
Lynch J M 1984 Interactions between biological processes, cultivation and soil structure. Plant Soil 76, 307–318.
Lynch J M and Bragg E 1985 Microorganisms and soil aggregate stability. Adv. Soil Sci. 2, 133–171.
Martin J P 1971 Decomposition and binding action of polysaccharides in soil. Soil Biol. Biochem. 3, 33–41.
Meyer J R and Linderman R G 1986. Selective influence on populations of rhizosphere or rhizoplane bacteria and actinomycetes by mycorrhizas formed by *Glomus fasciculatum*. Soil Biol. Biochem. 18, 191–196.
Miller R M and Jastrow J D 1990 Hierarchy of root and mycorrhizal fungal interactions with soil aggregation. Soil Biol. Biochem. 22, 579–584.
Molope M B, Grieve I C and Page E R 1982 Contributions by fungi and bacteria to aggregate stability of cultivated soils. J. Soil Sci. 38, 71–77.
Mullins C E, MacLeod D A, Northcote K H, Tisdall J M and Young I M 1990 Hardsetting soils: behaviour, occurrence, and management. Adv. Soil Sci. 11, 37–108.
Oades J M 1984 Soil organic matter and structural stability mechanisms and implications for management. Plant Soil 76, 319–337.
Oades J M and Waters A G 1991 Aggregate hierarchy in soils. Aust. J. Soil Res. 29, 815–828.
Ocampo J A and Hayman D S 1981 Influence of plant interactions on vesicular-arbuscular mycorrhizal infections. New Phytol. 87, 333–343.
Pojasak T and Kay B D 1990 Effect of root exudates from corn and bromegrass on soil structural stability. Can. J. Soil Sci. 70, 351–362.
Quirk J P and Williams B G 1974 The disposition of organic materials in relation to stable aggregation. Trans. 10th International Congress of Soil Science, Moscow. pp 165–172.
Reid J B, Goss M J and Robertson P D 1982 Relationship between the decreases in soil stability as affected by the growth of maize roots and changes in organically bound iron and aluminium. J. Soil Sci. 33, 397–410.
Schenck N C, Siqueira J O and Oliveira E 1989 Changes in the incidence of VA mycorrhizal fungi with changes in ecosystems. Dev. Soil Sci. 18, 83–91
Skinner M F and Bowen G D 1974 The uptake and translocation of phosphate by mycelial strands of pine mycorrhizas. Soil Biol. Biochem. 6, 53–56.
Thomas R S, Dakessian S, Ames R N, Brown M S and Bethlenfalvay G J 1986 Aggregation of a silty clay loam by mycorrhizal onion roots. Soil Sci. Soc. Am. J. 50, 1494–1499.
Tisdall J M and Oades J M 1979 Stabilization of soil aggregates by the root systems of ryegrass. Aust. J. Soil Res. 17, 429–441.
Tisdall J M and Oades J M 1980 The effect of crop rotation to stabilize aggregates of a red-brown earth. Aust. J. Soil Res. 18, 423–422.
Tisdall J M and Oades J M 1982. Organic matter and water-stable aggregates in soils J. Soil Sci. 33, 141–163.
Tisdall J M 1991 Fungal hyphae and structural stability of soil. Aust. J. Soil Res. 29, 729–743.
Verma L and Martin J P 1976 Decomposition of algal cells and components and their stabilization through complexing with model humic acid-type phenolic polymers. Soil Biol. Biochem. 8, 85–90.
Yeoh N S and Oades J M 1981 Properties of clays and soils after acid treatment. Aust. J. Soil Res. 19, 159–166.

Reprinted from *Plant and Soil*, Volume 159, No. 1 (1994), pp. 115–121.

Interactions between mycorrhizal fungi and other soil organisms

A.H. FITTER[1] and J. GARBAYE[2]
[1]*Department of Biology, University of York, York YO1 5DD, UK*
[2]*Laboratoire de Microbiologie Forestiere, INRA, 54280 Champenoux, France*

Key words: animals, bacteria, competition, mutualism, mycorrhizas, pathogens, soil organisms

Abstract

Mycorrhizal fungi interact with a wide range of other soil organisms, in the root, in the rhizosphere and in the bulk soil. These interactions may be inhibitory or stimulatory; some are clearly competitive, others may be mutualistic. Effects can be seen at all stages of the mycorrhizal fungal life-cycle, from spore population dynamics (predation, dispersal and germination) through root colonization to external hyphal growth. Two areas that seem likely to be of particular importance to the functioning of the symbiosis are the role of bacteria in promoting mycorrhiza formation and of soil animals in grazing the external mycelium. Mycorrhizal fungi also modify the interactions of plants with other soil organisms, both pathogens, such as root-inhabiting nematodes and fungi, and mutualists, notably nitrogen-fixing bacteria. These interactions are probably important both in natural ecosystems, where pathogens are increasingly recognized as playing controlling roles, and in agricultural systems, where mycorrhizas may be valuable in designing integrated systems of pest control and growth stimulation.

The rhizosphere: fauna and microbiota

Except in the surface layers of soils, where organic matter input from plant shoots is a major energy source, most biotic activity in soil occurs in the rhizosphere, simply because of the lack of food sources in the bulk soil. Mycorrhizal fungi are found inside roots (in the endorhizosphere, as it is sometimes termed), in the rhizosphere (or mycorhizosphere) itself and in the bulk soil. In all these zones they encounter and interact with other members of the soil biota. The internal mycelium interacts principally with the root itself, but there may be other organisms even in this part of the mycorrhizal fungus' habitat. Pathogenic fungi, for example, may invade roots and mycorrhizal fungi can alter host response to these pathogens (Perrin, 1991).

The external mycelium interacts with many more organisms, including bacteria, other fungi, protozoa, nematodes, arthropods and even large animals such as mammals. The interactions may be of various types, both beneficial and deleterious. In many cases it is not possible to define the interaction fully as yet; for example, mycorrhization helper bacteria (MHB), which are described in detail below, certainly improve the ability of mycorrhizal fungi to colonize plant roots, but whether the bacteria themselves gain from the interaction (in other words, whether the association is mutualistic) is unclear. Certainly, bacteria such as fluorescent pseudomonads may proliferate in the hyphosphere of mycorrhizal fungi (Klyuchnikov and Kozhevin, 1990). There may be a very large number of undescribed or unsuspected mutualisms in the interactions between microbes and mycorrhizal fungi.

In contrast, many documented interactions are apparently deleterious to the mycorrhizal fungus. Competitive interactions between the mycorrhizal fungus and both bacteria and other fungi have been observed, and there may be

A.D. Robson, L.K. Abbott & N. Malajczuk (eds.), Management of mycorrhizas in agriculture, horticulture and forestry, 123–132.

Table 1. A summary of the types of impact that members of the soil biota have on different stages of the life-cycle of mycorrhizal fungi

Life-cycle phase	Process	Effect	Responsible Group
spore population dynamics	grazing	dispersal death	small mammals invertebrates
	parasitism	death	chytrids, amoebae
germination colonization	chemical interactions	inhibition stimulation	bacteria bacteria
internal hyphal spread	uncertain	inhibition	fungi, nematodes
external hyphal growth	grazing	disconnection	invertebrates
sporulation	unknown	stimulation	bacteria

allelochemical interactions similar to antibiosis, which can however be either stimulatory or suppressive (Tylka et al., 1991). Pathogens of mycorrhizal fungi have not been widely described, yet certainly exist, while grazing of both spores and mycelium is now well documented. Grazing, however, may not always be deleterious; low levels of grazing may stimulate mycelial growth, possibly by recycling nutrients (Hanlon, 1981). Consumption of spores might seem an obviously damaging activity, but even this may result in spore dispersal (e.g. of truffles by small mammals) and so be in that sense beneficial.

The interactions between mycorrhizal fungi and other organisms in the soil are therefore complex and often poorly understood. In the following section we examine them from the standpoint of the fungal life-cycle (Table 1).

Impact of the soil biota on mycorrhizal fungi

Spore population dynamics

Surprisingly little is known of the ecology of mycorrhizal spores, despite the fact that for arbuscular fungi they are the only stage of the life-cycle where diagnostic features are expressed. It is an assumption that spores represent a resting stage in the life cycle, offering dispersal through time. They also disperse through space, and this dispersal can be promoted by animals. Fogel and Trappe (1978) emphasized the role of small mammals in dispersing spores of hypogeous ectomycorrhizal (ECM) fungi such as truffles, and many soil organisms have been shown to feed on spores of arbuscular mycorrhizal fungi (AMF), including earthworms (Lumbricidae), springtails (Collembola) and a range of other arthropods (see review by Fitter and Sanders, 1992). Spores can survive ingestion, since alfalfa roots can be infected with mycorrhizal fungi from woodlouse and millipede faeces (Rabatin and Stinner, 1988).

Dispersal and predation both therefore result from animals consuming mycorrhizal fungi. Whether either has an important impact on spore population dynamics is unknown, as are the effects of dispersal on the population structure of the fungi. The latter may be clarified with the advent of novel molecular techniques which will allow the genetic structure of populations to be investigated.

Spores are also vulnerable to parasitism from other soil organisms, principally chytridiaceous fungi (e.g. Sylvia and Schenck, 1983; Paulitz and Menge, 1986) and vampyrellid amoebae (e.g. Boyetchko and Tewari, 1991). Samples of field-collected spores often contain a high proportion colonized or damaged by these organisms, although some damaged spores may have died from other causes (Paulitz and Menge, 1984). Ross and Ruttencutter (1977) suggested that these parasites played an important role in regulating mycorrhizal spore population dynamics, but this remains to be tested fully.

Spore germination

Wilson, Hetrick and Kitt (1989) found that AMF spore germination could be suppressed in some non-sterile soils by an antagonistic microflora. However, Azcon-Aguilar et al. (1986) have shown that spore germination and hyphal growth of the AM fungus *Glomus mosseae* were enhanced by some rhizosphere bacteria or by the cell-free filtrates of their cultures, leading to a higher number of infection points on the roots. Mayo et al. (1986) obtained similar results with *Glomus versiforme* and spore-associated bacteria, as did Linderman and Paulitz (1990) with other systems. Tylka et al. (1991) found that three *Streptomyces* spp. had different effects on the germination of spores of three AM fungi, suggesting that specific interactions may occur. Many soil bacteria can therefore affect spore germination of AM fungi; some are inhibitors, while others, sometimes closely associated with the spores, promote germination and mycorrhizal development. In the case of ECM fungi, such strong experimental evidence has not been reported, but effects on germination may have been confounded with effects on colonization (see below).

Root colonization

During their pre-symbiotic phase, while growing from the propagule toward the root, mycorrhizal fungi compete with rhizobacteria, which may reduce both the development of the fungus and mycorrhiza formation. However, some bacteria may facilitate the establishment of the symbiosis. Mosse (1962) observed that mycorrhiza formation of alfalfa with *Glomus mosseae* was enhanced by bacteria on the spore surface, and Ames (1989) found some *Streptomyces* strains (and one *Nocardia*) to increase AM colonization of onion roots. Meyer and Linderman (1986) inoculated subterranean clover with unidentified arbuscular mycorrhizal fungi and a rhizosphere bacterium, *Pseudomonas putida*; the latter did not improve mycorrhizal infection, but amplified the effect of the symbiosis on the growth of the plant. This work is amongst the few which clearly shows that some bacteria can directly improve the functioning of the endomycorrhizal symbiosis without affecting its establishment. However, in these experiments, the effect of bacteria may have been on either spore germination or root colonization.

Clearer results have been obtained with ectomycorrhizas. Garbaye and Bowen (1987) inoculated *Pinus radiata* with *Paxillus involutus, Rhizopogon luteolus* and *Hebeloma crustuliniforme* in different sterilized soils, supplemented or not with various global soil microflora. Mycorrhiza formation was highly dependent upon the introduced microbial populations, with stimulatory as well as inhibitory effects. Similar results were obtained by McAfee and Fortin (1988) with larch. This was consistent with previous findings by Ridge and Theodorou (1972), who noted that the success of ECM inoculation was influenced by soil disinfection and reinfestation (independently of ECM competitors) or by bacterial contamination of the inoculum. For instance, inoculation of conifer seedlings with *Rhizopogon vinicolor* fails to result in mycorrhiza formation when a pure mycelial culture is used as inoculum, except in axenic conditions, but succeeds with a suspension of crushed sporocarps containing basidiospores and bacteria naturally present in the gleba (Castellano and Trappe, 1985). Thus, rhizosphere interactions with ECM fungi cannot be viewed only as competitive.

Bowen and Theodorou (1979) had shown that some bacterial strains isolated from the soil and the rhizosphere significantly interacted either positively or negatively with the growth of *Rhizopogon luteolus* mycelium along the root surface of *Pinus radiata* seedlings grown in vitro. With the same plant–fungus system, in both aseptic and non-aseptic conditions, Garbaye and Bowen (1989) found the same effects at different stages of the mycorrhization process; germination of propagules in the soil, growth along the root and primary infections. A higher proportion of bacterial isolates from the mycorrhizal mantle stimulated mycorrhizal development than from the soil, suggesting that these bacteria were symbionts of the mycorrhiza. This was the first demonstration that some bacteria closely associated with the fungus directly played a positive role in ectomycorrhiza formation. They were called MHB (mycorrhization helper bacteria).

MHB associated with the Douglas fir–*Laccaria laccata* symbiosis were studied in depth. Of 100 strains isolated from ectomycorrhizas and associated sporocarps of *L. laccata*, 14 consistently stimulated mycorrhiza formation in a range of environmental conditions as diverse as aseptic in vitro cultures, glasshouse experiments and several routine forest nurseries. They were all fluorescent pseudomonads or sporulating bacilli (Duponnois and Garbaye, 1991a), and were shown to be fungus-specific (they stimulate mycorrhiza formation by *L. laccata* and by the closely-related *L. bicolor* but inhibit mycorrhiza formation by all the other fungal genera tested), but not plant-specific, since they act as helpers of *L. laccata* with a range of angiosperm and gymnosperm ectomycorrhizal hosts (Garbaye and Duponnois, 1991). Therefore, these MHB associated with *L. laccata* selectively help a fungal species to form the symbiotic association by increasing its competitive ability against other potential symbionts. They act by improving the growth of the mycelium either indirectly, by detoxifying the substrate, or directly by producing organic acids used by the fungus as carbon sources and by emitting volatile substances responsible for the specific selectivity (Duponnois and Garbaye, 1991b; Garbaye and Duponnois, 1991).

The existence of MHB poses interesting new questions for mycorrhiza research; is the phenomenon general? Do MHB have an influence on the functioning of the symbiosis? Do they benefit from the presence of the fungus? What is their significance in terms of evolution of symbiotic systems? More generally, this evidence of positive interactions between rhizosphere bacteria and mycorrhizal fungi casts a new light on studies of PGPR (plant growth promoting rhizobacteria). PGPR are extensively studied for their potential applications in agriculture; they are commonly considered to stimulate plant growth by competing against root pathogens, producing plant-growth regulators and/or fixing free nitrogen (Chanway et al., 1991). However, the results reported above suggest that some of them may also promote plant growth by improving mycorrhizal symbiosis establishment and functioning. This hypothesis is further supported by the fact that most MHB and PGPR belong to the same taxonomic groups; fluorescent pseudomonads and sporulating bacilli. Unfortunately most work on PGPR so far has ignored mycorrhizas, and vice versa. A new joint field of research is obviously open.

Although the best-documented effects are with bacteria, there is also some evidence that soil animals may reduce colonization of roots by mycorrhizal fungi (Kaiser and Lussenhop, 1991), possibly simply by physical disturbance of the infection process, but most animal effects seem to be by grazing on the external mycelium (see below).

Internal hyphal spread

The growth of the internal mycelium of AM fungi is directly susceptible only to other organisms that themselves colonize the root tissue. Nematodes within roots may physically disrupt tissue (and probably cause physiological alterations too) and so impede the spread of mycelium (O'Bannon and Nemec, 1979). Other fungi in roots may have similar effects, but data on fungal interactions tend to emphasize the impact of the mycorrhizal fungus on the pathogen. For example, Muchovej et al. (1991) found that bean roots were colonized by *Glomus macrocarpum* more rapidly than by *Fusarium solani*; in apparent consequence, *F. solani* was only found in mycorrhizal root systems in areas where *G. macrocarpum* was absent, especially near the root tips, but the reverse effect was not found. Whether such interactions are physical or chemical in origin is unknown.

External hyphal growth

Many of the interactions discussed already, involving root colonization and the early growth after germination, presumably affect the development of the external mycelium as well. The established mycelium is, however, perhaps most vulnerable to grazing by the soil fauna (see Fitter and Sanders (1992) for a review). Soil animals are known to graze mycorrhizal mycelium, hyphal fragments having been identified in gut contents (Warnock et al., 1982). However, animals are automatically eliminated in experimental protocols involving soil sterilization, though the presence of animals (both collembola and nematodes have been used by various workers) in pot experiments can radically alter their outcome. Warnock et al. (1982) added 20 collembola (*Folsomia candida*) to pots containing individual leek seedlings infected or not with *Glomus* strain E3 (Rothamsted). In the presence of animals the mycorrhizal plants grew no better than the non-mycorrhizal plants, and significantly, the P inflow to the roots declined from 6.8 pmol m^{-1} s^{-1} in mycorrhizal plants with no collembola present, to 1.7 when collembola were present; the equivalent values for non-mycorrhizal plants were 2.3 and 1.0, but none of the lower three figures differed significantly. The reduction in P inflow can only be explained if P transport through the external mycelium was disrupted by the animals, and indeed hyphal fragments were found in animals' guts.

The most likely effect of grazing is either to limit the development of the external mycelium or to disconnect it from the internal mycelium by grazing in the rhizosphere on entry points (Fitter and Sanders, 1992). In the latter case, relatively low levels of hyphal predation could have dramatic effects on mycorrhizal function, since the transport of P to the roots would be prevented and the internal mycelium would remain as a significant resource drain on the host. The nature of the grazing interaction can only be elucidated by direct observation of grazing animals in the rhizosphere, a task that might be undertaken in a rhizotron.

Another possible result of grazing is a stimulus to fungal growth, as shown, for example, by Hanlon (1981) for saprotrophic fungi. Finlay (1985), in a series of experiments on the interaction between collembola and arbuscular mycorrhizal fungi, found that low densities of animals increased plant P concentration, but that high densities decreased it, and Harris and Boerner (1990) found the same non-linear impact of collembola on plant mass in *Geranium robertianum*. This non-linearity might be due to direct stimulus to fungal growth or simply to a balance of positive (increased rates of mineral cycling, for example) and negative effects, which is density-dependent.

Sporulation and sporophore production

Sporocarp production of *Agaricus bisporus* may be stimulated by bacteria, which might therefore have a similar effect on ECM. Wild sporocarps of *Laccaria bicolor* and *L. laccata* always contain large populations of bacilli and pseudomonads; these sporocarps form abundantly from mycorrhizal conifer seedlings in nurseries and pot experiments, but never when mycorrhizal synthesis is established aseptically in vitro. On the other hand, some fungi such as *Hebeloma cylindrosporum* do not contain bacteria in nature and form sporocarps in aseptic conditions. Bacteria may therefore be involved in basidiocarp initiation of *Laccaria* sp., especially since some of the bacteria naturally associated with the basidiocarps are specific and stimulate mycelial growth and mycorrhiza formation (MHB), as described above.

Influence of mycorrhizal fungi on other root-inhabiting microbes

Interactions with root pathogens

The effects of mycorrhizas on root-pathogenic bacteria are rather poorly documented in the literature, but some conclusive experimental results show that the damage due to *Pseudomonas syringae* on tomato is significantly reduced when the plants are endomycorrhizal (Garcia Garrido and Ocampo, 1989). In contrast, the protective effect of mycorrhizal symbioses against root pathogenic fungi has been attested by many authors (Caron, 1989); Perrin (1991) lists 91 examples, both AM and ECM, and on a wide range of plants, pathogens and symbionts, and in which a significant reduction of the disease occurs in 76% of the cases. Such effective protection is particularly common and intense with ectomycorrhizas of forest tree seedlings in nurseries and against common root pathogens such as *Phytophthora cinnamomi, Rhizoctonia solani, Fusarium oxysporum* and *Pythium* spp.

Much work has also been done with nematodes, some of which are detrimental obligate parasites. Most deals with the effect of AMF on *Meloidogyne* spp., particularly *M. incognita*, the root-knot nematode. Sikora and Schönbeck (1975) have shown that mycorrhizal infection of tobacco, tomato, carrot and oat by the fungus *Glomus mosseae* limited the impact of *M. incognita*. The larvae can penetrate mycorrhizal as well as non-mycorrhizal roots, but their development and reproduction is then inhibited (Schönbeck, 1979), suggesting that the antagonism takes place only inside the root. In contrast, Roncadori and Hussey (1977) found that cotton plants mycorrhizal with *Gigaspora margarita* grew better, but were more infected by the nematode; the mycorrhizal symbiosis appeared to induce plant tolerance to the disease but not plant resistance. Kellam and Schenck (1977) showed that gall formation by *M. incognita* in soybean roots was reduced only in the root portions mycorrhizal with *Glomus macrocarpum,* suggesting a direct short-range effect on the nematode similar to that on *Fusarium* (see above; Muchovej et al., 1991), and that the defence mechanisms of the plant are not modified by the symbiosis.

The mechanisms involved in these interactions include physical protection, chemical interactions and indirect effects (Table 2), but because good experimental control in field experiments is difficult to achieve, most of the studies have been performed in highly artificial conditions and the significance of the interactions in the field remains very unclear.

Interactions with nitrogen-fixing nodulating bacteria.

Both *Rhizobium* (and *Bradyrhizobium*) and *Frankia*-associated plant species are normally mycorrhizal. The mycorrhizal and nodule symbioses typically act synergistically, both on infection rate and on mineral nutrition and growth of the plant. Where nitrogen and phosphorus are limiting, arbuscular mycorrhizas may improve P uptake; the higher P concentration in the plant benefits the bacterial symbiont and the functioning of its nitrogenase, leading to increased N fixation, which in turn promotes root and mycorrhizal development. As a consequence, the interaction with AM fungi and either *Rhizobium* or *Frankia* is highly dependent on P and N fertilization. Moreover, when arbuscular mycorrhizas enhance nitrogen fixation, this can be through various responses of the bacterial symbiosis; the number of nodules per plant may be reduced, but either or both of their mean weight and specific nitrogenase activity may increase, as shown, for example, in legumes by Asimi et al. (1980) and in *Alnus* by Fragga-Beddiar and Le Tacon (1990).

Table 2. A simple classification of mechanisms described to date for the protective effect of mycorrhizal fungi against soil-borne pathogens

Mechanism	Fungal Type	Nature	References
Mechanical protection	ECM	especially those with dense mantles	Marx (1972)
Antibiotics	ECM	over 100 fungal species	Marx (1969) Trofast and Wickberg (1977)
Phytoalexins	AM	conflicting results	Morandi and Baylet (1984) Kellam and Schenk (1977)
	ECM	ethylene polyphenols terpenes	Graham and Linderman (1980) Krupa and Fries (1971) Sampangi et al. (1986)
Rhizosphere modification	ECM	altered competitive balance	Malajczuk (1988)
Host mineral nutrition	AM	improved P status	Graham (1987)

Implications of mycorrhizal interactions

Ecological implications

Mycorrhizal fungi interact with a wide range of other organisms at all stages of their life-cycle, and some of these interactions can have large effects on mycorrhizal function and behaviour. These interactions may account for the difficulty which many researchers have encountered in measuring mycorrhizal benefits to plants under natural conditions (Fitter, 1985). If grazing of external hyphae has a significant effect under normal field conditions, many mycorrhizal associations may be unable to perform what is normally taken to be their primary function, increasing P inflow.

One problem is logistical; the ideal experimental protocol involves removing mycorrhizal fungi from a system without disturbing the other organisms. No technique is available for doing this. Several workers have used fungicides to eliminate mycorrhizas and though these may be very effective in both laboratory (Fitter and Nichols, 1988) and field conditions (Finlay, 1985; Carey et al., 1992), as long as adequate penetration of fungicide into soil is achieved, they also attack other organisms, not only fungi, and disentangling direct and indirect effects may be difficult.

Nevertheless, one recent study, on the winter annual grass *Vulpia ciliata* ssp *ambigua* (Carey et al., 1992), suggests that mycorrhizal infection may offer benefits to plants that are not determined by P nutrition and are best explained by interactions of mycorrhizal and pathogenic fungi. Plant pathogens, such as *Fusarium* spp., are widespread in natural ecosystems (Alexander, 1990) and probably play a significant role in determining population dynamics and community structure. Their interaction with the equally widespread mycorrhizal symbiosis therefore merits urgent attention.

Practical implications

It may be possible to exploit these interactions in two ways. First, the effects of other organisms on mycorrhizal fungi may be modified so as to improve mycorrhizal function. For example, if hyphal grazing is important, then cultural techniques which limit the populations of fungivorous soil animals (nematodes, collembola, mites) might be advantageous. Similarly, mycorrhization helper bacteria (MHB) might be exploited to improve the colonization of roots by mycorrhizal fungi, especially in forest nurseries. Even with doses as low as 106 cfu per m^2 of soil, MHB provide higher rates of mycorrhizal infection with less inoculum; due to their selectivity, they limit the need for soil fumigation prior to inoculating by improving the competitiveness of the introduced fungus against the undesirable ectomycorrhizal competitors naturally present in the soil (Garbaye, 1991). The beneficial effect is consistent in various soil and climate conditions, and commercial development is under way. Another example where a practical application has been suggested is in the effect of pseudomonads on mycorrhizal fungi of hazel (Mamoun and Olivier, 1992); here the aim was to maximize infection by the truffle fungus *Tuber melanosporum*. Pseudomonads reduced infection of roots by all fungi initially, but *T. melanosporum* proved the best able to recolonize; in other words, the bacteria, although generally inhibitory to mycorrhiza formation, altered the competitive interaction between the various potential mycorrhizal colonists in the desired direction.

Second, the ability of mycorrhizal fungi to interfere with pathogens might be used as a form of biological control. However, contradictory experimental results are often obtained with the same plant–symbiont–parasite system in different soils or substrates, possibly because the efficiency of these mechanisms can be affected by the soil conditions. The two main soil factors involved are phosphorus availability (controlling both the phosphorus nutrition of the plant and the benefit of the mycorrhizal symbiosis) and microflora. Mycorrhizal infection does not always promote disease resistance; diseased plants are sometimes found to be more heavily mycorrhizal than healthy plants and mycorrhizal infection may even increase disease incidence (Caron, 1989). Furthermore, there are differences between genotypes of both mycorrhizal fungi and pathogens which determine the interaction. For example,

Laccaria bicolor prevented spread of *Fusarium oxysporum* in Douglas fir roots as a result of flavanoid wall infusions (Strobel and Sinclair, 1991b). In this case the control was only effective against non-aggressive strains; aggressive strains were unaffected, leading Strobel and Sinclair (1991a) to caution against suggestions of practical utility. It might, however, be possible to select strains of mycorrhizal fungi which were effective even against aggressive pathogens.

Another problem that may be encountered in using mycorrhizal fungi as biological control agents, is that disease susceptibility and mycorrhizal susceptibility may be positively correlated, as found by Toth et al. (1990). The practical application of mycorrhizas in this context, therefore, may require the explicit recognition of their role in plant selection programmes.

For these reasons, the practical application of controlled mycorrhiza formation as a deterrent of root diseases will be difficult and slow; to be a good candidate for such utilization, a mycorrhizal fungus must express its protective effect in a wide range of environmental conditions, be aggressive against pathogens and be able to colonize roots even of disease-resistant varieties. Nevertheless, the potential advantages of harnessing mycorrhizal fungi in integrated agricultural systems are large and should be pursued energetically.

References

Alexander H M 1990 Dynamics of plant–pathogen interactions in natural plant communities. *In* Pests, Pathogens and Plant Communities. Eds. J J Burdon and S R Leather. pp 31–48. Blackwell Scientific Publications, Oxford, UK.

Ames R N 1989 Mycorrhiza development in onion in response to inoculation with chitin-decomposing actinomycetes. New Phytol. 112, 423–427.

Asimi S, Gianinazzi-Pearson and Gianinazzi S 1980 Influence of increasing soil phosphorus level on interaction between vesicular-arbuscular mycorrhizae and *Rhizobium* in soybean. Can. J. Bot. 58, 2200–2205.

Azcon-Aguilar C, Diaz-Rodriguez R M and Barea J M 1986 Effect of soil microorganisms on spore germination of the vesicular-arbuscular mycorrhizal fungus *Glomus mosseae*. Trans. Br. Mycol. Soc. 86, 337–340.

Bowen G D and Theodorou C 1979 Interactions between bacteria and ectomycorrhizal fungi. Soil Biol. Biochem. 11, 119–126.

Boyetchko S M and Tewari J P 1991 Parasitism of spores of the vesicular-arbuscular mycorrhizal fungus *Glomus dimorphicum*. Phytoprotection 72, 27–32.

Carey P D, Fitter A H and Watkinson A R 1992 A field study using the fungicide benomyl to investigate the effect of mycorrhizal fungi on plant fitness. Oecologia 90, 550–555.

Caron M 1989 Problematique de l'utilisation des champignons endomycorhiziens commeagents de lutte biologique. Phytoprotection 70, 43–49.

Castellano M A and Trappe J M 1985 Ectomycorrhizal formation and plantation performance of Douglas-fir nursery stock inoculated with *Rhizopogon* spores. Can. J. For. Res. 15, 613–617.

Chanway C P, Turkington R and Holl F B 1991 Ecological implications of specificity between plants and rhizosphere microorganisms. Adv. Ecol. Res. 21, 122–170.

Duponnois R and Garbaye J 1991a Mycorrhization helper bacteria associated with the Douglas fir *Laccaria laccata* symbiosis: effects in vitro and in glasshouse conditions. Ann. Sci. For. 48, 239–251.

Duponnois R and Garbaye J 1991b Some mechanisms involved in growth stimulation of ectomycorrhizal fungi by bacteria. Can. J. Bot. 68, 2148–2152.

Finlay R D 1985 Interactions between soil micro-arthropods and endomycorrhizal associations of higher plants. *In* Ecological Interactions in Soil. Ed. A H Fitter. pp 319–332. Blackwell Scientific Publications, Oxford, UK.

Fitter A H 1985 Functioning of vesicular-arbuscular mycorrhizas under field conditions. New Phytol. 99, 257–265.

Fitter A H and Nichols R L 1988 The use of benomyl to control infection by vesicular-arbuscular mycorrhizal fungi. New Phytol. 109, 201–206.

Fitter A H and Sanders I R 1992 Interactions with the soil fauna. *In* Mycorrhizal Functioning. Ed. M F Allen. pp 333–354. Chapman and Hall, New York, USA.

Fogel R and Trappe J M 1978 Fungus consumption (mycophagy) by small mammals. Northwest Sci. 52, 1–31.

Fraga-Beddiar A and Le Tacon F 1990 Interactions between VA mycorrhizal fungi and *Frankia* associated with alder (*Alnus glutinosa*). Symbiosis 9, 247–258.

Garbaye J 1991 Biological interactions in the mycorhizospere. Experientia 47, 370–375.

Garbaye J and Bowen G D 1987 Effect of different microflora on the success of ectomycorrhizal inoculation of *Pinus radiata*. Can. J. For. Res. 17, 941–943.

Garbaye J and Bowen G D 1989 Stimulation of ectomycorrhizal infection of *Pinus radiata* by some microorganisms associated with the mantle of ectomycorrhizas. New Phytol. 112, 383–388.

Garbaye J and Duponnois R 1991 Specificity and function of mycorrhization helper bacteria (MHB) associated with the *Pseudotsuga menziesii–Laccaria laccata* symbiosis. Proc. International Symbiosis Congress, Jerusalem, Israel.

Garcia Garrido J M and Ocampo J A 1989 Effect of VA mycorrhizal infection of tomato on damage caused by *Pseudomonas syringae*. Soil Biol. Biochem. 21, 165–167.

Graham J H 1987 Non-nutritional benefits of VAM fungi. Do they exist? *In* Mycorrhizae in the Next Decade: Practical

Applications and Research Priorities, Proc. 7th NACOM. Ed. D M Sylvia, L L Hung and J H Graham. pp 237–239. Institute of Food and Agricultural Sciences, University of Florida, Gainesville, Florida.

Graham J H and Linderman R G 1980 Ethylene production by ectomycorrhizal fungi, *Fusarium oxysporum* f.sp. *pini* and by aseptically synthesized ectomycorrhizae and fusarium-infected Douglas-fir roots. Can. J. Microbiol. 26, 1340–1347.

Hanlon R D G 1981 Influence of grazing by Collembola on the activity of senescent fungal colonies grown on media of different nutrient concentration. Oikos 36, 362–367.

Harris K K and Boerner R E J 1990 Effects of below ground grazing by Collembola on growth, mycorrhizal infection, and P uptake of *Geranium robertianum*. Plant Soil 129, 203–210.

Kaiser P A and Lussenhop J 1991 Collembolan effects on establishment of vesicular-arbuscular mycorrhizae in soybean (*Glycine max*). Soil Biol. Biochem. 23, 307–308.

Kellam M K and Schenck N C 1977 The effect of initial sequence on the interaction between *Glomus macrocarpus* and *Meloidogyne incognita* on soybean. *In* Proc. 3rd NACOM, Georgia, USA. p 3.

Klyuchnikov A A and Kozhevin P A 1990 Dynamics of *Pseudomonas fluorescens* and *Azospirillum brasilense* populations during the formation of the vesicular-arbuscular mycorrhiza. Microbiology 59, 449–452.

Krupa S and Fries N 1971 Studies on ectomycorrhizae of pine. I—production of volatile organic compounds. Can. J. Bot. 49, 1425–1431.

Linderman R G and Paulitz T C 1990 Mycorrhizal–rhizobacterial interactions. *In* Biological Control of Soil-borne Plant Pathogens. Ed. D Hornby pp 261–283. Wallington, CAB International.

McAfee J and Fortin J A 1988 Comparative effect of the soil microflora on ectomycorrhizal inoculation of conifer seedlings. New Phytol. 108, 443–449.

Malajczuk N 1988 Interactions between *Phytophthora cinnamomi* zoospores and microorganisms on non-mycorrhizal and ectomycorrhizal roots of *Eucalyptus marginata*. Trans. Br. Mycol. Soc. 90, 375–382.

Mamoun M and Olivier J M 1992 Effect of soil pseudomonads on colonization of hazel roots by the ectomycorrhizal species *Tuber melanosporum* and its competitors. Plant Soil 139, 265–273.

Marx D H 1969 The influence of ectotrophic mycorrhizae fungi of pine roots to pathogenic infection. II—Production, identification and biological activity of antibiotics produced by *Leucopaxillus cerealis* var. *piceina*. Phytopathology 59, 411–417.

Marx D H 1972 Ectomycorrhizae as biological deterrents to pathogenic root infections. Ann. Revu. Phytopathol. 10, 429–454.

Mayo K, Davis R and Motta J 1986 Stimulation of germination of spores of *Glomus versiforme* by spore-associated bacteria. Mycologia 78, 426–431.

Meyer J R and Linderman R G 1986 Response of subterranean clover to dual inoculation with vesicular-arbuscular mycorrhizal fungi and a plant growth promoting rhizobacterium *Pseudomonas putida*. Soil Biol. Biochem. 18, 185–190.

Morandi D and Baylet J A 1984 Isoflavonoid accumulation in soybean roots infected with vesicular-arbuscular mycorrhizal fungi. Physiol. Plant Pathol. 24, 357–364.

Mosse B 1962 The establishment of vesicular-arbuscular mycorrhiza under aseptic conditions. J. Gen. Microbiology 27, 509–520.

Muchovej J J, Muchovej R M C and Goncalves E J 1991 Effect of kind and method of fungicidal treatment of bean seed on infections by the VA mycorrhizal fungus *Glomus macrocarpum* and by the pathogenic fungus *Fusarium solani*. 2. Temporal–spatial relationships. Plant Soil 132, 47–51.

O'Bannon J H and Nemec S 1979 The response of *Citrus limon* seedlings to a symbiont, *Glomus etunicatus* and a pathogen, *Radopholus similis*. J. Nematol. 11, 270–275.

Paulitz T C and Menge J A 1984 Is *Spizellomyces punctatum* a parasite or a saprophyte of vesicular-arbuscular mycorrhizal fungi? Mycologia 76, 99–107.

Paulitz T C and Menge J A 1986 The effects of a mycoparasite on the mycorrhizal fungus *Glomus deserticola*. Phytopathology 76, 351–354.

Perrin R 1991 Mycorhizes et protection phytosanitaire. *In* Les mycorhizes des arbres etplantes cultives. Ed. Strullu D G. pp 93–130. Lavoisier, Paris, France.

Rabatin S C and Stinner B R 1988 Indirect effects of interactions between VAM fungi and soil-inhabiting in vertebrates on plant processes. Agric. Ecos. Env. 24, 135–146.

Ridge E H and Theodorou C 1972 The effect of soil fumigation on microbial recolonization and mycorrhizal infection. Soil Biol. Biochem. 4, 295–305.

Roncadori R W and Hussey R S 1977 Interactions of the endomycorrhizal fungus *Gigaspora margarita* and root-knot nematode on cotton. Phytopathology 67, 1507.

Ross J P and Ruttencutter R 1977 Population dynamics of two vesicular-arbuscular endomycorrhizal fungi and the role of hyperparasitic fungi. Phytopathology 67, 490–496.

Sampangi R, Perrin R and Le Tacon P 1986 Disease suppression and growth promotion of Norway spruce and Douglas-fir seedlings by the ectomycorrhizal fungus *Laccaria laccata* in forest nurseries. *In* Mycorrhizae: Physiology and Genetic. Ed. V Gianinazzi. pp 799–806. Editions INRA, Paris, France.

Schönbeck F 1979 Endomycorrhiza in relation to plant diseases. *In* Soil-borne Plant Pathogens. Eds. Schippers B and Gams W. pp 271–280. Academic Press, New York, USA.

Sikora R A and Schönbeck F 1975 Effect of vesicular-arbuscular mycorrhiza (*Endogone mosseae*) on population dynamics of the root-knot nematodes (*Meloidogyne incognita* and *M. hapla*). VIII Int Congr. Plant Protect., Moscow. 5, 158.

Strobel N E and Sinclair W A 1991a Role of flavanolic wall infusions in the resistance induced by *Laccaria bicolor* to *Fusarium oxysporum* in primary roots of Douglas fir. Phytopathology 81, 420–425.

Strobel N E and Sinclair W A 1991b Influence of temperature and pathogen aggressiveness on biological control of *Fusarium* root rot by *Laccaria bicolor* in Douglas fir. Phytopathology 81, 415–420.

Sylvia D M and Schenck N C 1983 Soil fungicides for controlling chytridiaceous mycoparasites of *Gigaspora margarita* and *Glomus fasciculatum*. Appl. Environ.

Microbiol. 45, 1306–1309.

Toth R, Toth, D, Starke D and Smith D R 1990 vesicular-arbuscular mycorrhizal colonization in *Zea mays* affected by breeding for resistance to fungal pathogens. Can. J. Bot. 68, 1039–1044.

Trofast J and Wickenberg B 1977 Mycorrhizin A and chloromycorrhizin A, two antibiotics from a mycorrhizal fungus of *Monotropa hypopitys*. Tetrahedron 33, 875–879.

Tylka G L, Hussey R S and Roncadori R W 1991 Axenic germination of vesicular-arbuscular mycorrhizal fungi—effects of selected *Streptomyces* species. Phytopathology 81,754–759.

Warnock A J, Fitter A H and Usher M B 1982 The influence of a springtail *Folsomia candida* (Insecta, Collembola) on the mycorrhizal association of leek *Allium porrum* and the vesicular-arbuscular mycorrhizal endophyte *Glomus fasciculatus*. New Phytol. 90, 285–292.

Wilson G W T, Hetrick B A D and Kitt D G 1989 Suppression of vesicular-arbuscular mycorrhizal fungus spore germination by non-sterile soil. Can. J. Bot. 67, 18–23.

Reprinted from *Plant and Soil*, Volume 159, No. 1 (1994), pp. 123–132

The functioning of ectomycorrhizal fungi in the field: linkages in space and time

M. P. AMARANTHUS[1] and D. A. PERRY[2]
[1]*United States Department of Agriculture, Forest Service, Pacific Northwest Research Station, Portland, OR 97208-3890, USA*
[2]*Department of Forest Science, College of Forestry, Oregon State University, Corvallis, OR, 97331, USA*

Key words: disturbance, grass seeding, hyphal linkages, hardwoods, *Rhizopogon,* succession, site preparation

Abstract

Individual trees, either of the same or different species, can be linked spatially and temporally by the hyphae of ectomycorrhizal (ECM) fungi that allow carbon and nutrients to pass among them and promote forest establishment following disturbance. Spatial and temporal linkages between plants influence the function of ECM fungi in the field. Studies indicate that ECM linkages can reduce plant competition for resources, promote forest recovery, and influence the pattern of plant succession. The degree of influence depends on many factors, including the composition and arrangement of the vegetative community and soil and climatic conditions. Management practices that create intense disturbance and loss of organic matter or promote the introduction of non-ectomycorrhizal host species can decrease the ability of plants to form linkages with ECM fungi. Management practices that retain living trees and shrubs and input of organic matter provide the energy source and substrate necessary for ECM linkages. More research is needed to determine the degree to which ECM fungal linkages occur in the field and their role in ecosystem function and long-term health.

Introduction

The importance of ECM fungi to the survival and growth of individual trees in montane, boreal, and temperate forests is well established (Amaranthus and Perry, 1987, 1989; Harley and Smith, 1983). Ectomycorrhizal fungi affect field performance of trees through capture and uptake of nutrients (especially immobile nutrients such as phosphorous), protecting against pathogens and extending feeder root longevity (Harley and Smith, 1983). Ectomycorrhizal fungi are the primary interface between roots and the forest soil and perform a variety of important functions such as maintaining soil structure, buffering against moisture stress, and protecting plants from toxic heavy metals (Lynch and Bragg, 1985; Sutton and Sheppard, 1976).

Field study of ECM fungi has shifted fundamentally in recent years from the selection of beneficial fungal strains for inoculating seedlings to the ecological role of mycelium linking plants (Read, 1988) and effects on plant community development (Perry et al., 1989a). Many fungal species form mycorrhizae with more than one plant species within a community (Molina and Trappe, 1982; Molina et al., 1992) and new mycorrhizae often develop from established mycorrhizae (Read, 1988). Given this pattern of colonization one would expect plants to be linked. Laboratory and field studies demonstrate that roots of the same or dissimilar plants can be connected by commonly shared ECM fungi (Perry et al., 1989b; Read et al.,

A.D. Robson, L.K. Abbott & N. Malajczuk (eds.), Management of mycorrhizas in agriculture, horticulture and forestry, 133–140.

1985; Read, 1988).

Recognition of the importance of mycorrhizal mycelial linkages has been impeded by the difficulty of accessing intact mycelial networks. Although field studies suggest that some forest practices reduce spatial and temporal continuity provided by ECM fungal linkages among plants, to little is known on the degree to which plants form ECM linkages and the ecological function of the networking. The objective of this paper is to review what is known about the functional role of ECM fungi as spatial and temporal linkages and how such linkages might be impacted by forest management.

Spatial linkages

The idea that groups of plants might be linked spatially raises numerous questions about the function of ECM fungi. Studies have demonstrated that carbon and nutrients move between plants of the same and different species. Bjorkman (1960) showed that ^{14}C-labeled sugars injected into the phloem stream of conifers appeared in nearby *Monotropa* sp., a plant that is both achlorophyllos and ectomycorrhizal. He suggested the transfer was through mycorrhizal hyphae but did not demonstrate this. Read et al. (1985) fed ^{14}C to a lodgepole pine (*Pinus contorta* Dougl. ex Loud) surrounded by seedlings of various ectomycorrhizal and vesicular-arbuscular (VA) mycorrhizal tree species. Large amounts of tracer were transferred to nearby lodgepole pine seedlings, especially when those surrounding seedlings were shaded. Neighboring VA host plants accumulated only trace amounts or none at all. Studies where seedling mycorrhizal formation was controlled indicate that fungi enhance transfer of carbon and nutrients when participating plants share mycorrhizal fungi (Whittingham and Read, 1982; Francis and Read, 1984; Finlay and Read, 1986)

How might spatial linkages influence competition for resources by plants in the field? A study by Perry et al. (1989a) suggests that ECM fungi convert a negative interaction into either a neutral or positive one. Non-inoculated and inoculated Douglas-fir (*Pseudotsuga menziesii* (Mirb.) Franco) and ponderosa pine (*Pinus ponderosa* Dougl. ex Laws) seedlings were grown in a standard replacement series in pasteurized soils reconstructed by horizon. The study indicated that at least some types of mycorrhizal fungi reduce competition between the two species. Many factors could have produced this effect; (1) the ability of mycorrhizal fungi to exploit spatially distinct substrates such as humus vs mineral soil, (2) the ability of the fungi to be active during temporally changing conditions of moisture and temperature, and (3) a more even distribution of nutrients between plant species via hyphal connections (Perry et al., 1989a).

Numerous biotic and environmental factors can influence the spatial function of ECM fungi in native soil. One important factor is the matrix of neighboring plants and their ability to support fungi compatible between hosts. The ability of a particular plant to 'plug into' an existing network of fungal hyphae could allow for exploration of soil volume far beyond the root's capabilities. Thus, understanding the vegetative matrix is important for evaluating the potential for plants to share and benefit from linkages by ECM fungi. Hyphae supported by an adjacent plant or plants colonize and aid the establishment of a neighboring compatible plant (Amaranthus et al., 1990a). Energy supplied by an established host allow for young seedlings to form mycorrhizae and become linked to a fungal network of hyphae that would be far too expensive in terms of energy to maintain. Even without direct linkages an established adjacent ECM host provides a much higher level of fungal propagules that encourage mycorrhizal development with similar fungi. This mechanism also increases the eventual likelihood of the establishment of hyphal links.

Ectomycorrhizal fungi include thousands of fungal species that differ in their host associations, physiology and ecology. While many ECM fungi can form mycorrhizae with a wide variety of host plants, many others are family, genus or even species specific (Molina et al., 1992). For example, *Rhizopogon* species comprise a diverse and widespread group of hypogeous ECM fungi that are mostly restricted to Pinaceae. Massicote et al. (1993) In a study of

11 *Rhizopogon* species found that none had a broad host range, however a variety of responses occurred that ranged from intermediate to species specific. Interspecific linkages among host plants are unlikely if a site is contains a high proportion of ECM fungi with narrow host ranges. Yet even ECM fungi with narrow host ranges can greatly affect forest recovery and stability because hyphal links occur intraspecifically, and intraspecific linkages between overstory and understory plants may be quite common. Functional differences between interspecific and intraspecific linkages is poorly understood and research is needed.

The degree to which ECM fungi link plants spatially may differ as trees age and change mycorrhizal dependency. However, Massicote et al. (1993) found that most of the ECM present on four host plants grown in soils from a recent clearcut were also present on the same host plants grown in soil from an adjacent plantation and mature forest. Mycelium extending from the adjacent stands may aid rapid regeneration of small forest openings. Larger scale disturbances, such as clearcuting and site preparation, may require stabilization of the ECM community by compatible invading or sprouting hosts. Considerably more research is needed, however, before the degree to which ECM fungi are spatially linked can be determined solely by plant composition and distribution. Some ECM fungi have been shown to colonize herbs and grasses formerly thought to be strictly VA mycorrhizal and non-mycorrhzial (Trappe, 1988; Warcup, 1988) and even well studied species such as Douglas-fir, thought to be exclusively ectomycorrhizal forming have recently been found colonized by VA mycorrhizae (Cazares and Trappe, 1993).

Linkages in time

Forest ecosystems change over time. How they change is not perfectly predictable. The variables that affect ecosystems are many—fire, wind, floods, insects, pathogens, climate and management. During these periods of rapidly changing above-ground community structure and compostition, EM fungi beneath the soil surface colonize and aid establihment of successive ECM host species effectively linking old and new stands.

How are ECM fungi maintained through time and how do linkages between plants promote vegetative recovery and resilience? We are just beginning to understand how ECM fungi persist and are distributed temporally. It has been suggested that ECM fungi do not survive long in the absence of living hosts and supplies of essential photosynthate (Hacksaylo, 1973). The ability of early successional ECM hosts to rapidly occupy disturbed sites and allocate carbon resources into rhizospheres and mycorrhizospheres may facilitate the perpetuation of ECM fungi (Amaranthus and Perry, 1987, 1989; Perry et al., 1989a; Read, 1988).

In the Klamath Mountains of the Pacific Northwest United States sites that have been logged or burned are often rapidly invaded by woody hardwoods. These hardwoods, members of Ericaceae and Fagaceae, form ECM with many of the same members of the Pinaceae (Massicote et al., 1993; Molina and Trappe, 1982). It has been observed, for example, that Douglas-fir seedlings preferentially associate with the hardwood tree Pacific madrone (*Arbutus menziesii* Pursh) (Amaranthus et al., 1990b). We hypothesize that these hardwoods preserve ECM fungi during periods of rapidly changing above-ground community structure and that mycorrhizal links between hardwoods and conifers facilitate conifer establishment by providing a ready source of inoculum, nutrients and water.

Amaranthus and Perry (1989) planted Douglas-fir seedlings in adjacent areas that; (1) were cleared of manzanita (*Arctostaphylos viscida* Parry) and (2) a meadow cleared of annual grasses. Douglas-fir seedlings outplanted on the manzanita site contained significantly higher proportions of the ECM species *Rhizopogon* ($p < 0.01$) and were observed to form ECM more rapidly than seedlings at the adjacent cleared meadow. Total ECM formed on seedlings were not different between the two sites. By the end of the second growing season > 90% of the Douglas-fir planted in the manzanita stand had survived where as < 50% survived in the cleared meadow. Seedlings planted in the cleared manzanita had dramatically greater height and caliper growth ($p < 0.01$). By the end of the third growing season only one seedling remained in the meadow whereas survival in the manzanita stand was still > 90%. Other studies have shown that *Rhizopogon* species decrease seedling moisture stress (Parke et al., 1983) and increase outplanting performance (Castellano et al., 1985).

In the field, we examined the effect of greenleaf manzanita (*Arctostaphylos patula*) on the number and rapidity of Douglas-fir roots and ECM formation in southwest Oregon (Amaranthus et al., 1990a). The study site was a 20 year-old non-reforested clearcut that had been invaded largely by non-ECM annual plants and scattered clumps of manzanita, known to form mycorrhizae with some of the same fungi as Douglas-fir (Molina and Trappe, 1982). Non-mycorrhizal, 1 year-old containerized Douglas-fir seedlings were planted at 0.3, 1.2, and 3.6 m distances from manzanita clumps and sampled on three dates during the growing season. Douglas-fir seedlings nearest manzanita formed significantly greater number of root tips early in the growing season. Douglas-fir and manzanita mycorrhizal tips were greatest nearest manzanita. Soil moisture and temperature did not differ significantly with distance from the manzanita clumps.

Biological factors other than the presence of higher levels of inocula could be influencing ECM formation and conifer growth in hardwood soils. In the area studied by Borchers and Perry (1990), open soils had 10 times more colonies of the common soil actinomycete *Streptomycete* sp. than soils beneath hardwoods. These organisms are among the few that can degrade lignin and are relatively tolerant of environmental extremes; hence their abundance in open soils low in labile carbon and relatively harsh microclimatically is not surprising. *Streptomyces* are known to produce chemicals that inhibit plants and other microorganisms, including both mycorrhizal and pathogenic fungi. *Streptomyces* has been linked to reforestation failures in the Klamath Mountains (Friedman et al., 1989; Perry and Rose, 1983).

Additional factors could be responsible for the positive effect of hardwood soil. Mycorrhizal hardwood roots shape the composition of the surrounding rhizosphere community and influence exudates (Meyer and Linderman, 1986; Rambelli, 1973) that may account for observed early increases in root tip formation. Compared to non-mycorrhizal roots, mycorrhizal roots and rhizospheres have fewer pathogens, more nitrogen-fixers, and more bacteria that require complex nutritional factors from their environment. Amaranthus et al. (1990b) found 5 times greater nitrogenase activity in rhizospheres of 1-year-old Douglas-fir seedlings grown in cleared pure stands of *Arctostaphylos viscida* compared to seedlings grown in adjacent areas previously dominated by annual grass. The N_2 fixer, *Azospirillum*, was isolated from Douglas-fir mycorrhizospheres at the cleared *Arctostaphylos* site.

The importance of ECM temporal links between hardwoods and conifers depends on, among other things, how much one plant species depends on another to support below-ground mutualists in its absence. Ectomycorrhizal fungi may be able to successfully survive in the absence of a host plant by entering a resting stage, becoming saprophytic or maintaining effective inputs of propagules into a site. At least some ECM fungi have limited saprophytic capability, probably functioning as a means to extract nutrients from soil organic matter (Read, 1988). But it seems unlikely that all ECM fungi could compete successfully for long periods with saprophytes for carbon. Many ECM fungi produce abundant airborne spores and an important subset, the hypogeous fungi, are dispersed primarily by animals. How many spores are required to form abundant mycorrhizae in the field? No definite answer is available, but many thousand or millions maybe necessary in the field. Selected conifer seedlings in small containers with sterilized soil seedlings required at least 20,000 *Rhizopogon* spores each to be successfully colonized and seedlings in the field were not successfully colonized by the individual application of 1 million *Rhizopogon* spores (Amaranthus, unpublished results; Castellano et al., 1985). Spores alone appear unable to maintain the inoculation potential of most field sites, and the colonization potential of soils declines rapidly in the absence of hosts (Jasper et al., 1989a, b; Perry et al., 1989b).

Table 1. Height and basal area growth of 2 year-old Douglas-fir seedlings (from 6 families) with and without colonization with *Rhizopogon vinicolor* in the Oregon Coast Range. * indicates significantly greater than control ($p < 0.05$)

Family	Basal area growth (mm^2)		Height growth (mm)	
	Control	Colonized	Control	Colonized
1	20.4	19.9	16.3	48.1*
2	15.5	24.7*	38.8	53.8*
3	13.0	24.3*	17.5	45.0*
4	20.7	30.9*	32.9	41.9*
5	15.2	22.2*	38.8	50.0*
6	19.8	28.2*	42.6	53.3*

The importance of ECM fungi linking old and new stands is apparent from increased seedling performance in the field following nursery inoculation with native ECM fungi. We inoculated nursery-grown containerized Douglas-fir seedlings from 6 families with spores of *Rhizopogon vinicolor*. Colonized seedlings and uninoculated controls from all families were outplanted in the Oregon Coast Range. *Rhizopogon vinicolor* -colonized seedlings from all families had significantly greater height growth (6 of 6 families) and basal area growth (5 of 6 families) compared to controls (Table 1). Nursery inoculation with native ECM fungi can be an important tool to maintain linkages during periods of rapid change in above-ground community structure.

Severing the link

Many factors affect the ability of plants to maintain spatial and temporal linkages. Variation is likely determined by specific combinations of soil, biotic, and climatic factors. In Scandinavian forests, ECM spore and hyphal fragments remained metabolically active after 2 years, though the number of active fragments dropped rapidly over that time period (Ferrier and Alexander, 1985). Dry climates may limit the pesistence of ECM fungi by decreasing the length of time for spore production, germination, and optimal mycelial growth, which in turn can decrease the chances for seedlings in the field to become colonized. Moreover, seedlings in moist climates may be able to survive longer without forming ECM than those in dry climates thereby increasing the chances of the former to become colonized. Linkages by ECM fungi may be especially critical in cold climates where seedlings require rapid, early ECM formation to take advantage of the short growing season and obtain the nutrients and water to survive the long cold winter and early frosts. In studies in the Klamath Mountains, Amaranthus and Perry (1987, 1989) found that ECM formation most strongly influences seedling establishment on sites limited by moisture and temperature and that ECM linkages with early sucessional hosts can encourage seedling establishment.

Field studies indicate that the severity of a particular management activity may have the greatest impact on the persistence of ECM fungi. Wright and Tarrant (1958) found fewer ECM on Douglas-fir seedlings growing in burned clearcuts compared to unburned clearcuts. The greatest reductions were associated with the hottest burn. Parke (1982) compared ECM formation in soils from burned and unburned clearcuts of 36 'difficult to regenerate' sites in southwest Oregon and northwest California. Seedlings grown in soils from the unburned clearcuts formed 20% fewer ECM and seedlings grown in clearcut and severely burned soil formed 40% fewer ECM than seedlings grown in undisturbed forest soil. Amaranthus et al. (1987) grew Douglas-fir seedlings in soils from a clearcut and severely burned area and adjacent mature forest in southwest Oregon. They found 90% fewer ECM and 44% less basal area growth on seedlings grown in the clearcut and severely burned soil.

Not all studies report mycorrhizal reductions following clearcutting and prescribed fire (Pilz and Perry, 1984; Schoenberger and Perry, 1982). Some ECM inocula potential appears to persist in the first year on even the most catastrophically harvested and site prepared areas (D Luoma and M Amaranthus, unpublished results). Although it is difficult to generalize about effects of timber harvest and site preparation on the ability of ECM fungi to persist some impact in the Pacific Northwest United States has been associated with clearcutting in combination with intense prescribed fire.

The ability of ECM fungi to link plants spatially may depend on the structural and organic matter composition of soils in the field. Ectomycorrhizal fungi predominate in organic layers and are not distributed homogenously within the soil (Trappe and Fogel, 1977; Harvey et al., 1979). Thus, the degree of organic matter lost from a site can influence the spatial distribution of ECM fungi and potential to link plants. Harvey et al. (1979) found more than 87% of the active ECM in humus and decaying wood in a mature Douglas-fir/larch (*Larix*) forest. Some ECM fungi may be specific to special structural components such as rotten wood, and diversity of substrate may promote the temporal distribution of the ECM network; for example, in periods of adequate moisture, humus supports the highest level of ECM activity, but during periods of soil drought, soil wood becomes the most active sites in forests of the intermountain Western United States (Harvey et al., 1979). Rotting wood also may serve as a refuge for ECM fungi during periods of rapid community change. Amaranthus et al. (1989) found that decaying logs acted as reservoirs of moisture and ECM activity after drought and wildfire. Decayed logs contained 25 times more moisture than did surrounding soils and the ability of the logs to retain ECM fungi likely aids the transition between the previous stand and pioneering vegetation.

Temporal linkages by ECM fungi also may be impacted by invasion of sites by non-ECM plants. On sites invaded and dominated by VA mycorrhizal species or non-mycorrhizal species the ECM fungi may gradually disappear or be replaced. This may seriously affect linkages between old and new stands, especially in the case of ECM trees growing on difficult sites where seedlings must rapidly establish ECM to survive (Amaranthus and Perry, 1987, 1989). Some common practices on forest lands increase the potential to sever temporal ECM links. Attempting to mitigate unwanted post wildfire erosion effects forest managers in many parts of the Western United States routinely seed burned watershed with rapidly establishing annual ryegrass. Following extensive wildfires in 1987 Amaranthus, Perry and Trappe (unpublished) seeded 10 × 20 m blocks with 27 kg/ha of annual rye grass (*Lolium multiflorum*). One-year-old containerized, non-mycorrhizal sugar pine (*Pinus lambertianna* Dougl.) were planted in seeded and adjacent unseeded areas. The first year after planting, grass seeding reduced the number of ECM formed on the planted pine seedlings. Survival and growth were also greatly reduced. The seeding has had long-lasting effects on the composition and abundance of succeeding vegetation The introduction of grass, at least at these high rates, reduces the types and abundance of native shrub and hardwood species, many of which support the same ECM fungi as succeeding conifers. This may reduce the likelihood of temporal linkages between old and new stands.

Decline in ECM colonization potential is only one of several changes that might occur in soils where indigenous plants are weakened or removed. These changes include nutrient loss, moisture content, changes in other soil organisms, and degradation of soil physical structure (Jasper et al., 1989b; Meyer et al., 1988; Perry et al., 1989b). Soils altered in this way are less hospitable for establishing plants, thereby creating a positive feedback that can lead to rapid degradation (Perry et al., 1989b). Obviously plants have much more in common than shared ECM fungi, and as ecosystems are subjected to increased levels of antropogenic stresses, it is necessary to understand a number of interactions and their influence on fungal linkages between plants.

Disturbance is a major factor affecting the variation and diversity in the forest ecosystem. Events such as fire, winds, floods, and insect and pathogen activities shape the forest ecosystem through space and time. Disturbing the forest to remove commodities does not necessarily conflict with maintaining spatial and temporal linkages between plants via ECM fungi. After natural disturbance, such as fire in the western United States, biological remnants of the previous forest remain and become its basis for recovery. In most cases, this includes a various living trees and a wealth of organic inputs that provide the energy source and substrate for ECM fungal linkages. Studies in soil biology indicate that we cannot consider trees as crops in isolation from the community and ecosystem they are ecologically adapted to. Ectomycorrhizal fungi act as temporal and spatial connections in this highly interactive forest environment. Forest practices can mimic natural disturbance regimes and help assure that ECM fungal linkages among plants are not severed and that trees and soil ecosystem continue to operate as a coherent and dynamic unit.

References

Amaranthus M P and Perry D A 1987 Effect of soil transfer on ectomycorrhiza formation and the survival and growth of conifer seedlings in disturbed forest sites. Can. J. For. Res. 17, 944–950.

Amaranthus M P and Perry D A 1989 Interaction effects of vegetation type and Pacific madrone soil inocula on survival, growth and mycorrhiza formation of Douglas-fir. Can. J. For. Res. 19, 550–556.

Amaranthus M P, Perry D A and Borchers S L 1987 Reduction of native mycorrhizae reduce growth of Douglas-fir seedlings. *In* Mycorrhizae in the Next Decade: Practical Applications and Research Priorities, Proc. 7th NACOM. Eds. D M Sylvia and J H Graham. p 80. Institute of Food and Agricultural Sciences, University of Florida, Gainesville, Florida.

Amaranthus M P, Parrish D and Perry D A 1989 Decaying logs as moisture reservoirs following drought and wildfire. *In:* Stewardship of Soil, Air and Water Resources. Proc. Watershed 89, Juneau, Alaska. Ed. E Alexander. USDA Forest Service, Region 10. R10-MB-77.

Amaranthus M.P, Li C Y and Perry D A 1990a Influence of vegetation type and madrone-soil inoculum on associative nitrogen fixation in Douglas-fir rhizosperes. Can. J. For. Res. 20, 368–371.

Amaranthus M P, Molina R and Perry D A 1990b Soil organisms, root growth and forest regeneration. *In:* Forestry on the Frontier—Proc. National Society of American Foresters National Convention. Spokane, WA. pp 89–93.

Borchers S L and Perry D A 1990 Growth and ectomycorrhiza formation of Douglas-fir seedlings grown in soils collectcd at different distances from pioneering hardwoods in southwest Oregon clear-cuts. Can. J. For. Res. 20, 712–721.

Bjorkman E 1960 *Monotropa hypopitys* L. an epiparasite on tree roots. Physiol. Plant. 13, 308.

Cazares E and Trappe J M 1993 Vesicular-arbuscular mycorrhizae in the Pinaceae. New Phytol. (*in press*).

Castellano M A, Trappe J M and Molina R 1985 Inoculation of container-grown Douglas-fir seedlings with basidiospores of *Rhizopogon vinicolor* and *R. colossus:* effects of fertility and spore application rate. Can. J. For. Res. 15, 10–13.

Ferrier R C and Alexander I J 1985 Persistence under field conditions of excised fine roots and mycorrhizas of spruce. *In* Ecological interactions in soil. Eds. A H Fitter, D Atkinson, D J Read and M A Usher. pp 175–179.Blackwell Scientific Publications, Oxford.

Finlay R D and Read D J 1986 Translocation of CO_2-labelled carbon between plants interconnected by a common mycelium. New Phytol. 103, 143–156.

Francis R and Read D J 1984 Direct transfer of carbon between plants connectcd by vcsicular arbuscular mycorrhizal mycelium. Nature 307, 53–56.

Friedman J, Hutchins A, Li C Y and Perry D A 1989 Actinomycctes inducing phytotoxic or fungistatic activity in a Douglas-fir forest and in an adjacent area of repeated regeneration failure in southwestern Oregon. Biol. Plant. 31, 487–495.

Hacksaylo E. 1973. Carbohydrate physiology of ectomycorrhizae *In* Ectomycorrhizae: their Ecologyand Physiology. Eds. G C Marks and T T Kozlowski. pp 207–230. Academic Press, London.

Harley J L and Smith S E 1983 Mycorrhizal Symbiosis. Academic Press, London. 483 p.

Harvey A E, Larsen M J and Jurgensen M F 1979 Compariative distribution of ectomycorrhizae in soils of three western Montana forest habitat types. For. Sci. 25, 350–360.

Jasper D A, Abbott L K and Robson A D 1989a. Hyphae of a vesicular-arbuscular mycorrhizal fungus maintain infectivity in dry soil, except when the soil is disturbed. New Phytol. 112, 101–107.

Jasper D A, Abbott L K and Robson A D 1989b The loss of mycorrhizal infectivity during bauxite mining may limit the growth of*Acacia pulchella*.Aust. J. Bot. 37, 33–42.

Lynch J M and Bragg E 1985 Microorganisms and soil aggregate stability. Adv. Soil Sci 2, 133–171.

Massicote H, Molina R, Amaranthus M and Smith J 1993 Assessment of ectomycorrhizal fungal diversity and ecological specificity in association with 4 Pacific Northwest forest species. New Phytol. (*in press*).

Meyer J R and Linderman R G 1986 Selective influence on populations of rhizosphere or rhizoplane bacteria and actinomycetes by mycorrhizas formed by *Glomus fasciculalum*.Soil Biol. Biochem. 18, 191–196.

Meyer J, Schneider B, Werk K S, Oren R and Schulze E D 1988 Performance of two *Picea abies* stands at different stages of decline—Root tip and ectomycorrhiza development and their relation to above-ground and soil nutrients. Oecologia 77, 7–13.

Molina R and Trappe J M 1982. Lack of mycorrhizal specificity by the ericaceous hosts *Arbutus menziesii* and *Arctostaphylos uva-ursi*.New Phytol. 90, 495–509.

Molina R, Massicote H, and Trappe J M 1992 Specificity phenomena in mycorrhzial symbioses:community-ecological consequences and practical implications. *In* Mycorrhizal Functioning. Ed. M F Allen. pp 357–423. Chapman Hall, New York.

Parke J L 1982 Factors affecting the inoculum potential of VA and ectomycorrhizal fungi in forest soils of southwest Oregon and northern California. Ph.D. thesis. Oregon State University, Corvallis, OR, USA.

Parke J L, Linderman R, and Black C 1983 The role of ectomycorrhizas in drought tolerance of Douglas-fir seedlings. New Phytol 95, 83–95.

Perry D A and Rose S L 1983 Soil biology and forest productivity: opportunities and constraints. *In* IUFRO Symposium on Forest Site and Continuous Productivity. Eds. R Ballard and S P Gessel. pp 221–228. United States Department of Agriculture Forest Service General Technical Report PNW-163.

Perry D A, Margolis H, Choquette C, Molina R and Trappe J M 1989a Ectomycorrhizal mediation of competition between coniferous tree speeies. New Phytol. 112, 501–511.

Perry D A, Amaranthus M P, Borchers J, Borchers S and Brainerd R. 1989b Bootstrapping in Ecosystems. BioScience 39, 230–237

Pilz D P and Perry D A 1984 Impact of clearcutting and slash burning on ectomycorrhizal associations of Douglas-fir. Can. J. For. Res. 14, 94–100.

Rambelli A 1973 The rhizosphere of mycorrhizae. *In* Ectomycorrhizae: their Ecology and Physiology. Eds. A C Marks and T Kozlowski. pp 229–249. Academic Press, London.

Read D J 1988 Development and function of mycorrhizal hyphae in soil. *In* Mycorrhizae in the Next Decade, Practical Applications and Research Priorities, Proc. 7th NACOM. Eds. D M Sylvia, L L Hung and J H Graham. pp 176–180. Institute of Food and Agricultural Sciences, University of Florida, Gainesville, Florida.

Read D J Francis R and Finlay R D 1985 Mycorrhizal mycelia and nutrient cycling in plant communitics. *In* Ecological Interactions in Soil. Eds. A H Fitter, D J Read and M B Lusher. pp 193–217. Blackwell Scientific Publications, Oxford.

Schoenberger M M and Perry D A 1982 The effect of soil disturbance on growth and ectomycorrhizae of Douglas-fir and western hemlock seedlings: a greenhouse bioassay. Can. J. For. Res. 12, 343–353.

Sutton J C and Sheppard B R 1976 Aggregation of sand-dune soil by endomycorrhizal fungi. Can. J. Bot. 54, 326–333.

Trappe, J M 1988 Lessons from alpine fungi. Mycologia 80, 1–10.

Trappe J M and Fogel R 1977 Ecosystematic functions of mycorrhizae. *In* The Below-Ground Ecosystem. pp 205–214. Colorado State University, Range Science Department Scientific Series, Fort Collins, Colorado.

Warcup I H.1988 Mycorrhizal associations and seedling development in Australian Lobelioideae. Aust. J. Bot. 36, 461–472.

Whittingham J and Read D J 1982 Vesicular-arbuscular mycorrhizas in natural vegetation systems. Nutrient transfer between plants with mycorrhizal interconnections. New Phytol. 90, 277– 84.

Wright E and Tarrant R F 1958 Occurrence of mycorrhizae after logging and slash burning in the Douglas-fir forest type. Research Note PNW-160. Pacific Northwest Research Station, USDA Forest Service, Portland, OR, USA.

Reprinted from *Plant and Soil*, Volume 159, No. 1 (1994), pp. 133–140.

Research approaches to study the functioning of vesicular-arbuscular mycorrhizas in the field

I. JAKOBSEN
Plant Biology Section, Environmental Science and Technology Department, Risø National Laboratory, DK-4000 Roskilde, Denmark

Key words: compartments, field studies, P transport by hyphae, ^{32}P, ^{33}P, vesicular-arbuscular mycorrhizas

Abstract

Approaches suitable for the study on the functioning of vesicular-arbuscular mycorrhizas (VAM) in the field are discussed, with emphasis on those VAM-specific processes which may influence P nutrition of crops. These processes are related to hyphal P transport in particular, and conventional techniques based on comparisons between VAM-colonized plants and uncolonized controls are therefore of limited value in such studies. Novel methods are described, which make use of a two-compartment principle and isotopes to directly measure the P transport by the VAM mycelium under controlled conditions and in the field. The principle of isotopic equilibrium may be used for indirectly quantifying the relative amounts of P supplied by hyphae and by roots of VAM plants. The hyphal contribution to P uptake may also be assessed by comparing hyphal uptake of ^{32}P from a hyphal compartment with uptake of ^{33}P from a corresponding compartment containing both roots and hyphae. These methods are suitable for a functional characterization of VAM fungi and for identifying constraints to the functioning of VAM.

Introduction

Major research activities during the last two decades have provided much information on the occurrence of vesicular-arbuscular mycorrhizas (VAM) under field conditions (see Abbott and Robson, 1991). In contrast, our knowledge on the functioning of VAM in the field is limited (Fitter, 1985; 1991), although the importance of VAM as a growth factor is widely recognized (Mosse, 1986; Gianinazzi et al., 1989). Functioning is here defined as those VAM-specific processes which may influence mineral nutrition, stress tolerance and eventually also growth of the host plant. Mycorrhizas may also improve soil structure by their stabilizing effect on soil aggregates. Processes involved in the supply of P to plants will be emphasized in this paper, as the importance of VAM in crop production is primarily linked to its ability to alleviate P deficiency.

VAM-mediated P transport to plants results from the uptake and translocation of P by the external hyphae and the transfer of P across the symbiotic interface. The function of this hyphal transport system is affected by a range of biotic and abiotic environmental factors of the soil ecosystem. Conventionally, effects of VAM on nutrient transport and growth have been studied by comparing VAM-colonized plants with uncolonized controls. This approach is less suited for the quantification of processes and identification of constraints in VAM functioning than novel methods, which use a two-compartment principle to directly measure the nutrient transport by the VAM mycelium. The objectives of this paper are to discuss the limitations of the conventional methods and to describe the alternative methods and their advantages. The discussion will refer mainly to the functioning of VAM in cultivated systems, but the principles are applicable also for studying

A.D. Robson, L.K. Abbott & N. Malajczuk (eds.), Management of mycorrhizas in agriculture, horticulture and forestry, 141–147.

VAM in the more complex natural ecosystems.

Comparisons of mycorrhizal and non-mycorrhizal plants

The functioning of VAM has been studied primarily by comparing growth of plants colonized by indigenous or introduced VAM fungi with that of control plants grown in semi-sterile soil. Several factors complicate the evaluation of results from such studies under field conditions: confounding changes of nutrient availability and biological activity of soils, inefficient elimination of native VAM fungi, and difficulties in the attainment of previous levels of root colonization after reinoculation.

Soil fumigation

Undesirable growth responses to soil fumigation have been measured both in non-VAM hosts (Yost and Fox, 1979; Plenchette et al., 1983) and in cereals (Rovira, 1976; Jakobsen, 1983). The control of weak pathogens in the rhizosphere may play a role (Rovira, 1976) but increased levels of available P (5–25%) and in particular N (Fig. 1) mineralized from the soil microbial biomass are probably the major responsible factors (Jakobsen, 1983, 1987; Weber et al., 1992). The dramatic increases in ammonium may result from a rapid recolonization by ammonifying heterotrophs (Ridge, 1976) combined with a slow recolonization by nitrifying autotrophs (Rovira, 1976). The elevated ammonium levels in soil treated with 75 g dazomet m^{-2} persisted for at least 313 d (Fig. 1). The application of 100 g methyl bromide m^{-2} to a coarse, sandy soil resulted in 4–10 fold increases in ammonium levels (Jakobsen, 1983); consequently it can be assumed that the N contained in dazomet (17% by weight) contributes only marginally to the elevated ammonium levels.

The elimination of VAM propagules from field plots by fumigation may not be complete. Although root colonization in 0–20 cm soil depth is usually negligible during the first 40–50 days of crop growth, about 10% of the root length is most often mycorrhizal at crop maturity. Poor

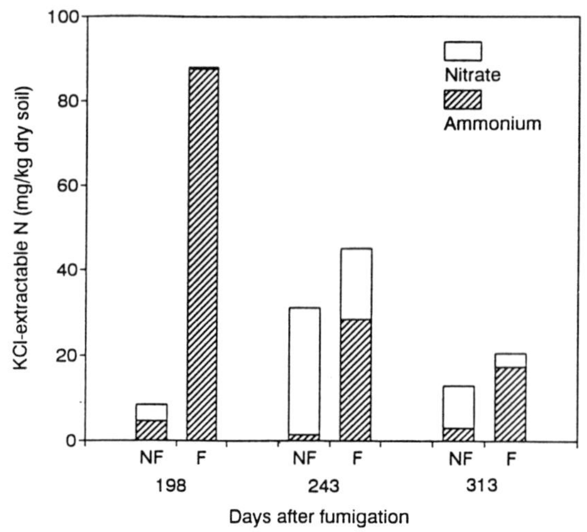

Fig. 1. Ammonium and nitrate in field plots sampled at different times after fumigation with 75 g dazomet m^{-2} (F) or not (NF) at 15 September. Untreated plots received 50 kg N ha^{-1} after 215 d.

penetration of the fumigant to deeper soil layers is probably responsible for this (Jakobsen, 1986; Weber et al., 1992).

Soil fumigation has been successfully used to demonstrate a marked dependency on VAM for a range of plant species grown at strong to moderate P deficiency (Plenchette et al., 1983; Jakobsen, 1983, 1987), but limited information is obtained from experiments done at one or two soil-P levels only. Phosphorus response curves for VAM and non-VAM plants are required to determine mycorrhizal effectiveness in terms of saved P fertilizer (Yost and Fox, 1979; Abbott and Robson, 1984). However, the use of fumigation for obtaining uncolonized controls is likely to underestimate the true VAM contribution to plant growth due to the increased nutrient levels in fumigated soil.

Fungicide application

Fungicides available may affect not only VAM fungi but also other soil microorganisms. However, these side effects may be less confounding than those caused by fumigation. Benomyl is the fungicide which has been most used in VAM research, and although it may effectively depress levels of root colonization (Boatman et al., 1978; Fitter and Nichols, 1988) and P inflow to roots of VAM plants in pot experiments (Hale and Sanders, 1982), reports on its function under field conditions are conflicting (see Fitter, 1991). Growth reductions in the field following the suppression of root colonization by benomyl have been demonstrated only twice (Lu and Miller, 1989; Bentivenga and Hetrick, 1991). Benomyl is strongly adsorbed to soil particles and its effect on VAM fungi may be enhanced if application is succeeded by mechanical tillage for a more even distribution in the soil (Lu and Miller, 1989). There is a requirement for a fungicide which is specific to VAM fungi and which may be transported to the roots via the phloem after shoot application.

Reinoculation after fumigation, fallowing, or pre-cropping with non-host

Comparison between uninoculated and inoculated plants grown in fumigated field plots are suitable for reducing the confounding effects of enhanced nutrient availability after fumigation. Inoculum may be applied to seedlings before transplanting to the field (Snellgrove and Stribley, 1986; Sasa et al., 1987) or directly to field plots before sowing (Jakobsen, 1983; Buwalda et al., 1985; Haas et al., 1987). Effects of VAM may also be studied by reinoculation of fallowed soil or soil pre-cropped with a non-host. Both treatments markedly reduce the number of propagules of the indigenous VAM fungi (Black and Tinker, 1979; Kucey and Paul, 1983; Buwalda et al., 1985; Thompson, 1987), and other soil organisms such as fungivorous animals are probably less affected by these treatments than by fumigation.

Although large growth responses may be demonstrated after reinoculation, this method cannot be used to measure the contribution to plant growth of the indigenous VAM fungi. Furthermore it remains uncertain whether a growth response to inoculation represents the optimal contribution from the introduced VAM fungus. This is because the rate and the maximum levels of root colonization is often lower in reinoculated than in untreated soil (Jakobsen, 1983; 1987).

Direct studies of nutrient uptake by VAM plants

The direct measurement of hyphal nutrient transport is superior to comparisons between VAM plants and their uncolonized controls, which are physiologically different entities. Information is required on amounts, distribution and longevity of the hyphae. Methods must be developed for the measurement of hyphal P transport and its variation between different isolates and populations of VAM fungi. Finally, the function of the hyphae must be evaluated under the influence of the natural biotic and physio-chemical soil environment.

The measurement of nutrient transport by VAM hyphae

The functioning of external VAM hyphae may be studied directly by means of two-compartment systems, similar in principle to the petri-dish systems used by Hattingh et al. (1973). A soil-filled hyphal compartment is separated from the rooting compartment by means of a fine mesh allowing for free passage of hyphae but not of roots. The critical mesh size will depend on the smallest root diameter of the plant species studied. Special multi-compartment systems were exploited by Schuepp et al. (1987) in a study of hyphal spread through different growth substrates and since then, hyphal compartments have been utilized in several experiments, which have improved our understanding of the functioning of VAM hyphae. This includes the hyphal incorporation of carbon by different VAM fungi (Jakobsen and Rosendahl, 1990; Pearson and Jakobsen, unpublished), differences between VAM fungi in total lengths and spread of hyphae (Jakobsen et al., 1992a), the depletion of different soil-P fractions and soil acidification by VAM hyphae (Li et al., 1991a), time-course studies of hyphal transport of ^{32}P by different VAM fungi (Jakobsen et al., 1992b), hyphal transport of N (Johansen et al., 1992) and hyphal transport of Cu (Li et al., 1991b) and Zn (Kothari et al., 1991).

Radioactive (Jakobsen et al., 1992b) and stable isotopes (Ames et al., 1983; Johansen et al., 1992) are probably the most powerful tools for directly demonstrating the nutrient transport by hyphae. The isotopes may either be supplied to the intact hyphal network as a solution (Hattingh et al., 1973), mixed into soil layers at defined positions from the root compartment (Jakobsen et al., 1992b) or uniformly mixed into the soil of the hyphal compartment (Johansen et al., 1992).

Quantification of contribution from VAM fungi

The relative importance of nutrient supply to VAM plants by hyphal transport and directly by root uptake is unknown. Estimates have been obtained for P inflow to roots via VAM hyphae, assuming that mycorrhizal and non-mycorrhizal roots absorb P at an equal rate (Sanders et al., 1977; Jakobsen, 1986; Jakobsen et al., 1992a). However, this assumption requires verification.

The two-compartment principle is useful for quantifying the VAM-hyphal contribution to P uptake by plants. Hyphal P transport from a hyphal compartment can be measured as the difference in P content between mycorrhizal plants and non-mycorrhizal plants, if the root density is high enough to prevent any differences in P uptake from the root compartment (Li et al., 1991a, 1991b; Jakobsen et al., 1992a). However, such growth conditions are highly artificial, and the hyphal P uptake measured in this way has to be considered as a potential contribution which cannot be expected under field conditions (Li et al., 1991a).

The principle of isotopic equilibrium may serve as an alternative for estimating the relative amounts of plant P derived from the hyphal compartment and from the root compartment in two-compartment systems. When ^{32}P applied to soil was allowed to equilibrate with the soil P for a certain period of time, the specific activity of ^{32}P in subsequently grown clover plants was similar to the specific activity of ^{32}P in soil extracts of different strength (Bolan et al., 1984). This relationship was observed irrespective of whether the plants were mycorrhizal or not.

The time needed for the major proportion of newly applied P to equilibrate with the native soil P will vary, but 3–4 weeks seem adequate (Bolan et al., 1984). If S_{HC} is the specific activity in an extract (e.g. 0.5 *M* $NaHCO_3$) of ^{32}P-labelled soil from a hyphal compartment then

$$S_{HC} = {}^{32}P_{HC} \div {}^{31}P_{HC} \quad \text{and} \quad {}^{31}P_{HC} = {}^{32}P_{HC} \div S_{HC} \qquad (1)$$

where the isotopes refer to plant content of P derived from the hyphal compartment (HC). Total plant P

$$^{31}P_{TOT} = {}^{31}P_{HC} + {}^{31}P_{RC} \qquad (2)$$

where $^{31}P_{RC}$ refers to P derived from the root compartment. Consequently

$$^{31}P_{RC} = {}^{31}P_{TOT} - ({}^{32}P_{HC} \div S_{HC}). \qquad (3)$$

S_{HC} should be measured both at initiation and termination of the growth period; changes with time may thus be partly accounted for by using a mean value in Eq. (3). The expected correspondence between the specific activity in soil extracts and the specific activity in plants should be tested in growth systems with only one

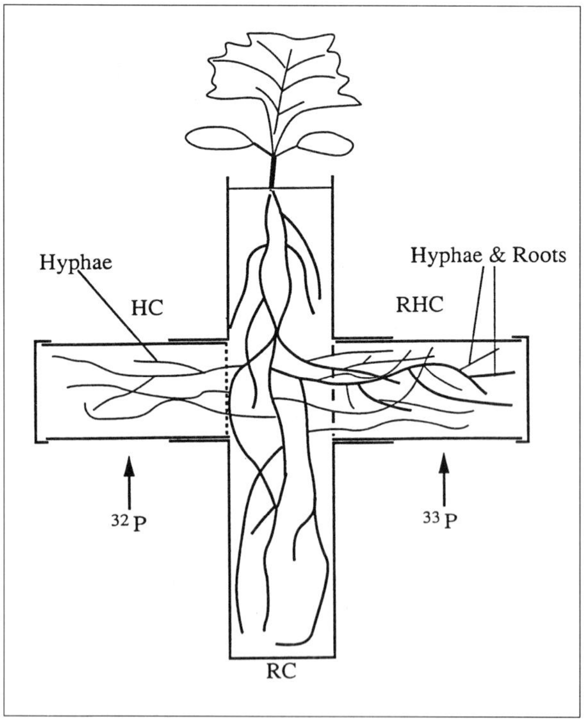

Fig. 2. Longitudinal section of three-compartment system for measuring hyphal contribution to P uptake by VAM plants.

compartment. The P uptake from the root compartment as estimated by Eq. 3 may be separated in a root and a hyphal component, if it is assumed that the hyphal uptake of ^{32}P is similar from compartments of similar size containing hyphae only or roots+hyphae. Whether this assumption is justified is discussed below.

A three-compartment system utilizing dual labelling with ^{32}P and ^{33}P makes it possible to directly compare the P uptake by hyphae alone and by roots and hyphae in combination. Plants are grown in PVC tubes, where a central compartment contains the major proportion of the root system (Fig. 2). Hyphae are allowed to grow through a 37 µm mesh into an adjacent compartment (HC) containing ^{32}P-labelled soil, while both roots and hyphae grow through a coarse mesh into an opposite compartment (RHC) of similar size. The soil in this compartment is labelled with the same quantity of ^{33}P as the ^{32}P supplied to the hyphal compartment. Radioacivity in the shoot originating from each isotope is distinguished during scintillation counting. The radioactivity of each isotope reflects the transport from each compartment, and the difference between levels of ^{33}P and ^{32}P in the plant would be expected to reflect the contribution from roots in RHC. It is a prerequisite that the hyphal density in RHC is not influenced by the presence of the roots. A non-mycorrhizal control should be included to correct for diffusion of ^{32}P into the main root compartment and for uptake of ^{32}P by root hairs growing through the mesh. A layer of unlabelled soil placed at the proximal end of HC helps to reduce these problems. Furthermore, the labelled soil may be incubated for some weeks to allow for as much isotopic exchange of ^{32}P and ^{33}P with the native soil P as possible. A mycorrhizal plant with only hyphae growing into both the ^{32}P and the ^{32}P labelled compartments may serve as a control to account for any isotopic discrimination in hyphal P transport. Attention should be paid to the maintenance of an adequate soil moisture content in RHC, where the water use will be higher than in HC. The absence of roots from HC results in a low activity of the general soil microflora as compared to levels measured in RHC (Olsson, Jakobsen and Söderström, unpublished results). The importance of this for the relative availability of the radioisotopes is unknown as yet.

Table 1. Content of ^{32}P and ^{33}P in 27 days old cucumber plants grown in association with VAM fungi or not in a three-compartment system. Compartments for hyphae only (HC) and for roots+hyphae (RHC) were filled with soil uniformly labelled with 10 µCi of ^{32}P and ^{33}P, respectively. HC and RHC were mounted 5 d after sowing (Pearson and Jakobsen, unpublished results)

Fungus	^{32}P (HC) (dpm in plants $\times 10^{-3}$)	^{33}P (RHC)
Control	64 (10)[a]	1332 (333)
Scutellospora calospora	182 (19)	2706 (365)
Glomus sp.	462 (105)	2235 (553)
Glomus caledonium	1823 (129)	1668 (107)

[a] Means (n = 4) with SE in brackets

The dual labelling method was used to demonstrate a major inter-fungal variation in hyphal P transport by three fungi in symbiosis with young cucumber plants (Table 1). The VAM fungal contribution to total P uptake varied accordingly: In associations with *Scutellospora calospora* the transport of ^{33}P from RHC was much higher than the transport of ^{32}P from HC, whereas the isotopes were present in similar amounts in plants colonized by *Glomus caledonium*, the fungus transporting most ^{32}P from HC. Root colonization in the central compartment and root length density in RHC were in the ranges 60–81% and 18–22 cm cm^{-3}, respectively and hyphal length densities were similar in HC and RHC. These results indicate that P uptake by roots may be strongly influenced by hyphal P transport to the plant. The P uptake by roots is regulated by the demand of the plant via negative feedback control (see Marschner, 1986), whilst the plant might have less control over the P supply from the fungus. A high fungal P supply by *G. caledonium* could in this way lead to a complete inhibition of root P uptake. It cannot be excluded, however, that overlapping of P depletion zones due to the high root densities in RHC may have reduced the amount of ^{33}P available to the hyphae. A combined use of hyphal compartments and split-root systems may prove useful for our further attempts to distinguish between P uptake by hyphae and roots.

Measurements in the field

The system described facilitates the measurement of the P uptake of the indigenous VAM population from disturbed or undisturbed field soils at controlled growth conditions. Ultimately, these measurements should be carried out in the field. Radiotracers have been widely used for studying root activity of field-grown crops (Abbott and Fraley, 1991) and Dighton et al. (1990) measured the P uptake by ectomycorrhizas of birch from an aqueous solution of ^{32}P injected into the soil. Mesh bags containing isotope-labelled soil may be placed in microplots in order to perform well-defined studies of hyphal P transport under field conditions. Pairs of bags of either fine or coarse nylon mesh are buried inside large PVC tubes (diameter = 30–40 cm) pushed into the soil. The large PVC tubes confine the uptake of isotopes to those plants growing inside the tubes. The mesh bags are easily positioned in or removed from the microplots at appropriate times, if they are contained in rigid cylinders with a perforated wall. PVC tubes (60 mm diameter, 250 mm length) with part of their wall removed have proven appropriate. Sequential harvesting and measurement of radioactivity in single leaves facilitate the study of the P contribution from hyphae within the context of the timing of P uptake which is important for crop productivity. The system has been successfully used to measure the hyphal uptake of ^{32}P and the uptake of ^{33}P by roots and hyphae into peas associated with an inoculant VAM fungus in fumigated soil or the indigenous VAM fungi in untreated soil.

Conclusions

The ultimate purpose of VAM research is to maximize the effectiveness of VAM in diverse ecosystems through an improved understanding of the functioning of mycorrhizas under field conditions. The novel approaches described for measuring hyphal nutrient transport seem adequate for the functional characterization of isolates and field populations of VAM fungi. This characterization should be combined with methods for describing the composition of VAM fungal populations outside and inside roots and used to establish a link between the dominating root colonizing fungi and mycorrhizal functioning. Eventually this will help us to understand the ecological fitness of a given VAM association.

The research approaches also facilitate the identification of constraints on the functioning of VAM and allow for a quantification of the impact of any potential constraint. Such quantification is needed for the development of models which may realistically predict nutrient uptake and growth of VAM plants.

References

Abbott M L and Fraley L Jr 1991 A review: Radiotracer methods to determine root distribution. Environ. Exp. Bot. 31, 1–10.

Abbott L K and Robson A D 1984 The effect of VA mycorrhizae on plant growth. *In* VA Mycorrhiza. Eds. C Ll Powell and D J Bagyaraj. pp 114–130. CRC Press, Inc. Boca Raton, Florida.

Abbott L K and Robson A D 1991 Factors influencing the occurrence of vesicular arbuscular mycorrhizas. Agric. Ecos. Environ. 35, 121–150.

Ames R N, Reid C P P, Porter L K and Cambardella C 1983 Hyphal uptake and transport of nitrogen from two ^{15}N-labelled sources by *Glomus mosseae*, a vesicular-arbuscular mycorrhizal fungus. New Phytol. 95, 381–396.

Bentivenga S P and Hetrick B A D 1991 Relationship between mycorrhizal activity, burning and plant productivity in tallgrass prairie. Can. J. Bot. 69, 2597–2601.

Black R and Tinker P B 1979 The development of endomycorrhizal root systems. II. Effect of agronomic factors and soil conditions on the development of vesicular-arbuscular mycorrhizal infection in barley and on the endophyte spore density. New Phytol. 83, 401–413.

Boatman N, Paget D, Hayman D S and Mosse B 1978 Effects of systemic fungicides on vesicular-arbuscular mycorrhizal infection and plant phosphate uptake. Trans. Br. Mycol. Soc. 70, 443–450.

Bolan N S, Robson A D, Barrow N J and Aylmore L A G 1984 Specific activity of phosphorus in mycorrhizal and non-mycorrhizal plants in relation to the availability of phosphorus to plants. Soil Biol. Biochem. 16, 299–304.

Buwalda J G, Stribley D P and Tinker P B 1985 Effects of vesicular-arbuscular mycorrhizal infection in first, second and third cereal crops. J. Agric. Sci. Camb. 105, 631–647.

Dighton J, Mason P A and Poskitt J M 1990 Field use of ^{32}P to measure phosphate uptake by birch mycorrhizas. New Phytol. 116, 655–661.

Fitter A H 1985 Functioning of vesicular-arbuscular mycorrhizas under field conditions. New Phytol. 99, 257–265.

Fitter A H 1991 Costs and benefits of mycorrhizas: Implications for functioning under natural conditions. Experientia 47, 350–355.

Fitter A H and Nichols R 1988 The use of benomyl to control infection by vesicular-arbuscular mycorrhizal fungi. New Phytol. 110, 201–206.

Gianinazzi S, Gianinazzi-Pearson V and Trouvelot A 1989 Potentialities and procedures for the use of endomycorrhizas with special emphasis on high value crops *In* Biotechnology of fungi for improving plant growth. Eds. J M Whipps and R D Lumsden. pp 41–54. Cambridge University Press.

Haas J H, Bar-Yosef B, Krikun J, Barak R Markovitz T and Kramer S 1987 Vesicular-arbuscular mycorrhizal fungus infestation and phosphorus fertilization to overcome pepper stunting after methyl bromide fumigation. Agron. J. 79, 905–910.

Hale K A and Sanders F E 1982 Effects of benomyl on vesicular-arbuscular mycorrhizal infection of red clover (*Trifolium pratense* L.) and consequences for phosphorus inflow. J. Plant Nutr. 5, 1355–1367.

Hattingh M J, Gray L E and Gerdemann J W 1973 Uptake and translocation of ^{32}P-labelled phosphate to onion roots by endomycorrhizal fungi. Soil Sci. 116, 383–387.

Jakobsen I 1983 Vesicular-arbuscular mycorrhiza in field-grown crops. II Effect of inoculation on growth and nutrient uptake in barley at two phosphorus levels in fumigated soil. New Phytol. 94, 595–604.

Jakobsen I 1986 Vesicular-arbuscular mycorrhiza in field-grown crops. III Mycorrhizal infection and rates of phosphorus inflow in pea plants. New Phytol. 104, 573–581.

Jakobsen I 1987 Effects of VA mycorrhiza on yield and harvest index of field-grown pea. Plant Soil 98, 407–415.

Jakobsen I and Rosendahl L 1990 Carbon flow into soil and external hyphae from roots of mycorrhizal cucumber plants. New Phytol. 115, 77–83.

Jakobsen I, Abbott L K and Robson A D 1992a External hyphae of vesicular-arbuscular mycorrhizal fungi associated with *Trifolium subterraneum* L. 1. Spread of hyphae and phosphorus inflow into roots. New Phytol. 120, 371–380.

Jakobsen I, Abbott L K and Robson A D 1992b External Hyphae of vesicular-arbuscular mycorrhizal fungi associated with *Trifolium subterraneum* L. 2. Hyphal transport of ^{32}P over defined distances. New Phytol. 120, 509–516.

Johansen A, Jakobsen I and Jensen E S 1992 Hyphal transport of ^{15}N-labelled nitrogen by a vesicular-arbuscular mycorrhizal fungus and its effect on depletion of inorganic soil N. New Phytol.122, 281–288.

Kothari S K, Marschner H and Römheld V 1991 Contribution of the VA mycorrhizal hyphae in acquisition of phosphorus and zinc by maize grown in a calcareous soil. Plant Soil 131, 177–185.

Kucey R M N and Paul E A 1983 Vesicular arbuscular mycorrhizal spore populations in various Saskatchewan soils and the effect of inoculation with *Glomus mosseae* on faba bean growth in greenhouse and field trials. Can. J. Soil Sci. 63, 87–95.

Li X-L, George E and Marschner H 1991a Phosphorus depletion and pH decrease at the root–soil and hyphae–soil interfaces of VA mycorrhizal with clover fertilized with ammonium. New Phytol. 119, 397–404.

Li X-L, Marschner H and George E 1991b Acquisition of phosphorus and copper by VA-mycorrhizal hyphae and root-to-shoot transport in white clover. Plant Soil 136, 49–57.

Lu S and Miller M H 1989 The role of VA mycorrhizae in the absorption of P and Zn by maize in field and growth chamber experiments. Can. J. Soil Sci. 69, 97–109.

Marschner H 1986 Mineral Nutrition of Higher Plants. Academic Press, London. 674p.

Mosse B 1986 Mycorrhiza in a sustainable agriculture. Biol. Agric. Hortic. 3, 191–209.

Plenchette C, Fortin J A and Furlan V 1983 Growth responses of several plant species to mycorrhizae in a soil of moderate P-fertility. Plant Soil 70, 199–209.

Ridge E H 1976 Studies on soil fumigation—II Effects on bacteria. Soil Biol. Biochem. 8, 249–253.

Rovira A D 1976 Studies on soil fumigation—I Effects on ammonium, nitrate and phosphate in soil and on the growth, nutrition and yield of wheat. Soil Biol. Biochem. 8, 241–247.

Sanders F E, Tinker P B, Black R L B and Palmerley S M 1977 The development of endomycorrhizal root systems: I Spread of infection and growth-promoting effects with four species of vesicular-arbuscular endophyte. New Phytol. 78, 257–268.

Sasa M, Zahka G and Jakobsen I 1987 The effect of pre-transplant inoculation with VA mycorrhizal fungi on the subsequent growth of leeks in the field. Plant Soil 97, 279–283.

Schüepp H, Miller D D and Bodmer M 1987 A new technique for monitoring hyphal growth of vesicular-arbuscular mycorrhizal fungi through soil. Trans. Br. Mycol. Soc. 89, 429–435.

Snellgrove R C and Stribley D P 1986 Effects of pre-inoculation with a vesicular-arbuscular mycorrhizal fungus on growth of onions transplanted to the field as multi-seeded peat modules. Plant Soil 92, 387–397.

Thompson J P 1987 Decline of vesicular-arbuscular mycorrhizae in long fallow disorder of field crops and its expression in phosphorus deficiency of sunflower. Aust. J. Agric. Res. 38, 847–867.

Weber E, Saxena M C, George E and Marschner H 1993 Effect of vesicular-arbuscular mycorrhiza on vegetative growth and harvest index of chickpea grown in Northern Syria. Field Crops Res. 32, 115–128.

Yost R S and Fox R L 1979 Contribution of mycorrhizae to P nutrition of crops growing on an oxisol. Agron. J. 71, 903–908.

Reprinted from *Plant and Soil*, Volume 159, No. 1 (1994), pp. 141–147.

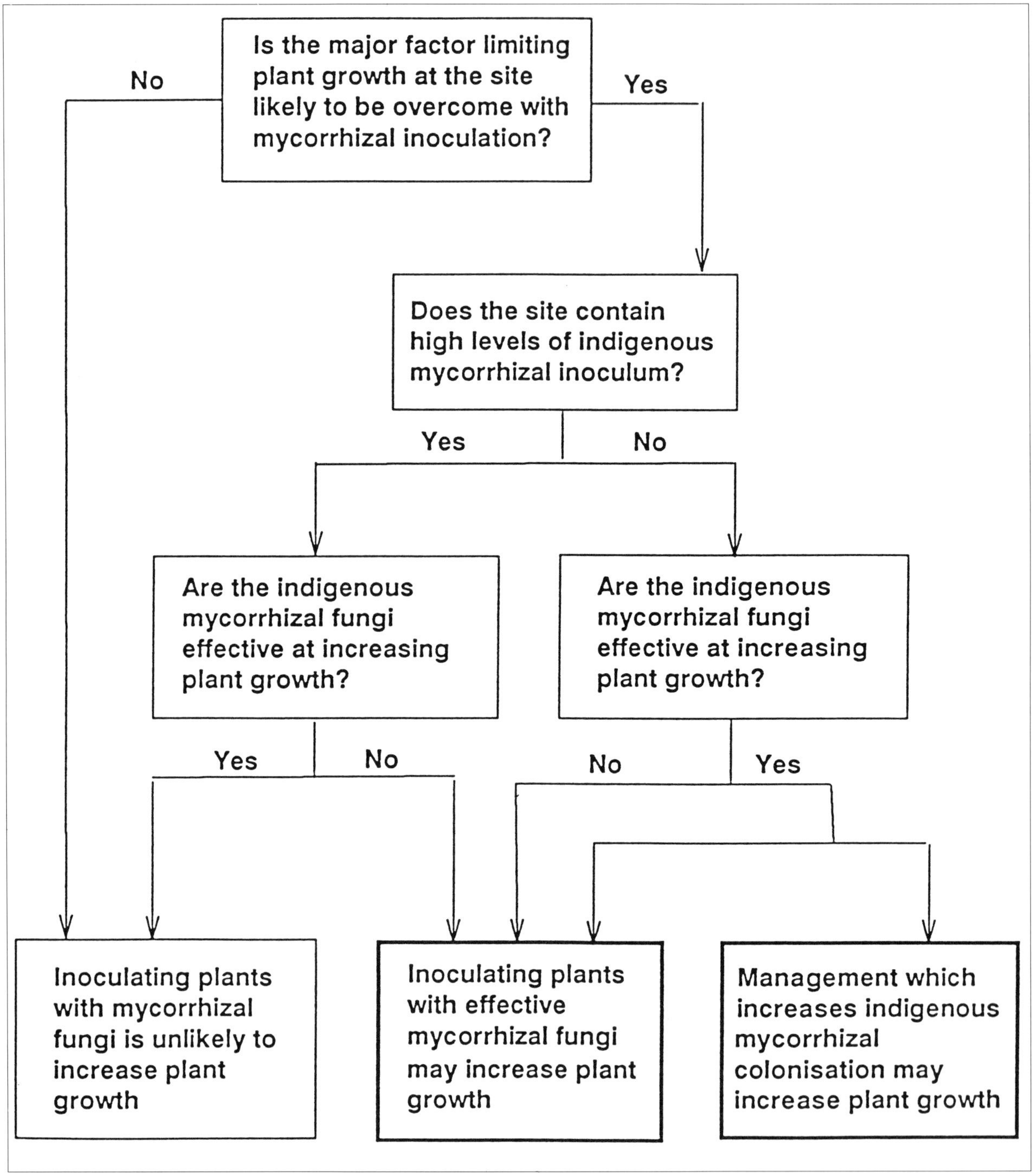

Fig. 1. Identifying responsive sites for mycorrhizal inoculation.

diversity. These additional benefits of the mycorrhizal symbiosis have widened the scope for the use of mycorrhizal inoculants.

Most natural field soils and non-sterile nursery soils contain indigenous mycorrhizal fungi. A decision to introduce mycorrhizal fungi under these circumstances will also depend on the 'effectiveness' of these indigenous fungi [i.e. "their ability to benefit plant growth" (Abbott et al., 1992), through whatever mechanism] when compared with possible inoculant fungi (Fig. 1). The 'colonization capacity' of a mycorrhizal fungus [i.e. "the amount of root colonized in a certain time or the time taken to colonize a certain amount of root by a measured dose of inoculum" (Tommerup, 1992)], will generally help to determine the role of the indigenous mycorrhizal population. Where the indigenous fungi have a low colonization capacity, or where they have a high colonization capacity but are ineffective, inoculating with effective mycorrhizal fungi may increase plant growth. Where the indigenous mycorrhizal fungi have a low colonization capacity but prove to be effective when root colonization is increased, then the use of management practices to increase the abundance of fungal populations may be more appropriate (Dodd et al., 1990a, b; Sieverding, 1991). Management practices could also be used to decrease the colonization capacity of ineffective indigenous mycorrhizal populations to allow a bigger initial niche for the inoculant fungus (or fungi). For instance, fallowing decreases the colonization capacity of AM fungi in the northern areas of the Australian grain belt (Thompson, 1987) and clear-felling and burning of forests decreases levels of ECM fungi inoculum (Parke et al., 1984).

The potential for management of indigenous populations of mycorrhizal fungi is discussed more fully by Thompson (Chapter 18), Chang (Chapter 19), Grove and Malajczuk (Chapter 20) and Jasper (Chapter 21). The greatest immediate potential for the use of mycorrhizal inoculants will be in areas where indigenous mycorrhizal populations have been eradicated or drastically reduced by human influence or natural disturbance. These areas include: fumigated soils (Krikun et al., 1990), mine sites (Jasper et al., 1988), badly eroded sites (Dodd et al., 1991), reforestation zones (Jeffries and Dodd, 1991; Parke et al., 1984) and low-input agricultural systems (Sieverding, 1991). ECM fungi can also be host-specific (Molina and Trappe, 1982), so that where trees are planted as exotics at sites, the indigenous population of ECM fungi may be unsuitable. The early failures of pine plantations in Australia and Africa, for example, were attributed to the lack of suitable mycobionts (Kessel, 1927). Also, the best results from inoculation of conifers (Douglas Fir, Norway Spruce, Sitka Spruce and Scots Pine) with ECM FUNGI in Europe have been obtained with trees planted as exotics in previously cultivated soils and natural broadleaved forests (Le Tacon et al., 1991).

Site characterization

Having identified an opportunity for benefits from inoculation with a suitable mycorrhizal fungus (or fungi), the initial stage in the selection of isolates involves defining the edaphic and climatic conditions where the fungi will be introduced. Where a specific tolerance to environmental extremes is sought, e.g. salinity, polluted sites, root pathogens etc., then ideally the isolation of mycorrhizal fungi should begin in such soils. There is already considerable evidence, for example, that few AM and ECM fungi are effective across the soil pH range found commonly in agricultural and forestry ecosystems. Indeed, there are distinct groups found commonly in either acid or alkaline soils (Sieverding, 1991; Trappe, 1977). To minimize the likelihood of incompatibility between fungus and host or between fungus and soil it seems more and more likely that inoculant fungi should be selected from the soil in which inoculated seedlings are to be planted. This does not exclude the collection of fungi from locations with similar edaphic and climatic conditions to the site targeted for inoculated seedlings. The types of the data that should be recorded when collecting soil and fruiting bodies for the isolation of inoculant fungi include: site location, climate and vegetation, soil pH, soil P, soil texture, soil electrical conductivity, and site management history (Abbott et al., 1992; Sieverding; 1991).

Collection and isolation of fungi

Fig. 2 outlines possible routes for the isolation of

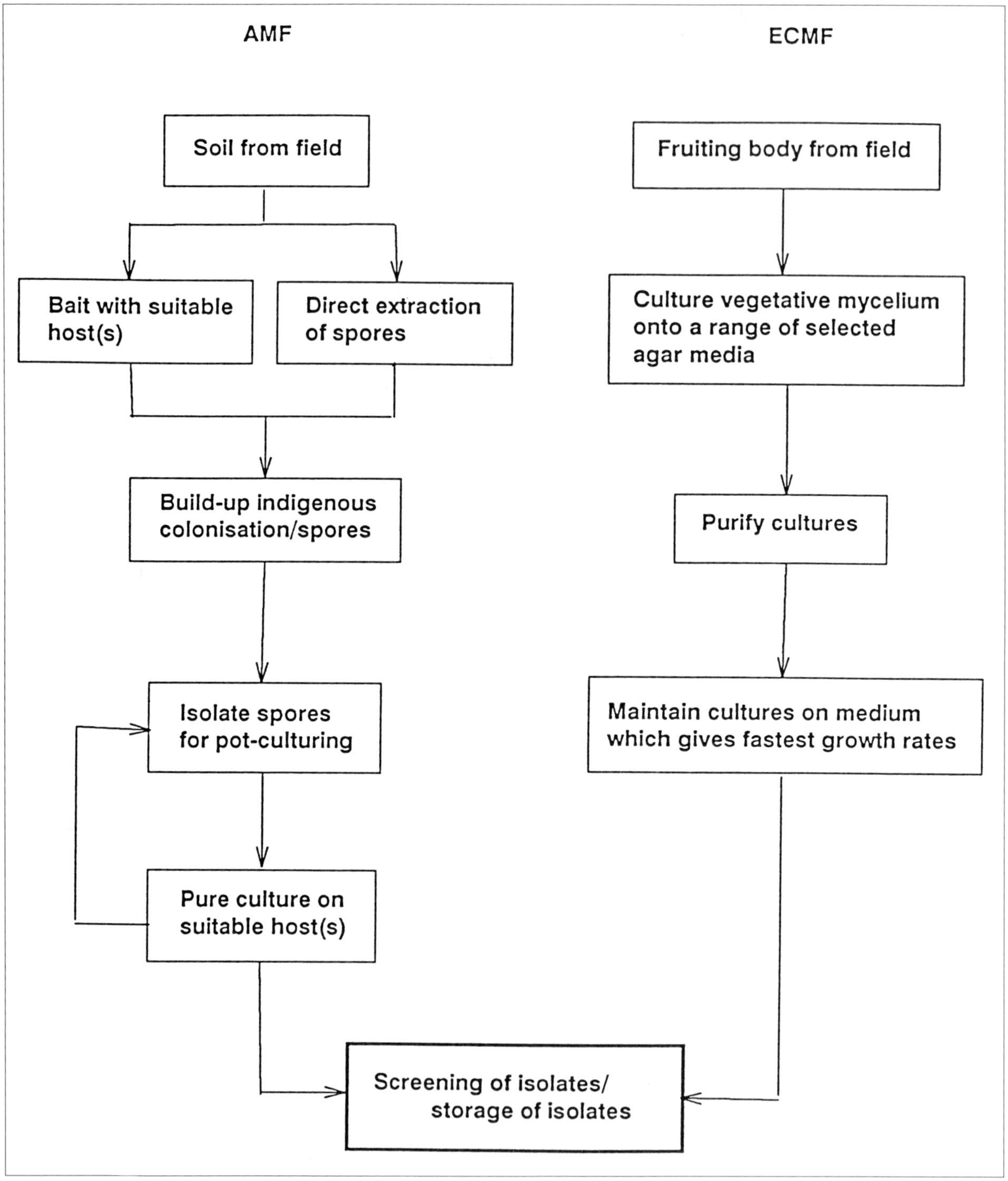

Fig. 2. Isolating mycorrhizal fungi.

AM and ECM fungi from the field. Mycorrhizal fungi can be difficult to isolate and maintain in pure culture, so that this isolation procedure also forms part of the screening of fungi (Fig. 3).

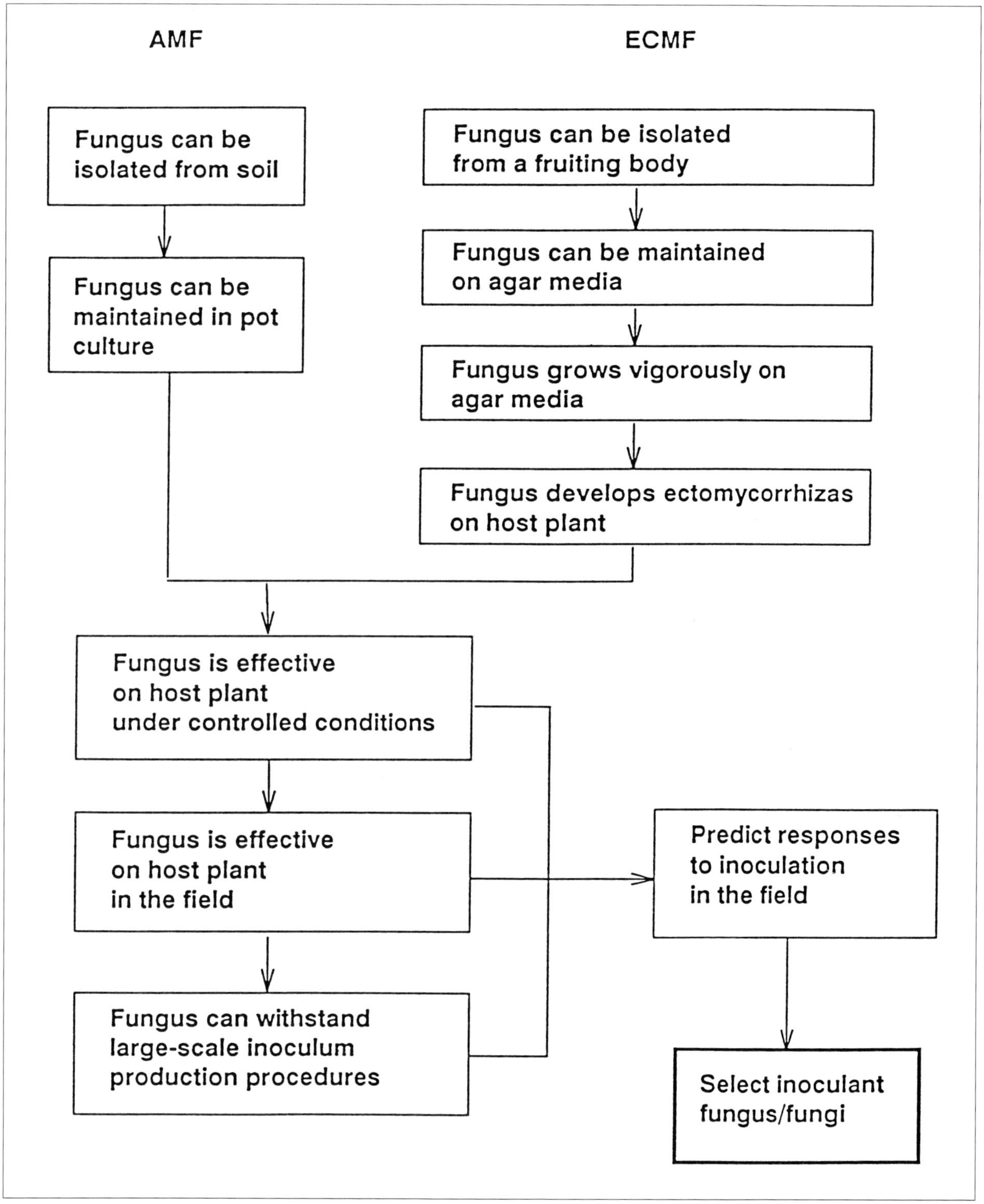

Fig. 3. Screening mycorrhizal fungi.

In collecting AM fungi from field sites, soil samples should be taken from the rhizosphere of native or crop plants at a soil depth where most root proliferation occurs. If the plant sampled is also the plant which is to be inoculated, then sampling should concentrate on the various stages of development of this species. It has been observed, for example, that different populations of spores of AM fungi are found at different sampling times in various ecosystems (Gemma et al., 1989). Depending on the amount of soil available and the spore numbers in the samples, the next step is to 'bait' the soils by planting a suitable mycotrophic host directly into the soil to induce sporulation of indigenous AM fungi (trap cultures). Spores can be extracted from the soil and used to inoculate host roots at the same time, although this may not be as successful as using spores produced in trap cultures (INVAM, 1992). Unsuccessful reproduction of indigenous AM fungi is often the result of the incorrect selection of host and soil medium as well as unsuitable environmental conditions. Ideally, two different hosts should be used to increase the likelihood of stimulating the sporulation of the full complement of indigenous AM fungi.

Trap cultures may need to be grown for 4–6 months to allow sporulation of *Acaulospora, Scutellospora*, and *Gigaspora* species (Sieverding, 1991), whereas *Glomus* spp. generally sporulate earlier. Pure cultures of the AM fungi can be initiated from these spores. Spores can be characterized based on their wall morphologies, at least to the level of genus, and compared with spores extracted from the original field soil to confirm the presence of the original AM fungi. If mixed cultures result, spores can be surface-sterilized and reinoculated onto a host plant to obtain a pure culture. Pure cultures should be maintained on living plants until the conditions for successful storage of inoculum have been investigated. Cultures can be stored at room temperature [although there appears to be little problem in storing cultures at 4°C if desired (Dodd and Jeffries, pers. obs.)], as air-dried soil containing colonized roots and spores.

ECM fungi, in contrast to AM fungi, can be grown on agar media in the absence of a host plant. Isolation of ECM fungi from the field is therefore a simpler process and is achieved primarily from fruiting bodies. These are favoured over mycorrhizal roots because they can easily be identified, at least to the level of genus, are easy to collect from the field and pure cultures are comparatively easy to obtain. Ideally, fruiting bodies should be collected (i) from beneath the tree species which is to be inoculated or a related tree species, since ECM fungi can be host-specific (Molina and Trappe, 1982), (ii) at different stages of plant development, because ECM fungi can colonize roots at different stages of tree development (Mason et al., 1987), and (iii) from soil types which closely approximate those of the soil in which inoculated seedlings are to be planted.

Isolation from fruiting bodies can be achieved by inoculating vegetative mycelium onto a range of selected agar media (see Molina and Palmer, 1982). Spores are infrequently used for isolation because of their poor germination and genetic variability. Spores are, however, widely used as sources of inoculum (Kropp and Langlois, 1990) but because of the large quantities required, this form of inoculum is restricted to only a few fungal species e.g. *Pisolithus tinctorius, Scleroderma* spp. and *Rhizopogon* spp. Purified ECM fungi isolates can be maintained on the agar medium which gives the fastest growth rates. Comparatively fast growing cultures are preferred because they are considered to be physiologically more active and, therefore, better sources of inoculum. However, growth rates on agar may bear little relationship to the ability of the fungal isolate to increase plant growth (see below). Fungal cultures can be stored on agar at 3–5°C (Molina and Palmer, 1982), although this can result in loss of vigour for some fungi (Thomson et al., 1991). Other forms of storage warrant further investigation.

Screening of fungi

Having collected and isolated a range of mycorrhizal fungi which are likely to be more or less adapted to a given set of ecological conditions, the next stage of the selection process is to screen these isolates for 'effectiveness'. In the majority of screening programmes, this will mean searching for isolates which increase nutrient uptake and therefore plant growth. Criteria used in the selection of mycorrhizal fungi for the uptake and transport of P from soil to host have recently been reviewed by Abbott et al. (1992). In recent years, however, there has been an increased awareness of the additional benefits resulting from the presence of an effective fungal partner. In these cases, different criteria may be used to select inoculant mycorrhizal fungi which are unrelated to the ability to increase rates of growth or nutrient uptake. For instance, increased and rapid development of the extramatrical hyphal network of AM and ECM fungi may be of more ecological value in stabilizing eroded soils and sand dune ecosystems (Bethlenfalvay et al., 1988). The ability of a mycorrhizal fungus to function efficiently on a wide range of plant species may also be a more desirable attribute in increasing the diversity of mycotrophic plants in restoration programmes such as revegetating mine sites (Jasper et al., 1988).

The selection of ECM fungi should begin with an initial screening of isolates to determine whether they form mycorrhizas with the selected host (Fig. 3) because these fungi can be host-specific (Molina and Trappe, 1982). AM do not display the same degree of host-specificity. Synthesis of ECM fungi can be quickly assessed in axenic culture in the laboratory. ECM fungi can also be screened for other characteristics in vitro in the laboratory including their growth rates on different media, the response to different temperatures, pH and moisture tensions, antagonism against selected pathogens and production of secondary metabolites. The value of pure culture growth rates as an indication of an isolates suitability as an inoculant fungus requires further study.

AM and compatible ECM fungi can be screened under controlled conditions for their effectiveness on the selected host (Fig. 3). Growth responses to inoculation with mycorrhizal fungi can be affected by the quantity and distribution of inoculum in the growth medium (Abbott and Robson, 1984). Thus, experiments comparing the effectiveness of different mycorrhizal isolates should use saturating amounts of inoculum, such that further increases in the quantity of the inoculum has little or no effect on root colonization (Smith and Walker, 1981). Isolates should initially be compared in sterile soil/growth media chosen as being representative of, or the same as, the field environment in which the isolates will be introduced. Where isolates are being selected for stress situations such as saline soils, polluted sites etc. this will mean choosing a medium which is representative of these sites or which can be altered to approximate conditions occurring at these sites. In many cases, however, isolates can be screened in a sterilized P-deficient sand/soil.

Effective mycorrhizal isolates can then be tested in natural soils under controlled conditions (Fig. 3) for their competitive ability with indigenous mycorrhizal populations and other soil microorganisms. Isolates may need to be screened in undisturbed soil collected from the field because disturbance can decrease the colonization capacity of indigenous mycorrhizal populations (Jasper et al., 1991). Programmes involving the out-planting of nursery-raised seedlings or the control of nursery pathogens may need to proceed straight to the nursery because inoculant fungi must be able to withstand nursery inoculation practices. Several factors have been identified as affecting mycorrhizal development in the nursery: substrate pH, soil microbiological activity, the toxicity of certain pesticides and substrate fertility (Marx and Cordell, 1987). Trappe (1977) and Kropp and Langlois (1990) suggested selecting broadly adapted isolates so that they are able to cope with the changes which occur in environmental conditions and soils from nursery to field and over plantation areas. Currently, it is also unrealistic to expect nursery managers to incorporate a number of different 'site-specific' mycorrhizal inoculants into their nursery programmes, particularly when inoculation methods and the specific physiological requirements of each fungus may differ.

Isolates of AM and ECM fungi which prove to be effective in competition with indigenous mycorrhizal fungi must ultimately be tested for effectiveness in the field (Fig. 3). These fungi must not only be able to compete with the indigenous mycorrhizal population under field conditions, but in some cases (e.g. reforestation and revegetation programmes) must also be able to persist and spread on roots over several years. Isolates will need to be able to colonize soil and plant roots away from points of inoculation so that they will be present in an extensive volume of soil for colonization of roots in the following season. Many ECM fungi have an efficient means of spread from spore inoculum. Despite this, however, several studies have demonstrated an increase in colonization by indigenous ECM fungi and a decrease in colonization by inoculant ECM fungi in the field with time (Garbaye et al., 1988; Hatchell and Marx, 1987). Colonization by inoculant AM in the field can also decrease with time (Dodd et al., 1990a, b). Our current understanding of the longer-term ecological role of AM and ECM fungi inoculants in the field has been restricted by our inability to accurately distinguish between inoculant and indigenous mycorrhizal fungi. However, recent advances in the development of genetic markers (see Martin et al., Chapter 22) should facilitate further research in this area.

An important challenge for future mycorrhizal research is to develop screening procedures which reduce the need for extensive field testing of isolates (which can be time-consuming and expensive). Current laboratory and glasshouse procedures can give a poor prediction of plant growth responses to inoculation in the field (Le Tacon et al., 1988; Grove et al., 1991). These procedures would be improved by simulating more closely in the glasshouse the biotic and abiotic conditions of the field. A better understanding of the physiological and morphological characteristics of mycorrhizal fungi which contribute to mycorrhizal 'effectiveness' would also help in identifying a few well-defined fungal characteristics which relate to field performance of isolates. For example, the ability of a mycorrhizal fungus to decrease plant disease may be related to the production of specific antibiotics, and these could be quantified in the laboratory. Also, the ability of a mycorrhizal fungus to increase P uptake by the plant and to stabilize soil may be related to the formation of external hyphae, and this could be quantified in the glasshouse. By screening isolates for specific fungal characteristics at an early stage of the screening procedure, it should be possible to reduce the total number of isolates which need to be tested in the field.

Inoculum production

The production of inoculum of selected AM and ECM fungi will also pose constraints on the use of mycorrhizal fungi as inoculants (Fig. 3) (Abbott et al., 1992; Jeffries and Dodd, 1991; Malajczuk et al., 1993). The reliable large-scale production of pathogen-free inoculum will be interlinked with the ability of the selected fungi to tolerate conditions during and after inoculum preparation. A rapid loss of viability of inoculum in the chosen carrier over a short period of time may negate its usefulness as an inoculant. The production and quality control of inoculum is considered by Kuek (Chapter 23) and Sylvia and Jarstfer (Chapter 24).

There is currently an increasing interest in producing 'mixed' mycorrhizal inocula containing more than one mycorrhizal isolate. This interest has stemmed from revegetation and rehabilitation studies (Jasper et al., 1991; Sylvia, 1990), where the introduction of a single inoculant fungus may be inappropriate for a broad range of hosts. Several AM or ECM fungi inoculants are more likely to allow the re-establishment of a greater diversity of mycotrophic plant species, similar to that found in the undisturbed native ecosystem. With the succession of ECM fungi on trees as the forest matures (Mason et al., 1987) and the changes which occur in AM fungi communities with different crop rotations (Dodd et al., 1990b) and with secondary plant succession (Johnson et al., 1991), mixed mycorrhizal inocula may also ensure the persistence of inoculant fungi at field sites. Furthermore, mixed inocula may be an alternative to screening for broad range fungi adapted to conditions in the nursery as well as in the field or to field sites which experience extremes in environmental conditions.

Discussion

In the 1970's and 1980's, the aim of many

mycorrhizal screening programmes was to select several 'superstrains' of fungi which could be used in the high-input agriculture/forestry systems prevalent in 'developed' countries. Many applied investigations concentrated on a few well-documented and easily-cultured fungi, with scant regard for their origins or suitability for the experimental conditions. Not surprisingly, these mycorrhizal programmes had mixed success in demonstrating the potential for increasing plant productivity with mycorrhizal inoculation. Happily, there now appears to be a revived interest in gaining more biological and ecological data from the field which will aid in selecting suitable mycorrhizal partners for plants growing in different soils and climatic zones (Allen, 1991; Brundrett, 1991; Sieverding, 1991). Our present knowledge of the functioning of the mycorrhizal symbiosis in the field is limited to several particular ecosystems or soils where long-term funding has enabled productive studies to evolve. If we are to learn more about the development of suitable inoculants or the potential for manipulation of mycorrhizal fungi then long-term ecological studies in the field in diverse ecosystems are required. Mycorrhizal fungi are, and will continue to be, a major component of 'sustainable plant production systems' and funding should therefore parallel these long-term aims. Mycorrhizal fungi are an environmentally-sound biological resource which can be used for the maintenance of plant diversity and productivity.

Acknowledgments

Dr J C Dodd is currently funded by the Baring Bank Foundation and Dr B D Thomson by Bunnings Treefarms Pty. Ltd. We thank Drs P. Jeffries, T S. Grove and N. Malajczuk for helpful comments on an earlier draft of the manuscript.

References

Abbott L K and Robson A D 1984 The effect of root density, inoculum placement and infectivity of inoculum on the development of vesicular-arbuscular mycorrhizas. New Phytol. 97, 285–299.

Abbott L K, Robson A D and Gazey C 1992 Selection of inoculant vesicular-arbuscular mycorrhizal fungi. *In* Methods in Microbiology—Vol. 24, Techniques for the study of mycorrhiza. Eds. J R Norris, D J Read and A K Varma. pp 1–21. Academic Press, London.

Allen M F 1991 The Ecology of Mycorrhizae. Cambridge University Press, Cambridge. 184 p.

Bethlenfalvay G J, Thomas R S and Dakessian S 1988 Mycorrhizae in stressed environments: Effects on plant growth, endophyte development, soil stability and soil water. *In* Arid Lands: Today and Tomorrow. Eds. E E Whitehead, C F Hutchinson, B N Timmerman and R G Varady. pp 1015–1029. Westview Press, Boulder, Colorado.

Brundrett M 1991 Mycorrhizas in natural ecosystems. Adv. Ecol. Res. 21, 171–313.

Dodd J C, Arias I, Koomen I and Hayman D S 1990a The management of vesicular-arbuscular mycorrhizal populations in acid–infertile soils of a savanna ecosystem. I—The effect of pre-cropping and VAM fungi inoculation on plant growth and nutrition in the field. Plant Soil 122, 229–240.

Dodd J C, Arias I, Koomen I and Hayman D S 1990b The management of vesicular-arbuscular mycorrhizal populations in acid-infertile soils of a savanna ecosystem. II—The effects of inoculation and pre-crops on the native VAM fungi spore populations. Plant Soil. 122, 241–247.

Dodd J C, Williams S and Jeffries P 1991 Mycorrhiza and Vetiver—Rehabilitating Degraded Lands. Newsletter of the Vetiver Information Network of the World Bank 7, 3–4.

Garbaye J, Delwaulle J C and Diangana D 1988 Growth response of *Eucalyptus* in the Congo to ectomycorrhizal inoculation. For. Ecol. Manag. 24, 151–157.

Gemma J N, Koske R E and Carreiro M 1989 Seasonal dynamics of selected species of V-A mycorrhizal fungi in a sand dune. Mycol. Res. 92, 317–321.

Grove T S, Malajczuk N, Burgess T, Thomson B D and Hardy G 1991 Growth responses of plantation eucalypts to inoculation with selected ectomycorrhizal fungi. *In* Proc. IUFRO Symp. Intensive Forestry: The Role of Eucalypts. pp 86–93. Durban, South Africa.

Hatchell G E and Marx D H 1987 Response of longleaf, sand and loblolly pines to *Pisolithus* ectomycorrhizal and fertilizer on a sandhills site in South Carolina. For. Sci. 38, 301–315.

Jasper D A Robson A D and Abbott L K 1988 Revegetation in an iron-ore mine—Nutrient requirements for plant growth and the potential role of vesicular-arbuscular mycorrhizal fungi. Aust. J. Soil Res. 26, 497–507.

Jasper D A Robson A D and Abbott L K 1991 The effect of soil disturbance on vesicular-arbuscular mycorrhizal fungi in soils from different vegetation types. New Phytol. 118, 471–476.

Jeffries P and Dodd J C 1991 The use of mycorrhizal inoculants in agriculture. *In* Handbook of Applied Mycology. Vol. I—Soil and Plants. Eds. D K Arora, B Rai, K G Mukerji and G R Knudsen. pp 155–186. Marcel Dekker, Inc., Madison Ave., New York.

Johnson N C, Zak D R, Tilman D and Pfleger F L 1991 Dynamics of vesicular-arbuscular mycorrhizae during old field succession. Oecologia 86, 349–358.

Kessel S L 1927 Soil organisms. The dependence of certain pine species on a biological soil factor. Emp. For. J. 6, 70–74.

Krikun J, Haas J H, Dodd J C and Kinsbursky R 1990 Mycorrhizal dependence of four crops in a P-sorbing soil. Plant Soil 122, 213–217.

Kropp B R and Langlois C-G 1990 Ectomycorrhizae in reforestation. Can J. For. Res. 20, 438–451.

LeTacon F, Garbaye J, Bouchard D, Chevalier G, Olivier J M, Guimberteau J, Poitou N and Frochot N 1988 Field results from ectomycorrhizal inoculation in France. *In* Canadian Workshop on Mycorrhizae in Forestry. Eds. M Lalonde and Y Piché. pp 51–74. Universite Lavel, Ste-Foy.

Le Tacon F, Alvarez I F, Bouchard D, Henrion B, Jackson R M, Parlade J I, Pera J, Stenström E, Villeneuve N and Walker C 1991 Variations in field response of forest trees to nursery ectomycorrhizal inoculation in Europe. *In* Mycorrhizas in Ecosystems—Structure and Function, Abstracts 3rd ESM, Sheffield.

Malajczuk N, Grove T S, Tommerup I C Bougher N L, Thomson B D, Dell B and Kuek C 1993 Ectomycorrhizas. *In* Microorganisms that Promote Plant Productivity. Eds. P Dart and J Dawson (*in press*).

Marx D H and Cordell C E 1987 Ecology and management of ectomycorrhizal fungi in regenerating forests in the eastern United States. *In* Proceedings of the 7th NACOM. Eds. D M Sylvia, L L Hung and J H Graham. p. 69. Gainesville, Florida.

Mason P A, Last F T, Wilson J, Deacon J W, Fleming L V and Fox F M 1987 Fruiting and successions of ectomycorrhizal fungi. *In* Fungal Infection of Plants. Eds. G F Pegg and P G Ayres. pp 253–268. Cambridge University Press, Cambridge.

Molina R and Palmer J G 1982 Isolation, maintenance, and pure culture manipulation of ectomycorrhizal fungi. *In* Methods and Principles of Mycorrhizal Research. Ed. N C Schenck. pp 115–129. The American Phytopathological Society, Minnesota, USA.

Molina R and Trappe J M 1982 Patterns of ectomycorrhizal host specificity and potential among Pacific North West conifers and fungi. For. Sci. 28, 423–458.

Parke J L, Linderman R G and Trappe J M 1984 Inoculum potential of ectomycorrhizal fungi in forest soils of southwest Oregon and northern California. For. Sci. 30, 300–304.

Sieverding E 1991 Vesicular-Arbuscular Mycorrhiza Management in Tropical Agrosystems. Deutsche Gesellsschaft fur Technische Zusammenarbeit (GTZ) GmbH, Federal Republic of Germany. 371 p.

Smith S E and Walker N A 1981 A quantitative study of mycorrhizal infection in *Trifolium*: separate determination of the rates of infection and mycelial growth. New Phytol. 89, 225–240.

Sylvia D M 1990 Inoculation of native woody plants with vesicular-arbuscular mycorrhizal fungi for phosphate mine land reclamation. Agric. Ecos. Environ. 31, 253–261.

Thompson J P 1987 Decline of vesicular-arbuscular mycorrhizae in long fallow disorder of field crops and its expression in phosphorus deficiency of sunflower. Aust. J. Agric. Res. 38, 847–867.

Thomson B D, Malajczuk N and Grove T S 1991 Increasing the infectivity and effectiveness of ectomycorrhizal isolates of *Eucalyptus* spp. be re-isolating them from root infection and fruiting bodies. *In* Mycorrhizas in Ecosystems–Structure and Function, Abstracts 3rd ESM, Sheffield.

Tommerup I C 1992 Methods for the study of the population biology of vesicular-arbuscular mycorrhizal fungi. *In* Methods in Microbiology—Vol. 24, Techniques for the study of mycorrhiza. Eds. J R Norris, D J Read and A K Varma. pp 1–21. Academic Press, London.

Trappe J M 1977 Selection of fungi for ectomycorrhizal inoculation in nurseries. Ann. Revu. Phytopathol. 15, 203–222.

Reprinted from *Plant and Soil*, Volume 159, No. 1 (1994), pp. 149–158.

Genetics of ectomycorrhizal fungi: progress and prospects

F. MARTIN[1], I.C. TOMMERUP[2] and D. TAGU[1]
[1] *Laboratoire de Microbiologie Forestière, Centre de Recherches Forestières de Nancy, INRA, F-54280 Champenoux, France*
[2] *CSIRO Division of Forestry, Private Bag, P.O. Wembley, WA 6014, Australia*

Key words: DNA-mediated transformation, DNA polymorphisms, ectomycorrhizal fungi, genetics, polymerase chain reaction, RAPD, symbiosis-related genes

Abstract

The variability within and among ectomycorrhizal species provides a substantial genetic resource and the potential to increase forest productivity and environmental sustainability. Two parallel and interacting approaches, classical and molecular genetics, are being developed to acquire the genetic information underpinning selection of improved ectomycorrhizal strains. Determining the genetic traits of the fungi which contribute to symbiosis and plant function are being followed using natural variability combined with classical and molecular genetic manipulations. Classical and molecular manipulations for breeding rely on key information including sexual and parasexual reproduction, post-meiotic nuclear behaviour, mating-types and vegetative incompatibility mechanisms. Progress in the manipulation of genomes of ectomycorrhizal fungi will depend on efficient methods for gene cloning and DNA transformation. Gene transfer into fungal cells have been shown to be successful and include treatment of protoplasts and intact mycelium with naked DNA in the presence of polyvalent cations, electroporation, and microbombardment. The merits and limitations of these methods are discussed. Using this technology the expression of foreign DNA, the functional analysis of fungal DNA sequences, as well as molecular exploitation for commercial purposes can be carried out. This review concentrates on these aspects of fungal molecular biology and discusses the applications of the experimental systems that are currently available to ectomycorrhizal fungi. As it is essential to be able to define the traits which a breeder is seeking to improve, availability of genetically defined strains that are isogenic for a character or differ only in one character and a thorough knowledge of the biochemistry of the symbiosis will be necessary before any genetic manipulation be carried out. Genetic variability of ectomycorrhizal strains has been assessed by DNA fingerprinting. This approach allows the evaluation of DNA variability and the exchange of genetic information in natural populations, the identification of species and isolates by DNA polymorphisms, and tracking the environmental fate of the introduced fungi to determine their survival, growth, and dissemination within the soil.

Introduction

Cost effective forest tree production as plantations or as regenerated clear felled forests and tree establishment in degraded sites are major problems on all the continents and in all climatic regions. Most forest regrowth and plantation sites have less than optimal mineral nutrition. Ectomycorrhiza increase tree growth primarily due to increased supply to the trees of P and N and on some rehabilitation sites by amelioration of, for example, heavy metal toxicity (Grove and Le Tacon, 1993; Grove and Malajczuk, Chapter 20). The variability within and among ectomycorrhizal species provides a substantial genetic resource and the potential to

A.D. Robson, L.K. Abbott & N. Malajczuk (eds.), Management of mycorrhizas in agriculture, horticulture and forestry, 159–170.

develop wide scale inoculation as a management option to increase forest productivity and profitability with current inputs or to maintain productivity with decreased inputs. Two parallel and interacting approaches, classical and molecular genetics, are being developed to acquire the genetic information which underpins selection and breeding of ectomycorrhizal strains effective for increasing tree survival and productivity when the prime purpose may be for environmental management or commercial timber production. Assessing and understanding the genetic traits of the fungi which interact with and contribute to plant function are being followed using natural variability combined with classical and molecular genetic manipulations. Four key questions relating to current and future management of ectomycorrhizal fungi are: How widely has the existing genetic pool been examined? Is genetic variability within populations likely to be large? Can fungi with desirable characteristics relating to formation and function be bread or engineered? Can fungi that do not form ectomycorrhiza be genetically manipulated to be mycorrhizal?

Classical genetics for breeding ectomycorrhizal fungi

Before any breeding involving molecular or classical manipulation can be carried out key information includes: sexual and parasexual reproduction, post-meiotic nuclear behaviour, mating-types and vegetative incompatibility mechanisms. Classical genetic studies are in progress for about ten ectomycorrhizal species, this being approximately of the same order as for saprophytic or plant pathogenic filamentous fungi. Many ectomycorrhizal fungi can be cultured on defined media and some can be successfully stored at low temperatures. Techniques have been developed for several fungi to produce homonucleate mycelia either from spores or by protoplasting, to produce natural or induced mutants, to induce mating-type dikaryons and fertile mycelia to produce sporocarps and to recover some of the progeny. Others because of a high proportion of inbreeding have less readily manipulable homonucleate life-cycle stages. Several types of DNA markers (see section *Identification of genetic variability of ectomycorrhizal fungi and molecular ecology*) are possible for many other mycorrhizal species, although classical experimental genetics is not presently a practical reality because they have not been induced to produce sexual sporocarps under controlled conditions, or their spores have proved to be impossible to germinate or have erratic, low germination, or the vegetative phase does not grow or grows poorly in axenic culture.

Sexual reproduction and breeding systems

The taxonomy of ectomycorrhizal fungi implies substantial general information about breeding mechanisms including sexual compatibility mechanisms, regulators of sporocarp development and sexual spore dispersal (see Tommerup and Malajczuk, 1993). Breeding systems are an important factor in determining the proportion of genetic variation found between subpopulations relative to the total genetic diversity. Breeding systems promoting outcrossing have evolved in ascomycetes and basidiomycetes but they are particularly well developed in the latter. A major difference, of significance to genetic variation, between ascomycetes and basidiomycetes is that asexual nuclear fusion and segregation has a greater role in genetic structuring in basidiomycetes. In basidiomycete ectomycorrhizal species inbreeding may be more common than in non-mycorrhizal species however, this is probably secondary inbreeding due to the superimposed control exerted by mating-type genes.

Intersterility groups and vegetative incompatibility. In basidiomycete ectomycorrhizal fungi failure to recognise incompatibility products of mating-type genes and/or failure to recognise compatibility products of putative intersterility genes may be the genetic basis for subspecific divergence. Intersterility groups are not absolute in demarcation and groups are not always demarcated by geographic regions. Intersterile groups occur in all species examined for them. They include outbreeding species of *Laccaria* and *Suillus* (Fries and

Mueller, 1984; Dahlberg and Stenlid, 1990; Sen, 1990a, b; Mueller and Gardes, 1991). Predominantly inbreeding species such as *Laccaria fraterna* may essentially be genetically isolated subspecies (Tommerup and Malajczuk, 1993).

Mating-types. Sexual compatibility usually involves nonself recognition, and the morphogenetic events imply that a series of gene products are involved. Detailed mating-type analysis of ectomycorrhizal ascomycetes is lacking. Three patterns of behaviour are known for ascomycetes: two mating-type alleles at a single locus; several genes may contribute to mating competency; mating-type switching (Glass and Lorimer, 1991).

Mating in homobasidiomycetes may be based either on a single locus, termed a unifactorial (bipolar, with multiple alleles) incompatibility system or two unlinked mating factors, A and B, each with multiple specificities, termed a bifactorial (tetrapolar) incompatibility system (Casselton and Economou, 1985). For compatibility (i.e. fertile mycelium), the monokaryons must have different factors. In the bifactorial type (e. g. *Coprinus cinereus*) functionally redundant genes determine specificity of the A factor and they are separated into two subcomplexes $A\alpha$ and $A\beta$ (Kües and Casselton, 1992). The specificity genes encode proteins which are predicted to interact and trigger A regulated sexual development. Many conifer and eucalypt ectomycorrhizal isolates are bifactorial with multiple A and B alleles e.g., *Laccaria laccata, Pisolithus tinctorius, Hebeloma cylindrosporum, H. westraliense, Suillus granulatus,* and *S. luteus* and (Fries and Mueller, 1984; Gay and Debaud, 1987; Fries and Neumann, 1990; Doudrick et al., 1990; Lamhamedi et al., 1990; Bougher et al., 1991; Tommerup et al., 1991). Several other quadrisporic and bi-quadrisporic *Laccaria* spp., mono-bisporic species such as *Hydnangium carneum* and *L. fraterna,* and fungi having six to eight sterigmata per basidium such as *Pisolithus* spp., and *Scleroderma* spp. are bifactorial suggesting this is a dominant characteristic for these genera (Fries and Mueller, 1984; Tommerup, 1992; Tommerup and Malajczuk, 1993). One of the consequences of sexual hyphal fusion is that mitochondrial and other cytoplasmic inheritance factors including plasmids may be exchanged. These aspects have not yet been addressed in ectomycorrhizal species but there are well described examples to provide analogies from non-mycorrhizal taxa (Casselton and Economou, 1985).

In/out breeding, post meiotic nuclear behaviour and spore development. Outcrossing, due to mating-type factors, is a dominant feature of ectomycorrhizal basidiomycete species and primary homothallism has not been found. Outbreeding is controlled principally by the allocation, after meiosis, of single mating-types to spores, and inbreeding by the allocation of mating-type compatible nuclei (Tommerup, 1992). The spore genome is determined at an early stage of spore development. Mature spores are often bi- or multi-nucleate due to mitosis in delimited spores. Sexual spores are the main long distance dispersal agent. The significances to ecological success of mating-type allocations involve considerations of the relative merits of outbreeding compared to changing heterozygosity in general, versus inbreeding which may promote successful traits. The importance to mycorrhiza of heterozygosity is indicated by ectomycorrhiza formation of several outbreeding basidiomycetes being more vigorous for mating-type compatible strains than the component mating types alone (Debaud and Gay, 1987; Kropp et al., 1987; Debaud et al., 1988; Gardes et al., 1990; Lamhamedi et al., 1990; Tommerup, 1988; Tommerup et al., 1991). Common A or common B dikaryon mycorrhiza are also more vigorous than either of the component homonucleate strains derived from a sporocarp. Heterozygosity is probably important to inbreeding species like *L. fraterna* and *H. carneum* due to more vigorous mycorrhiza development of dikaryons compared to common A or common B phenotypes. High levels of deleterious recessives or lethals in populations favours directional selection due to elimination of progeny homozygous for these recessives and in inbreeding populations, can effectively promote high levels of heterozygosity. A tightly inbreeding species, *L. fraterna,* may have death of 60–99% of spores during or soon after germination depending on the strain.

Asexual reproduction and breeding systems

The asexual cycle is to a large extent controlled by vegetative compatibility mechanisms and sexual incompatibility mechanisms, discussed above, because they control whether or not new genetic introgression from another mycelium is possible. In nature ecological factors may also have a role. Mycelial fusion and nuclear exchange which can occur in axenic cultures appear undetectable in nature as shown by studies of *Suillus*. Mating-types of basidiocarps, intersterility groups and isozyme patterns have used to define individual asexual mycelia in several species of *Suillus* (Dahlberg and Stenlid, 1990; Sen, 1990a, b). Results support the notion that different individual mycelia are long lived and apparently rarely successfully somatically interbreed even with their own sexual progeny, although they coexist in the same habitat. However, new genetic input via asexual recombination or mutation is possible and this is being experimentally tested using DNA markers (see section *Identification of genetic variability of ectomycorrhizal fungi and molecular ecology;* Tommerup, 1992). Analysis of mating-types in single protoplasts (and single basidiospores) supports the hypothesis of recombination during vegetative growth of known tetrapolar isolates of *L. laccata* and *H. westraliense* in axenic cultures or as mycorrhizal associations of trees grown in pots in a glasshouse (Tommerup and Malajczuk, 1993).

Host-range relationships

Plant–ectomycorrhizal fungus interactions involve basic compatibility interactions, plant-beneficial and pathogenic relationships, non-host incompatibility and no obvious interaction. In compatible interactions the fungus apparently has the tools necessary to achieve successful colonization. Basic compatibility appears to be controlled by many genes encoding factors enabling the fungus to establish a colony in host tissue and complete the whole or a portion of its life-cycle (Kropp et al., 1987; Tommerup and Malajczuk, 1993). Hence for ectomycorrhizal fungi, it seems unlikely that they will gain access to a new host (at or above species level) through a few mutations. There are a very large number of ectomycorrhizal fungus species–plant species associations (Tommerup and Malajczuk, 1993). A few, including pine or eucalyptus and *Pisolithus, Laccaria* and *Hebeloma* are currently being developed as model systems for genetic analysis of mycorrhiza-capacity and host-range (Malajczuk et al., 1990; Martin and Hilbert, 1991). Mycorrhiza-capacity, like many other fitness traits, are 'polygenic' (Kropp et al., 1987) or 'quantitative' and their phenotypes may involve modifier or epistatic genes and/or linkage disequilibria. Fitness in both the host and fungus may have directionally selected for the association. Alternatively, hosts may have had negligible evolution of resistance genes. The nature and role of symbiosis-related genes are discussed in Section *Genetic engineering of mycorrhizal fungi.*

Breeding ectomycorrhizal fungi to increase tree growth

Improved productivity is the result of growing superior genotypes in an environment allowing them to express their superiority. A major emphasis of ectomycorrhiza programmes has been selecting for superior individual isolates which cause increased growth of trees in the field. This is a natural first stage because there is a very taxonomically diverse population of ectomycorrhizal fungi and from amongst them certain species and in some cases genera have proven to be more valuable than others as inoculant fungi (see references in Grove and Le Tacon, 1993; Dodd and Thomson, Chapter 15; Grove and Malajczuk, Chapter 20). Present knowledge indicates that variation among isolates of a species is as much as among species in some genera, including for the capacity to increase tree growth in the field. Recognition of variation in economic traits is directing attention to breeding to combine useful levels of traits into an organism.

Programmes aimed at developing a systematic and highly reproducible breeding strategy for superior ectomycorrhizal fungi have two aspects. They are the development of efficient methods for trait identification and selection, and breeding involving classical techniques and protoplast production, protoplast regeneration, DNA marker production and an efficient transformation system. Trait selection may be for fungal characteristics increasing mycorrhiza function (Durand et al., 1992) and plantation tree growth or tree survival, characters increasing the competitive ability of the fungus, or agronomic characteristics of sporocarps if mushrooms or truffles are the final product. Use of recombinant DNA techniques and DNA markers to analyse simple and quantitative traits are discussed in sections *Genetic engineering of ectomycorrhizal fungi* and *Identification of genetic variability of ectomycorrhizal fungi and molecular ecology.*

Preservation of isolates and strains as genetic resources. Loss of vigour in culture and of capacity to rapidly form mycorrhiza commonly occurs when ectomycorrhizal fungi have been axenically cultured for extended periods. This highlights the need to preserve cultures for consistent and comparable isolate performance and excellent preservation of some mycorrhizal fungi is possible in liquid N_2 and the techniques should be generalizable.

Molecular genetics for breeding ectomycorrhizal fungi

Considerable effort is currently directed towards the genetic improvement of trees using genetic engineering techniques. However, success with this approach has been limited to a few species (e.g. poplar, chestnut), due to the lack of an efficient transformation system for introducing foreign genes and the inability to regenerate plants from transformed callus (e.g. many *Eucalyptus*). The barriers to the genetic improvements of major tree crops such as conifer and eucalypt species may partly be overcome by the genetic engineering of their fungal ectomycorrhizal associates. Recombinant DNA techniques presently are a sound addition to the array of methods used for improving the productivity of fungal species. An essential step in applying recombinant DNA technology to a filamentous fungus is to develop a DNA-mediated transformation system.

Fungal DNA transformation procedures

They are several essential features required to set up an efficient DNA-mediated transformation procedure and to design selectable markers and vectors to allow the selection of the transformed fungal cells and the development of procedures for DNA introduction into the genome.

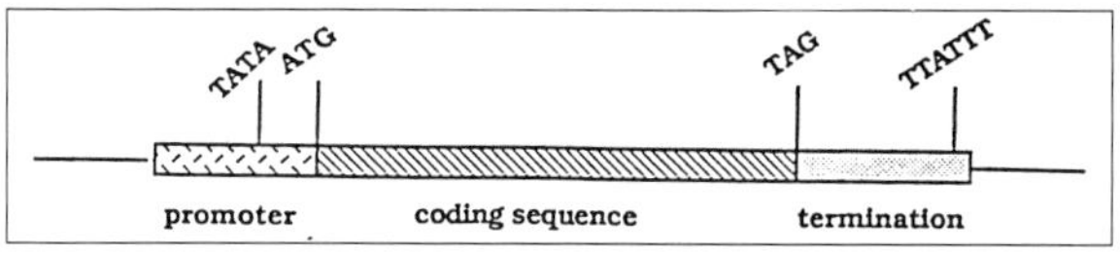

Fig. 1. Diagram of a chimeric gene. The coding sequence is generally from a prokaryotic origin and can confer, for instance, a resistance to an antibiotic (hygromycin). The promotor and the termination signals are isolated often from fungi such as Aspergillus or Neurospora.

Selectable genes and vectors. The genetic markers set the criteria by which a cell is judged to be transformed by providing unique assayable properties or enzymatic activities. Generally, the genetic markers that have been developed for use in filamentous fungi fall into three categories: complementation of a metabolic activity absent from a mutant strain (e. g. nitrate reductase, arginase), introduction of a gene encoding for an antibiotic or fungicide resistance (hygromycin, benomyl), and insertion of a non selectable gene encoding for an enzyme lacking in the fungal cells and easy to detect in the transformed cells (β-glucuronidase, β-galactosidase). The use of the complementation approach requires an excellent knowledge of the genetics of the fungus and a large collection of auxotrophic mutants which are not yet available for ectomycorrhizal fungi In contrast, a selection protocol based on the use of fungicide dominant markers is appropriate for ectomycorrhizal fungi since information on the genetics of the fungus is not necessary.

Nature of the promoters. Sequences 5′ to the coding regions of the gene contain promoters, a combination of several sequence motifs, which are responsible for the quantitative and qualitative (i.e. tissue-specific and/or developmentally regulated) control of gene expression. These regulatory sequences are likely to control RNA polymerases. Since the majority of the genetic markers encoding for antibiotic resistance are cloned from bacteria, they are fused to fungal promoters before DNA-mediated transformation. To ensure a high level of gene expression, the use of a homologous promoter isolated from the species which is submitted to the transformation procedure is recommended. Schründer et al. (1991) reported the occurrence of extrachromosomal DNA (linear double strand DNA) in *Heleboma circinans*. Sequences isolated from these plasmids (replicative origins) may be useful to set up transformation vectors.

Transformation procedures. Different procedures have been used to transform fungal cells. In the majority of these studies, protoplasts have been used as competent cells for transformation. The application of electroporation and high velocity microprojectile bombardment have been successfully used to transform intact mycelium of some species.

Transformation of protoplasts. This procedure is the most widely used DNA-mediated transformation system because protoplasts are ideal for gene transfer; the readily accessible plasmalemma assures that genes can reach and enter each and every protoplast; the enzymatic isolation procedures increase the proportion of competent cells; the foreign genes can enter every competent cell, and gene transfer does not require a biological vector (Fincham, 1989). A combination of polyethylenglycol (PEG) and Ca^{2+} is used to direct the DNA into protoplasts. PEG presumably induces the fusion of the plasmalemma and binds the highly negatively charged DNA to the plasmalemma. Yields for many fungal species have been consistently low (Fincham, 1989) and the preparation of protoplasts entails a lengthy procedure comprising thorough monitoring of the various steps of the protocol (Kropp and Fortin, 1986; Hebraud and Fevre, 1988; Tommerup, 1988; Barrett et al., 1989; Lemke et al., 1991).

Electroporation. Discharge of capacitors across cell populations induces transient gaps in the plasmalemma which facilitate entry of DNA molecules into cells. The transformation yield for fungal cells (Goldman et al., 1990; Chakraborty and Kapoor, 1990) is within the same range as the PEG method, but electroporation is easier to handle and is more reproducible (Fig 2).

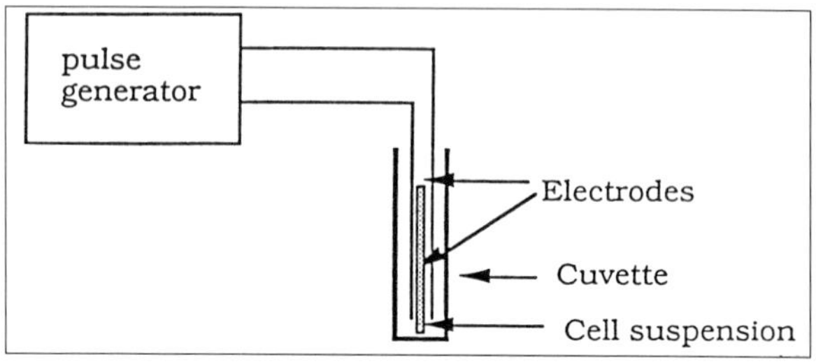

Fig. 2. Diagram of electroporation apparatus for gene transfer. Protoplasts and DNA are mixed together and disposed into the electroporation cuvette and submitted to a brief electrical discharge.

Biolistic. The second procedure which allows the transformation of intact cells has been developed and optimized for plant tissues (Klein et al., 1987). Tungsten or gold microprojectiles coated with DNA are shot into cells by

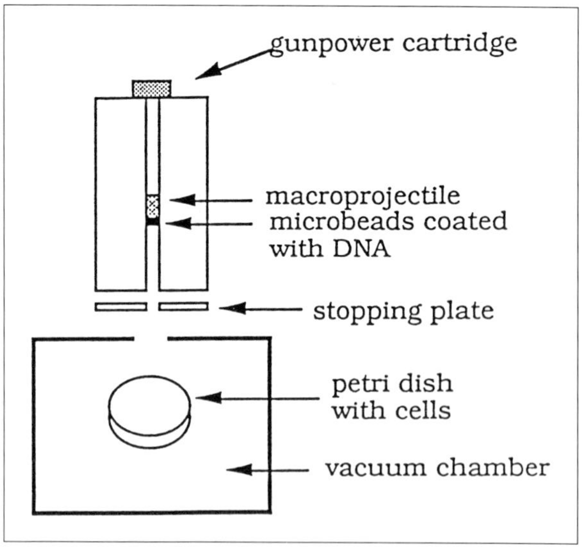

Fig. 3. Diagram of a biolistic device (from Butow and Fox, 1990). The DNA is coated to tungsten microbeads. The macroprojectile is accelerated and allows the microbeads to bombard the cells in the vacuum.

acceleration to a high velocity with subsequent insertion of the foreign DNA into the genome (Fig 3). Some expected that this technique, so-called 'biolistic', would solve all gene transfer problems. Indeed, this procedure provides a biological vector-independent DNA delivery system into a great variety of cells. It is easy to handle and particle guns are not very expensive. However, a careful optimization of the parameters of the technique is required. This microbombardment procedure has been applied to fungal cells (Armaleo et al., 1990). Biolistic should be the method of choice for direct gene transfer in ectomycorrhizal mycelium.

Transformation of ectomycorrhizal fungi. Barrett et al. (1990) and Marmeisse et al. (1992a) have used protoplasts to develop transformation procedures of the mycorrhizal fungi *L. laccata* and *H. cylindrosporum.* The work of Barrett et al. (1990) on the transformation of *L. laccata* is the first report of a transgenic ectomycorrhizal fungus. Protoplasts were isolated and transformed with the *Escherichia coli* hygromycin resistance gene fused to the *gpd* (glyceraldehyde phosphate dehydrogenase gene) regulatory sequences from Aspergillus. A combination of PEG and Ca^{2+} was used to direct the DNA into dikaryotic protoplasts. The transformation yield ranged from 5 to 50 transformed cells per μg of DNA. The phenotype was stable even after 4 to 6 weeks of growth in the absence of the antibiotic. Southern blot analysis demonstrated that the foreign DNA integrated into the chromosomal DNA. The transformed mycelium retained the ability to form mycorrhizas on *Pinus taeda.* A similar procedure has been used for the transformation of a homokaryotic strain of *H. cylindrosporum.* The transformation yield was low (1 to 5 transformants per μg of DNA) and the majority of the transformants integrated several copies of the plasmid into the chromosome. The transformed mycelium was able to form mycorrhizas. Despite the low transformation yield, the mycelium was co-transformed with plasmids carrying either genes of the tryptophan biosynthesis pathway or the NADP-glutamate dehydrogenase gene from *Coprinus cinereus.* This study demonstrated that co-transformation can be applied to ectomycorrhizal fungi. Several ectomycorrhizal fungi are naturally resistant to high levels of hygromycin and it may not be an appropriate genetic marker for widespread use.

Genetic engineering of ectomycorrhizal fungi

The genetic engineering of ectomycorrhizal fungi is a realistic proposition and this approach may be the potential for improving agriculture and forest sustainability worldwide. However, mycorrhiza is an incompletely understood system and, at present, it is not possible to predict with certainty the precise consequences of many of the genetic modifications envisaged. As it is essential to be able to define the traits which a breeder is seeking to improve, a thorough knowledge of the physiology and molecular processes of the symbiosis will be necessary before any genetic manipulation can be carried out. Candidate genes for manipulation are those encoding: factors involved in the mycorrhiza development; enzymes involved in the plant and fungus nutrition; resistance to fungicides and pesticides; substances repellent to nematode or collembola predation; metabolites conferring osmotolerance to alleviate water stress (see also Lemke et al., 1991).

An important conceptual advance, due to gene isolation and transformation methodologies, is the capacity to analyse complex traits (e. g., symbiosis-related genes).

Manipulation of the ectomycorrhiza development. There is compelling biochemical evidence that ectomycorrhiza development is determined by genes in the fungus and root (Martin and Hilbert, 1991). Thus, molecular cloning of symbiosis-related genes has been carried out by differential screening (Martin et al., 1992). $PolyA^+$ mRNA were isolated from eucalypt ectomycorrhiza in the early stages of the symbiosis differentiation and used to construct a cDNA library. Two hundred fungal cDNAs were randomly selected, amplified by the polymerase chain reaction (PCR) and screened either for clones with preferential hybridization to mycorrhiza RNA (mycorrhiza-stimulated genes) or with a high level of expression in the free-living mycelium (mycorrhiza-down regulated genes). Once several fungal symbiosis-related genes have been cloned, it would be possible to alter their expression in the mycelium through efficient DNA-transformation systems to determine their function in the development of the symbiotic tissues.

Manipulation of plant nutrition. P and N are essential, but limited, nutrients for tree growth and development. Given these circumstances, it is not surprising that both soil microorganisms and roots have evolved highly efficient mechanisms for absorbing these minerals from very dilute solutions. In addition, some bacteria and fungi have evolved an enhanced capability to solubilize soil organic P, hence increasing the external Pi concentration. It is widely accepted that the mycorrhizal symbiosis improves the growth of the host-plant by increasing the availability of soil minerals. Excreted enzymes (e.g. phosphatases and proteases) presumably play a prominent role in these processes. An obvious genetic change that would hopefully give rise to a rapid and significant improvement of the mycorrhizal symbiosis would be the modification of the copy number of the gene loci encoding for these enzymes. For example, the highly efficient mineral phosphate solubilizing (MPS) trait of the bacteria *Erwinia herbicola* has been cloned and transferred to *E. coli* via molecular cloning technology (Goldstein and Liu, 1987). The overexpression of the genes controlling the MPS activity in an ectomycorrhizal fungus or the transfer of the MPS trait from an efficient strain (or species) to fungal isolates used in large scale inoculation in a phosphate-limited environment could have obvious ecological and commercial advantages. In addition, this might allow determination of the physiological and biological importance of each of the enzyme activities.

Fungicide resistance. Since fungicides applied against pathogenic fungi of tree nurseries often lack selectivity and are therefore detrimental to symbiotic fungi, efforts should be devoted to modifying inoculated ectomycorrhizal fungi to become resistant to broad-spectrum fungicides. Toward this task, different approaches can be applied; overproduction of the target protein, expression of an altered form of the target protein that is less sensitive to the fungicide or introduction of a gene encoding an enzyme for the detoxification or degradation of the fungicide.

Once transformed ectomycorrhizal mycelium are selected under laboratory conditions, it will be essential to analyse the expression of the new trait in symbiosis produced in the 'real world' where the symbiotic fungus is in competition with the endogenous soil microflora.

Identification of genetic variability of ectomycorrhizal fungi and molecular ecology

A current application of molecular biology which is rapidly developing is the characterization of the genetic variability in the mycorrhizal fungus genomes by DNA fingerprinting. This approach allows the evaluation of the genetic variability of DNA sequences (mutations, chromosome rearrangements) in natural populations of fungi, the identification of species and isolates by using the observed DNA polymorphisms, and tracking the environmental fate of the introduced fungi to determine their survival, growth, and dissemination within the soil. The primary technique for constructing DNA fingerprints is RFLP analysis which detects length mutations and alterations in base sequences (i.e. restriction sites) (Armstrong et al., 1989; LoBuglio et al., 1991; Martin et al., 1991; Marmeisse et al., 1992b). However, the inability to detect DNA on a single mycorrhiza and utilization of radioactive isotopes prevented the spread of conventional RFLPs in identifying ectomycorrhizal isolates for characterization of community structures.

PCR/RFLP. The recently developed PCR (Saiki et al., 1988) provided an in vitro method for increasing the number of copies of a target DNA sequence, thereby allowing increased sensitivity in detecting DNA sequences present in small amounts in samples (spores, mycorrhizal tip) with DNA from other rhizospheric microbe populations. This technique consists of the enzymatic amplification of DNA fragments spanning between terminal sequences recognized by specific oligonucleotide primers. Amplified products are then digested by endonucleases giving rise to DNA fingerprints. PCR probes with high specificity and high sensitivity are based on DNA sequences which have a low degree of conservation and are highly repeated. PCR amplification of various nuclear rDNA sequences could serve as target sites and provide useful criteria for identification of genera, species, and isolates of ectomycorrhizal fungi. Length polymorphism and restriction endonuclease analysis of amplified rDNA demonstrated that genetic variation occurs in the targeted rRNA genes within local geographic population of *Laccaria* species, as well as among isolates from a worldwide distribution (Gardes et al., 1991a; Henrion et al., 1992). Use of other target sites (mitochondrial DNA and tRNA genes) and different endonuclease combinations will extend the scored RFLPs. The ability to determine RFLPs and sequences from crude DNA extracts of small amount of fungal material and single mycorrhizal tips enables identification and tracking of introduced mycorrhizal isolates in nurseries and forest ecosystems without employing microbiological methods.

RAPD. An alternative way of obtaining polymorphisms based on PCR is the random amplified polymorphic DNA (RAPD) technique (Williams et al., 1990; Welsh and McClelland, 1990). Short oligonucleotide primers of arbitrary sequences anneal to complementary sequences occurring randomly in the genome. Successfully amplified templates are separated by electrophoresis to reveal DNA fingerprints. The advantages of this technique are that it is quick, does not require target sequence information, and can provide markers in genome regions inaccessible to PCR/RFLP analysis. It has been successfully applied to distinguish isolates of *Laccaria* and *Hydnangium* (Tommerup, 1992; Tommerup et al., 1992) and Tuber (Wyss and Bonfante, 1992). RAPD is therefore applicable to routine analysis of breeding populations and practical strain typing of a wide range of ectomycorrhizal fungi.

Routine applications of PCR/RFLP and RAPD will be facilitated by automation of the analysis of the amplified products using DNA sequencers (Tommerup et al., 1992), capillary electrophoresis (Martin and Vairelle, unpublished results) and appropriate software.

Chromosome and gene mapping. Molecular techniques for producing markers and relating these by linkage analysis or to chromosome-sized pieces of DNA will enable statistical and physical map construction. DNA markers can be found in far greater numbers than visible, isozyme or other types. A more complete and therefore informative map covering all chromosomes of an organism can be constructed using DNA markers together with any others. DNA markers enable map construction and marker association with traits when there is little genetic knowledge of an ectomycorrhizal fungus. By knowing the 'map position' of a gene, a nearby DNA marker can be used to diagnose the presence of a gene without waiting for the gene to be expressed. This predictive value enables informed decisions to be made at an early stage in breeding. It offers an efficient means of screening and accelerating some steps in classical breeding programmes. It also offers new approaches to some objectives which are difficult with classical techniques.

Biodiversity in ecosystems: a genetic resource. Techniques used for producing markers for breeding are equally applicable to studies of diversity in natural populations which are a major source of future genetic stocks. Genetic markers have provided a glimpse of the nature of change underlying diversity, for example, DNA markers and mating-type analysis of taxonomic rank in *Laccaria* and *Hydnangium* (Gardes et al., 1991a, b; LoBuglio et al., 1991; Henrion et al., 1992; Tommerup et al., 1992), DNA sequence analysis of *Rhizopogon* and *Suillus* (Bruns et al., 1989) and isozymes and mating-types in *Suillus* (Sen, 1990 a, b).

Future prospects

This brief review of the application of molecular genetics and recombinant DNA technology for ectomycorrhizal fungi indicates that many DNA markers and DNA-transformation systems have been successfully applied. Further progress in these new developing technologies concerned with gene identification, genetic manipulation and gene transfer postulate a more thorough understanding of the genome structure and gene expression of the ectomycorrhizal symbionts. Without this knowledge it will not be possible to bring about the precisely directed changes that breeders of mycorrhizal fungi and trees aspire to achieve in economically important tree crops.

Potential targets for mycorrhiza improvement include fungicide resistance, drought tolerance, heavy metal detoxification, and increased utilization of soil nutrients. The ability to overexpress or mutate a gene for a rate-limiting enzyme (or transporter) will also allow the metabolic consequences of these changes to be investigated. Finally, as the detailed mechanisms responsible for the fungal alterations in the host-plant gene transcription are better understood, it should be possible to tailor the metabolism of the transgenic partners of the symbiosis in order to study the impact of changing individual regulatory parameters on the functioning of interacting nutritional processes.

In view of the fast progress that is being made in crop plant and industrial fungal biotechnologies, it is likely that some ectomycorrhizal fungi and their host-plants will be amenable to crop improvement by genetic engineering within the next decade. However, such an optimistic projection should be restrained by the fact that firstly, success requires more than occasional gene transfer into experimentally well suited strains of some species. It also requires routine and efficient gene transfer into any desired strains of any given species, changing the genome only by the addition of one defined gene with predictable biological consequences. Secondly, non-scientific issues such as the number of scientists working on these new approaches to mycorrhizal research will be decisive in future design and field-testing of mycorrhizal fungi genetically engineered using recombinant DNA technology. Thirdly, the cost, in both time and money, of manipulating the ectomycorrhizal genomes will always present limitations to the adoption of this technology. Finally, the sensibilities of the environmentalists and the statutory bodies that exist to protect the environment will have to be carefully considered.

Acknowledgements

Work in the authors laboratories has been supported by grants from the Eureka-Eurosilva Tree Physiology programme and the District de l'Agglomération Nancéienne (to FM) and Rural Industries Development Corporation (to IT). We would like to acknowledge the collaboration of C Di Battista, B Henrion, F Le Tacon and M Python.

References

Armaleo D, Ye G N, Klein T M, Shark K B, Sanford J C and Johnston S A 1990 Biolistic nuclear transformation of *Saccharomyces cerevisiae* and other fungi. Curr. Genet. 17, 97–103.

Armstrong J L, Fowles N L and Rygiewicz P T 1989 Restriction fragment length polymorphisms distinguish ectomycorrhizal fungi. Plant Soil 116, 1–7.

Barrett V, Dixon R K and Lemke P A 1990 Genetic transformation of a mycorrhizal fungus. Appl. Microbiol. Biotech. 33, 313–316.

Barrett V, Lemke P A and Dixon R K 1989 Protoplast formation from selected species of ectomycorrhizal fungi. Appl. Microbiol. Biotech. 30, 381–387.

Bougher N L, Tommerup I C and Malajczuk N 1991 Taxonomy, nuclear behaviour in the basidiomes, and ectomycorrhizas of *Hebeloma westraliense* sp. nov. Mycol. Res. 95, 683–688.

Bruns T D, Fogel R, White T J and Palmer J D 1989 Accelerated evolution of a false-truffle from a mushroom ancestor. Nature 339, 140–142.

Casselton L A and Economou A 1985 Dikaryon formation. *In* Developmental Biology of Higher Fungi. Eds. D Moore, L A Casselton, D A Wood and J C Frankland. pp 213–229. Cambridge University Press, Cambridge.

Chakraborty B N and Kapoor M 1990 Transformation of filamentous fungi by electroporation. Nucl. Acid. Res. 18, 6737.

Dahlberg A and Stenlid J 1990 Population structure and dynamics in *Suillus bovinus* as indicated by spatial distribution of fungal clones. New Phytol. 115, 487–493.

Debaud J C and Gay G 1987 In vitro fruiting under controlled conditions of the ectomycorrhizal fungus *Hebeloma cylindrosporum* associated with *Pinus pinaster*. New Phytol. 105, 429–435.

Debaud J C, Gay G, Prevost A, Lei J and Dexheimer J 1988 Ectomycorrhizal ability of genetically different homokaryotic and dikaryotic mycelia of *Hebeloma cylindrosporum*. New Phytol. 108, 323–328.

Doudrick R L, Furnier G R and Anderson N A 1990 The number and distribution of incompatibility alleles in *Laccaria laccata* var. *moelleri* (Agaricales). Phytopathol. 80, 869–872.

Durand N, Debaud J C, Casselton L A and Gay G 1992 Isolation and preliminary characterization of 5-fluoroindole-resistant and IAA-overproducer mutants of the ectomycorrhizal fungus *Hebeloma cylindrosporum* Romagnesi. New Phytol. 121, 545–553.

Fincham J R S 1989 Transformation in fungi. Microbiol. Rev. 53, 148–170.

Fries N and Mueller G M 1984 Incompatibility systems,

cultural features, and species circumscriptions in the ectomycorrhizal genus *Laccaria* (Agaricales). Mycologia 76, 633–642.

Fries N and Neumann W 1990 Sexual incompatibility in *Suillus luteus* and *S. granulatus*. Mycol. Res. 94, 64–70.

Gardes M, Fortin J A, Mueller G M and Kropp B R 1990 Restriction fragment length polymorphisms in the nuclear ribosomal DNA of four Laccaria spp.: *L. bicolor, L. laccata, L. proxima,* and *L. amethystina*. Phytopathology 80, 1312–1317.

Gardes M, White T J, Fortin J A, Bruns T D and Taylor J W 1991a Identification of indigenous and introduced symbiotic fungi in ectomycorrhizae by amplification of nuclear and mitochondrial ribosomal DNA. Can. J. Bot. 69, 180–190.

Gardes M, Mueller G M, Fortin J A and Kropp B R 1991b Mitochondrial DNA polymorphisms in *Laccaria bicolor, L. laccata, L. proxima* and *L. amethystina*. Mycol. Res. 95, 206–216.

Gay G and Debaud J C 1987 Genetic study on indole-3-acetic acid production by Hebeloma species: inter- and intra-specific variability in homo- and di-karyotic mycelium. App. Microbiol. Biotech. 26, 141–146.

Glass N L and Lorimer I A J 1991 Ascomycete mating types *In* More Gene Manipulations in Fungi. Eds. J W Bennett and L L Lasure. pp 193–216. Academic Press, San Diego.

Goldman G H, Van Montagu M and Herrera-Estrella A 1990 Transformation of *Trichoderma harzianum* by high-voltage electric pulse. Curr. Genet. 17, 169–174.

Goldstein A H and Liu S T 1987 Molecular cloning and regulation of a mineral phosphate solubiliqing gene from *Erwinia herbicola*. Biotechnol. 5, 72–74.

Grove T S and Le Tacon F 1993 Mycorrhiza in plantation forestry. *In* Mycorrhiza Synthesis. Advances in Plant Pathology Vol 9. Ed. I C Tommerup. pp 191–227 Academic Press, London.

Hebraud M and Fevre M 1988 Protoplast production and regeneration from mycorrhizal fungi and their use for isolation of mutants. Can. J. Microbiol. 34, 157–161.

Henrion B, Le Tacon F and Martin F 1992 Rapid identification of genetic variation of ectomycorrhizal fungi by amplification of ribosomal RNA genes. New Phytol. 122, 289–298.

Klein T M, Wolf E D, Wu R and Sanford J C 1987 High-velocity microprojectiles for delivering nucleic acids into living cells. Nature 327, 70–73.

Kropp B R and Fortin J A 1986 Formation and regeneration of protoplasts from the ectomycorrhizal basidiomycete *Laccaria bicolor*. Can. J. Bot. 64, 1224–1226.

Kropp B R, McAfee B J and Fortin J A 1987 Variable loss of ectomycorrhizal ability in monokaryotic and dikaryotic cultures of *Laccaria bicolor*. Can. J. Bot. 65, 500–504.

Kües U and Casselton L A 1992 Homeodomains and regulation of sexual development in basidiomycetes. TIG 8, 154–155.

Lamhamedi M S, Fortin J A, Kope H H and Kropp B R 1990 Genetic variation in ectomycorrhiza formation by *Pisolithus arhizus* on *Pinus pinaster* and *Pinus banksiana*. New Phytol. 115, 689–697.

Lemke P A, Barrett B and Dixon R K 1991 Procedures and prospects for DNA-mediated transformation of ectomycorrhizal fungi. *In* Methods in Microbiology, Vol. 23. Eds. J R Norris, D J Read and A K Varma. pp 281–293. Academic Press, London.

LoBuglio K F, Rogers S O and Wang C J K 1991 Variation in ribosomal DNA among isolates of the mycorrhizal fungus *Cenococcum geophilum*. Can. J. Bot. 69, 2331–2343.

Malajczuk N, Lapeyrie F and Garbaye J 1990 Infectivity of pine and eucalypt isolates of *Pisolithus tinctorius* on roots of *Eucalyptus urophylla* in vitro. I. Mycorrhiza formation in model systems. New Phytol. 114, 627–631.

Marmeisse R, Gay G, Debaud J C and Casselton L A 1992a Genetic transformation of the ectomycorrhizal fungus *Hebeloma cylindrosporum*. Curr. Genet. 22, 41–45.

Marmeisse R, Debaud J C and Casselton L A 1992b DNA probes for species identification in the ectomycorrhizal fungus *Hebeloma*. Mycol. Res. 96, 161–165.

Martin F and Hilbert JL 1991 Morphological, biochemical and molecular changes during ectomycorrhiza development. Experientia 47, 321–331.

Martin F, Zaiou M, Le Tacon F and Rygiewicz P 1991 Strain specific differences in ribosomal DNA from the ectomycorrhizal fungi *Laccaria bicolor* (Maire) Orton and *Laccaria laccata* (Scop ex Fr) Br. Ann. Sci. For. 48, 133–142.

Martin F, Python M, Tagu D 1992 Molecular cloning of symbiosis-related genes in the eucalypt ectomycorrhiza. *In* The International Symposium on Management of Mycorrhizas in Agriculture, Horticulture and Forestry, The University of Western Australia, Nedlands (Abstracts). p 157.

Mueller G M and Gardes M 1991 Intra- and inter-specific relations within *Laccaria bicolor* sensu lato. Mycol. Res. 95, 592–601.

Saiki R K, Gelfand D H, Stoffel S, Scharf S J, Higuchi R, Horn G T, Mullis K B and Erlich H A 1988 Primer-directed enzymatic amplification of DNA with a thermostable DNA polymerase. Science 239, 487–491.

Schründer J, Debaud J C and Meinhardt F 1991 Adenoviral-like genetic elements in *Hebeloma circinans*. *In* Mycorrhizas in Ecosystems—Structure and Function, Abstracts 3rd ESM Sheffield, UK.

Sen R 1990a Intraspecific variation in two species of *Suillus* from Scots pine (*Pinus sylvestris* L.) forests based on somatic incompatibility and isozyme analysis. New Phytol. 114, 607– 616.

Sen R 1990b Isozymic identification of individual ectomycorrhizas synthesized between Scots pine (*Pinus sylvestris* L.) and isolates of two species of *Suillus*. New Phytol. 114, 617–622.

Tommerup I C 1988 Tissue culture technique to improve ectomycorrhizal fungi for increasing forestry production. Inst. For. Aust. Newsletter 29, 10–11.

Tommerup I C 1992 Genetics of eucalypt ectomycorrhizal fungi. *In* International Symposium on Recent Topics in Genetics, Physiology and Technology of Basidiomycetes. Eds. M Miyaji, A Suzuki and K Nishimura. pp 74–79. Chiba University, Chiba, Japan.

Tommerup I C and Malajczuk N 1993 Genetics and Molecular Genetics of Mycorrhiza. *In* Mycorrhiza Synthesis. Advances in Plant Pathology Vol 9. Ed. I C Tommerup. pp 103–134. Academic Press, London.

Tommerup I C, Bougher N L, and Malajczuk N 1991 *Laccaria fraterna* in south Western Australia, a common ectomycorrhizal fungus with mono- and bi-sporic basidia: its nuclear behaviour and taxonomy, and *Hydnangium carneum*, a secondarily homothallic secotoid relative. Mycol. Res. 95, 689–698.

Tommerup I C, Barton J E and O'Brien P A 1992 RAPD fingerprinting of *Laccaria, Hydnangium* and *Rhizoctonia* isolates. *In* The International Symposium on Management of Mycorrhizas in Agriculture, Horticulture and Forestry, The University of Western Australia, Nedlands (Abstracts). p 161.

Welsh J and McClelland M 1990 Fingerprinting genomes using PCR with arbitrary primers. Nuc. Acid. Res. 18, 7213–7218.

Williams J G K, Kubelik A R, Livak K J, Rafalski J A and Tingey S V 1990 DNA polymorphisms amplified by arbitrary primers are useful genetic markers. Nucl. Acid. Res. 18, 6531–6535.

Wyss P and Bonfante P 1992 Identification of mycorrhizal fungi by DNA fingerprinting using short arbitrary primers *In* The International Symposium on Management of Mycorrhizas in Agriculture, Horticulture and Forestry, The University of Western Australia, Nedlands (Abstracts). p 154.

Reprinted from *Plant and Soil*, Volume 159, No. 1 (1994), pp. 159–170.

Genetic manipulation in vesicular-arbuscular mycorrhizal fungi

Y. PICHÉ, L. SIMON and A. SÉGUIN
Centre de Recherche en Biologie Forestière, Département des Sciences Forestières, Faculté de Foresterie et de Géomatique, Université Laval Sainte-Foy, Québec, Canada G1K-7P4

Key words: *Agrobacterium*, *Gigaspora margarita*, *Glomus intraradix*, PCR, mycorrhizae, rDNA gene, root organ culture

Abstract

The symbiosis between vesicular-arbuscular mycorrhizal (VAM) fungi and host plants develops after successful interactions between both partners. These interactions probably involve signal molecules produced by the host plant, by the fungi, or by both. So far the biotrophic status of VAM fungi has hampered the understanding of the processes regulating their physiology. However, among different methods for co-cultivating VAM fungi, root organ cultures (ROC) appear to be a useful technique for studying VAM development. This system has been useful in defining the nutritional requirements of VAM fungi in the precolonization stage and in obtaining axenic fungal material in various developmental stages.

The work discussed here focuses on the application of Polymerase Chain Reaction (PCR) technology and the potential of promoting hyphal growth in the absence of the plant. These techniques are being used to study VAM fungi in two main areas. The first concerns the determination of the DNA sequences coding for the SSU ribosomal RNA of two VAM fungi. This approach has allowed the design of specific primers for the rapid identification and quantification of VAM fungi. The second area of research concerns the potential use of PCR technology to study selective expression of specific genes during fungal spore development in defined in vitro conditions. The achievement of this future prospect depends on the ability to prepare PCR-based cDNA libraries from small amounts of fungal material after stimulation of hyphal growth with CO_2 and plant flavonols.

Introduction

Biotrophic status of vesicular-arbuscular mycorrhizal fungi

Terrestrial plants appeared during the siluro-devonian era (Foster and Gifford, 1974) along with ancestral vesicular-arbuscular mycorrhizal (VAM) fungi (Pirozynski, 1981). It is hypothesized by Pirozynski and Malloch (1975) that terrestrial plants arose as a result of the union between eukaryotic ancestral algae and primitive VAM fungi. This integrated vesicular-arbuscular mycorrhizal system has persisted in the majority of plants and has facilitated adaptation to many hostile soil habitats. This ancient, mutually beneficial relationship between plants and VAM fungi has probably been kept during evolution because it facilitates plant nutrient uptake and provides an unique microniche for the endosymbiont.

Among evolutionary concepts and theories involving genomic mutations and rearrangements, the serial endosymbiosis theory (SET), best explained the evolution of the eukaryotic cell as a result of symbiotic associations between prokaryotic cells (reviewed by Margulis, 1981). By this theory,

A.D. Robson, L.K. Abbott & N. Malajczuk (eds.), Management of mycorrhizas in agriculture, horticulture and forestry, 171–178.

mitochondria and plastids of eukaryotic cells are former free-living organisms which have been integrated into a host cell, giving rise to the actual eukaryotic nucleocytoplasm (Ahmadjian and Paracer, 1986). VAM fungi, like those ancient bacteria which became symbiotically integrated into eukaryotic cells, have developed a dependency on their hosts. By analogy with the impossibility of propagating mitochondria and plastids alone, it is not surprising that VAM fungi have been resistant to most, if not all, attempts to be cultivated in pure culture.

The growth of VAM fungi in pure culture in the absence of host roots is considered by many researchers as a prerequisite to fully understanding and manipulating VAM plant symbioses. Recently, Williams (1992) reviewed this subject and suggested that the isolation and axenic culturing of VAM fungi may be accomplished by applying procedures used in culturing other fungal biotrophs, such as rust fungi. It has been found that rust fungi have a better capacity to regenerate in pure cultures when intercellular mycelia from infected tissues are used as inoculum (Williams, 1984). This approach toward the culture of VAM fungi has received little attention; only Strullu and Romand (1986), Mosse (1988) and Williams (1990) have shown fungal regrowth from surface-sterilized VAM root pieces. Surprisingly, it seems that this problem has already been solved according to Barrett (1947) and Janardhanan et al. (1990) when they claimed that they had isolated and axenically propagated VAM fungi alone in defined media. Unfortunately, Gerdemann (1968) and our group at Laval University have been unable to verify these results using same living materials and procedures. Thus, to our knowledge, the growth of VAM fungi in long term axenic cultures has not been demonstrated and the question remains whether this constant failure reflects an insurmountable obstacle.

Available in vitro co-culture systems

Driven by the need to understand the physiology of VAM fungi during early plant–fungus interactions, different methods for co-cultivating VAM fungi have been developed (reviewed in Norris et al., 1992). Among these techniques, root organ cultures (ROC) appear well suited for advancing the knowledge of the processes regulating mycorrhizal development (Bécard and Piché, 1992). ROC methods were first used by Mosse and Hepper (1975) to obtain typical in vitro VAM colonizations with *Glomus mosseae*. Later Mugnier and Mosse (1987), building on the work of Tepfer and Yacoub (1986), reported a ROC method using isolated roots transformed by *Agrobacterium rhizogenes* as a host partner for *Glomus* spp. This method has been improved and extended to other VAM fungi including *Gigaspora margarita* (Bécard and Fortin, 1988) and *Glomus intraradix* (Chabot et al., 1992a). Clonal Ri T-DNA-transformed carrot roots have been useful in showing both the spore dependent and exclusively root dependent modes of hyphal growth of *G. margarita* (Bécard and Piché, 1989a). Moreover, it has been shown that hyphal growth stimulation during the prepenetration stage involves the synergistic action of carbon dioxide and root exudates (Bécard and Piché, 1989b). These results suggest that molecular signals produced by roots control different stages of in vitro development of *G. margarita*. This hypothesis is consistent with the fact that Ri T-DNA transformed sugar beet roots (a non-mycorrhizal host) do not secrete exudates that stimulate *G. margarita* hyphal growth (Bécard and Piché, 1990).

Purpose of the present paper

The use of Ri T-DNA-transformed roots described above has clarified the essential nutritional requirements of *G. margarita* in the precolonization stage. Also, this in vitro system serves as a source of fungal material at all stages of VAM development, including axenic and viable daughter spores. This system, which is controllable and reproducible, presents for the first time an opportunity to investigate the molecular changes associated with hyphal growth stimulation and fungal spore production. The remainder of this paper discusses the potential of Polymerase Chain Reaction (PCR) derived methods when only minute amounts of fungal material are available. A description of a recent application of such an approach to obtain DNA sequence information from VAM fungi and the subsequent design of specific molecular markers is followed by an outline of a strategy that we are pursuing to study gene expression in VAM fungi.

Amplification of VAM fungal genes by the polymerase chain reaction

All VAM fungi (≈130 spp.) are of the family Glomaceae, division Zygomycota (Morton and Benny, 1990). Systematics of the VAM fungi has always been difficult and is currently restricted to microscopic observations of spores as a means of species identification. Over the last decade there has been a great interest in using molecular techniques to identify VAM species by comparing their genomic DNA (Millner, 1989). However, the methods needed to isolate and separate nuclear and mitochondrial DNAs (Garber and Yoder, 1983), which can be tedious, especially for intractable species such as VAM fungi, have discouraged the use of molecular biology in VAM research. The sole method reported for the isolation of VAM fungal DNA requires hundreds of thousands of spores (Cummings and Wood, 1989). The recent introduction and application of DNA amplification by PCR (Mullis and Faloona, 1987; White et al., 1989) has reduced the constraints of isolating quantitative amounts of DNA from biological materials. The PCR method can amplify target DNA sequences from minute amounts of samples, such as a single fungal spore (Lee and Taylor, 1990). PCR is based on the exponential repeated duplication of a specific DNA fragment, using a thermostable DNA polymerase and specific flanking primers. PCR, involving 20–40 successive cycles of denaturation, primer annealing and primer extension leads to yields of microgram amounts of amplified DNA product(s).

In the first application of PCR to VAM fungi, Simon et al. (1992a) reported the amplification of the nuclear genes coding for the small subunit ribosomal RNA from *Glomus intraradix* and *Gigaspora margarita*. This was achieved using universal rDNA primers that anneal to highly conserved regions within these ubiquitous genes (Fig. 1). The use of axenic spores produced in vitro on Ri-T-DNA transformed roots was critical to the success of this work, since homologous genes from non-VAM fungi and other organisms usually present in standard spore preparations would have also been amplified by the universal primers. The direct sequencing of the amplified rDNA genes gave the first sequence data reported for VAM fungi. It allowed the design of a VAM fungal-specific primer (VANS1, Fig. 1) that could be used to amplify rDNA genes from other VAM fungal species. Since this primer is VAM-specific it

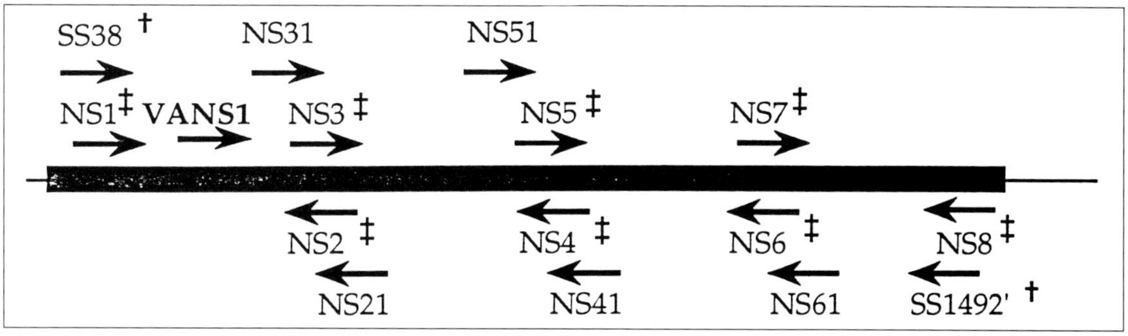

Fig. 1. Diagram of the relative location of the SSU primers. Locations of the universal primers and taxon specific primer (VANS1) are shown on a schematic representation of the nuclear gene coding for the small subunit ribosomal RNA (SSU). Thin lines represent flanking intergenic spacers. Sequences of the primers have been described previously–†(Bousquet et al., 1990), ‡(White et al., 1990)–with the exception of the following:
VANS1: GTCTAGTATAATCGTTATACAGG; **NS21**: AATATACGCTATTGGAGCTGG;
NS31: TTGGAGGGCAAGTCTGGTGCC; **NS41**: CCCGTGTTGAGTCAAATTA; **NS51**: GGGGGAGTATGGTCGCAAGGC;
NS61: TCAGTGTAGCGCGCGTGCGGC
Reprinted from Simon et al. (1992a) with permission.

allows the direct amplification of VAM-rDNA from preparations of pot collected spores or even from colonized roots.

This VAM fungal-specific primer has led to the development of a quantitative PCR assay to assess the amount of VAM fungi present in samples such as colonized roots (Simon et al., 1992b). This assay is based on the simultaneous amplification of a 550 bp fragment of the fungal rDNA gene and of a 430 bp fragment from an engineered internal standard. Known amounts of internal standard are added to different aliquots of a sample. Both fragments are then amplified by the same VANS1-NS21 primer pair in a single reaction, and the measured ratio of the two amplified fragments is used to calculate the initial amount of VAM fungi in the sample.

The VANS1 primer has also been useful in obtaining rDNA sequences from species of VAM fungal genera including *Acaulospora, Entrophospora,* and *Scutellospora* in addition to *Glomus* and *Gigaspora*. This will allow the design of DNA probes designated to detect and identify these VAM fungi, even when axenic spores are not available.

Investigation of symbiotic gene expression in VAM fungi

Hyphal growth stimulation (S+) during the preinfection stage of *G. margarita* involves the synergistic action of carbon dioxide and root exudates (Bécard and Piché, 1989b). We hypothesize that the S+ condition depends initially on molecular signals present in root exudates.

Flavonoids and other secondary metabolites produced by the host plant have been shown to be involved in the establishment of several pathological and symbiotic relationships, a phenomenon well reviewed by Siquiera et al. (1991a). Phenolic signals have been shown to be involved in the induction of virulence (vir) genes in *Agrobacterium* (Stachel et al., 1985) and nodulation (nod) genes in *Rhizobium* and *Bradyrhizobium* (Long, 1989). Similarly, various host secondary metabolites have been shown to promote haustoria formation in *Striga* (Lynn and Chang, 1990) and even spore germination of the ectomycorrhizal fungus *Suillus* (Fries et al., 1987).

Following from these studies, commercially available flavonoids were shown to enhance spore germination and hyphal growth of the VAM fungus *G. margarita* (Gianinazzi-Pearson et al., 1989). Naturally released host flavonoids were shown to promote hyphal growth of the VAM fungus *Glomus etunicatum* (Tsai and Phillips, 1991) and to increase VAM colonization (Nair et al., 1991; Siqueira et al., 1991b). These studies on hyphal growth promotion by flavonoids were not tested with CO_2 enrichment. Subsequently, Bécard et al. (1992) and Chabot et al. (1992b) have shown that carbon dioxide (2%), in combination with flavonoids (quercetin, kaempferol, myricetin at 10 μM), is an absolute requirement for substantial enhancement of *G. margarita* hyphal growth.

The elucidation of these in vitro conditions makes possible the molecular investigation of gene expression associated with VAM hyphal growth stimulation. Yet the small colonies produced by the VAM fungi have continued to be a limiting factor in obtaining enough biological material to use with standard molecular techniques. For this reason, the procedures developed for studies of *Saccharomyces cereviseae, Neurospora crassa* and *Aspergillus nidulans* (reviewed by, among others, Bennett and Lasure, 1985; Timberlake and Marshall, 1989), including the two successful transformations for ectomycorrhizal fungi *Laccaria laccata* (Barrett et al., 1990) and *Hebeloma cylindrosporum* (Marmeisse et al., 1992) may not be applicable for research on VAM fungi. Indeed, Hirsch (1985) and Beringer et al. (1987) in reviews of the applicability of such systems in manipulations of VAM fungi concluded that the use of molecular techniques will be possible only after pure cultures of VAM fungi are available. We propose however, that with ROC techniques and the powerful tools of PCR technology, molecular genetic analysis of VAM fungi is now possible. We are initiating studies comparing S+ and S- (as described above) messenger RNA expression. It is our belief that PCR derived techniques will be an indispensable tool in compensating for the low amount of S+ and S- fungal material available,

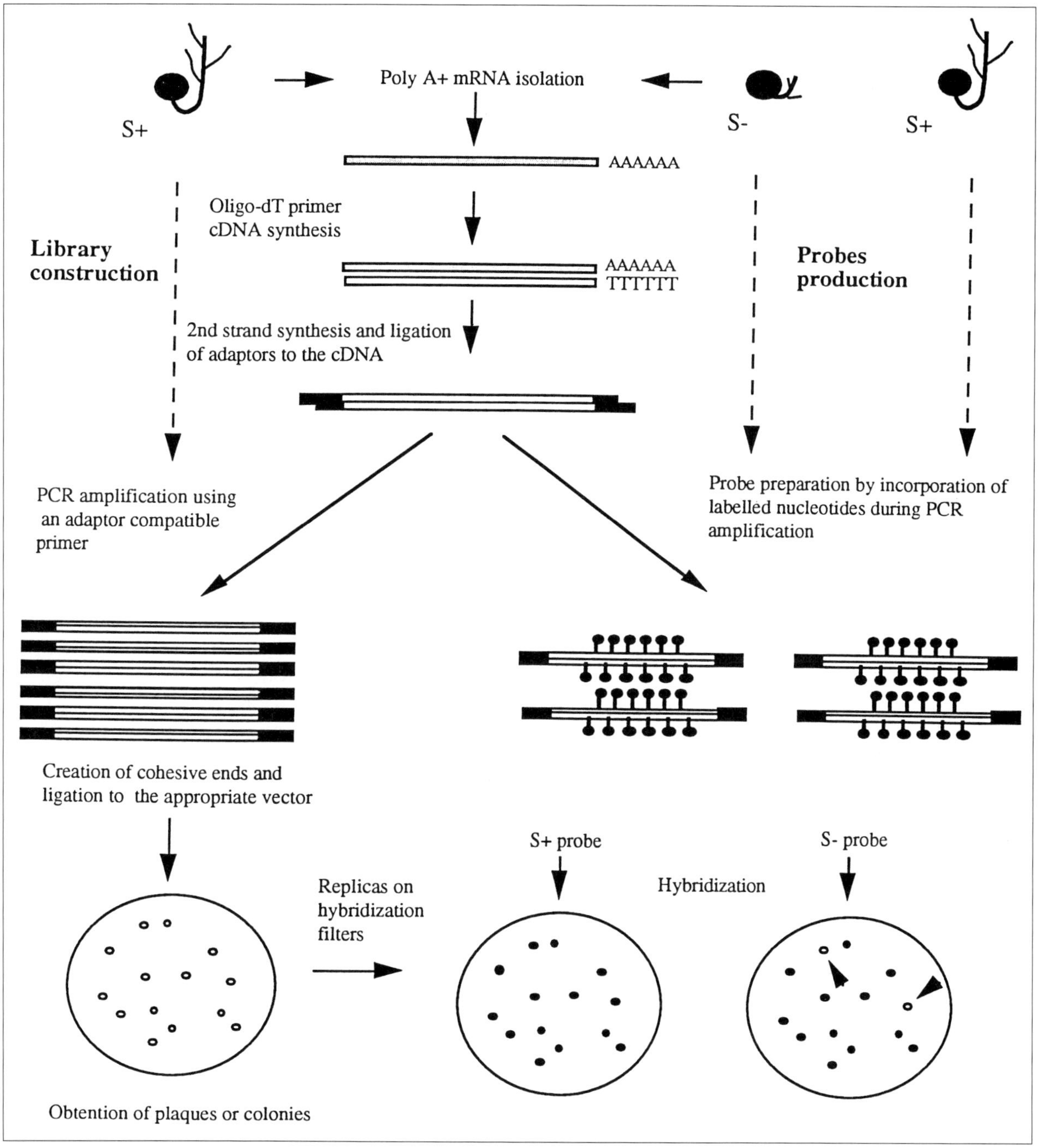

Fig. 2. Scheme of PCR-based library production and differential screening (adapted from Jepson et al., 1991). The double-strand cDNA is synthesized by a conventional procedure (left side of the figure). Following ligation of adaptors to cDNAs, amplification is performed with a single primer compatible with the adaptor, and containing an appropriate restriction enzyme recognition sequence, using standard PCR procedure. Cohesive ends on the cDNAs are obtained by digestion with the restriction enzyme prior to ligation with the cloning vector (bottom left). The preparation of S- and S+ specific probes follow the same initial steps as for the obtention of the cDNA (right side of the figure). The labelled probes are then obtained by replacing one nucleotide in the PCR reaction by a radiolabelled or chemically modified (e.g. biotin or digoxigenin) nucleotide. Standard protocols are available for the differential screening of cDNA librairies with such probes. Hypothetical clones that would be differentially expressed in S+ developing VAM hyphae are indicated by a short arrow (bottom left of the figure).

and will allow us to study gene expression in VAM fungi.

Different procedures for the synthesis and cloning of cDNA from very small amounts (< 50 mg of tissue or a few cells) of material have been already published (Belyavsky et al., 1989; Welsh et al., 1990 and Jepson et al., 1991). Using these approaches, representative cDNA libraries containing over 10^6 independent recombinant clones have been obtained by these authors. A general overview is presented in Fig. 2. More information on cDNA cloning is available elsewhere (Sambrook et al., 1989). To identify the genes expressed during the S+ in vitro conditions, an S+ cDNA library can be screened by differential hybridization with probes obtained from mRNA preparations isolated from S+ and S- in vitro conditions. These probes are obtained using the same procedure as for the cDNA reaction, but include a labelled nucleotide during the PCR amplification. Subsequently, the two cDNA probes are then used to screen replicas of the cDNA library constructed from mRNA extracted from S+ hyphae. The clones that hybridize preferentially to the S+ probes are chosen for further analysis. This approach will provide an effective method to identify genes that are induced during the S+ in vitro condition. In addition, a similar screening of a S- cDNA library may lead to the identification of genes that are repressed during S+ in vitro induction. The analysis of the DNA sequences of these genes may also give new insights into the physiology of hyphal development of VAM fungi.

Although host genotypes in VAM are not covered here (see Graham and Eissenstat, Chapter 24) it is important to mention the observation by Duc et al. (1989) and Bradbury et al. (1991) that nodulation capacity and mycorrhizal formation are phenotypically related. This suggests that similar molecular signals might be involved in both root symbioses. A research program to understand this genetic similarity between the mycorrhizal (myc) and nodulation (nod) characters is well underway at the INRA, Dijon (Gianinazzi-Pearson and Gianinazzi, 1989; Gianinazzi-Pearson et al., 1991). Moreover, evidence that nodulin gene expression is indifferently activated by rhizobial or mycorrhizal infections has also been obtained (Wyss et al., 1990). It should be possible to find the fundamental mechanisms regulating VAM development and these may overlap with those responsible for the activation of regulatory genes in rhizobial symbiosis.

Conclusion

Since genetic manipulations have revolutionized the understanding of early infection and microbial development in *Rhizobium*–legume symbiosis (Hennecke and Verma, 1991), it is important for VAM researchers to initiate the identification and characterization of symbiotic genes. Even if pure cultures of VAM fungi are not yet achievable, the combination of root organ cultures and molecular techniques should permit a coherent strategy for addressing many fascinating questions concerning the biology of these heterokaryotic symbionts.

The applicability of PCR technology to the study of root symbioses has already been demonstrated on ectomycorrhiza (Gardes et al., 1991; Henrion et al., 1992), and more recently on VAM fungi (Simon et al., 1992a; 1992b). Transformed root systems, PCR methods and gene cloning technologies are undoubtedly powerful tools that will be employed to study specific genes controlling the mycorrhizal formation as well as developmental processes such as hyphal growth stimulation.

References

Ahmadjian V and Paracer S 1986 Symbiosis, an Introduction to Biological Associations. University Press of New-England, London. 212 p.

Barrett J T 1947 Observations on the root endophyte, *Rhizophagus*, in culture. Phytopathology 37, 359–360.

Barrett V, Dixon R K and Lemke P A 1990 Genetic transformation of a mycorrhizal fungus. Appl. Microbiol. Biotech. 33, 313–316

Bécard G and Fortin J A 1988 Early events of vesicular-arbuscular mycorrhiza formation on Ri T-DNA transformed roots. New Phytol. 108, 211–218.

Bécard G and Piché Y 1989a New aspects on the acquisition of biotrophic status by a VAM fungus, *Gigaspora margarita*. New Phytol. 112, 77–83.

Bécard G and Piché Y 1989b Fungal growth stimulation by CO_2 and root exudates in vesicular-arbuscular mycorrhizal symbiosis. Appl. Environ. Microbiol. 55, 2320–2325.

Bécard G and Piché Y 1990 Physiological factors determining vesicular-arbuscular mycorrhizal formation in host and non-host Ri T-DNA transformed roots. Can. J. Bot. 68, 1260–1264.

Bécard G and Piché Y 1992 Establishment of vesicular-arbuscular mycorrhizal in root organ culture: review and proposed methodology. *In* Methods in Microbiology: Techniques for the Study of Mycorrhiza. Eds. J R Norris, D J read and A K Varma. pp 89–108. Academic Press, London, U K.

Bécard G, Douds D D and Pfeffer P E 1992 Extensive in vitro hyphal growth of vesicular-arbuscular mycorrhizal fungi in the presence of CO_2 and flavonols. Appl. Environ. Microbiol. 58, 821–825.

Belyavsky A, Vinogradova T and Rajewsky K 1989 PCR-based cDNA library construction: general cDNA libraries at the level of a few cells. Nucl. Acid. Res. 17, 2919–2932.

Bennett J W and Lasure L 1985 Genetic Manipulation of Fungi. Academic Press, London. 552p.

Beringer J E, Burggraaf A J P, Reddell P and Turner G 1987 The role of mycorrhizas in crop growth and prospects for producing modified strains of mycorrhizal fungi. *In* Genetics and Plant Pathogenesis. Eds. P R Day and J L Jellis. pp 91–100. Blackwell Scientific Publications, Oxford, U K.

Bousquet J, Simon L and Lalonde M 1990 DNA amplification from vegetative and sexual tissues of trees using polymerase chain reaction. Can. J. For. Res. 20, 254–257.

Bradbury S M, Peterson R L and Bowley S R 1991 Interactions between three alfalfa nodulation genotypes and two *Glomus* species. New Phytol. 119, 115–120.

Chabot S, Bécard G and Piché Y 1992a Life cycle of *Glomus intraradix* in root organ culture. Mycologia 84, 315–321.

Chabot S, Bel-Rhlid R, Chênevert R and Piché Y 1992b Hyphal growth promotion in vitro of the VA mycorrhizal fungus *Gigaspora margarita* Becker and Hall by the activity of structurally specific flavonoid compounds under CO_2-enriched conditions. New Phytol. 122, 461–467.

Cummings B and Wood T 1989 A simple and efficient method for the isolating genomic DNA from endomycorrhizal spores. Genet Anal. Techn. 6, 89–92.

Duc G, Trouvelot A, Gianinazzi-Pearson V and Gianinazzi S 1989 First report of non-mycorrhizal plant mutants (Myc-) obtained in pea (*Pisum sativum*) and fababean (*Vicia Faba* L). Plant Sci. 60, 215–222.

Foster A S and Gifford E M Jr 1974 Comparative Morphology of Vascular Plants. Freeman W H and Co., San Francisco. 751p.

Fries N K, Serck-Hanssen K, Dimberg L H and Theander O 1987 Abietic acid, an activator of basidiospore germination in ectomycorrhizal species of the genus *Suillus* (Boletaceae). Exp. Mycol. 11, 360–363.

Garber R C and Yoder O C 1983 Isolation of DNA from filamentous fungi and separation into nuclear, mitochondrial, ribosomal, and plasmid components. Anal. Biochem. 135, 416–422.

Gardes M, White T J, Fortin J A, Bruns T D and J W Taylor 1991 Identification of indigenous and introduced symbiotic fungi in ectomycorrhizae by amplification of nuclear and mitochondrial ribosomal DNA. Can. J. Bot. 69, 180–190.

Gerdemann J W 1968 Vesicular-arbuscular mycorrhiza and plant growth. Annu. Rev. Phytopathol. 6, 397–418.

Gianinazzi-Pearson V and Gianinazzi S 1989 Cellular and genetical aspects of interactions between hosts And fungal symbionts in mycorrhizae. Genome 31, 336–341.

Gianinazzi-Pearson V, Branzanti B and Gianinazzi S. 1989 In vitro enhancement of spore germination and early hyphal growth of a vesicular-arbuscular mycorrhizal fungus by host root exudates and plant flavonoids. Symbiosis 7, 234–255.

Gianinazzi-Pearson V, Gianinazzi S, Guillemin J P, Trouvelot A and Duc G 1991 Genetic and cellular analysis of resistance to vesicular-arbuscular (VA) mycorrhizal fungi in pea mutants. *In* Advances in Molecular Genetics of Plant–Microbe Interactions. Eds. H Hennecke and D P S Verma. pp 336–342. Kluwer Academic Publishers, Dordrecht, The Netherlands.

Hennecke H and Verma D P S 1991 Advances in Molecular Genetics of Plant–Microbe Interactions. Kluwer Academic Publishers, Dordrecht, The Netherlands. 482 p.

Henrion B, Le Tacon F and Martin F 1992 Rapid identification of genetic variation of ectomycorrhizal fungi by amplification of ribosomal RNA genes. New Phytol. 122, 289–298.

Hirsch P R 1985 Gene cloning and its potential application to mycorrhizal fungi. *In* Mycorrhizae: Physiology and Genetics. Eds. V Gianinazzi-Pearson and S Gianinazzi. pp 147–157. INRA, 1st ESM, Dijon, France.

Janardhanan K K, Gupta M L and Husain A 1990 Axenic culture of a vesicular-arbuscular mycorrhizal fungus. Curr. Sci. 59, 509–513.

Jepson I, Bray J, Jenkins G, Schuch W and Edwards K 1991 A rapid procedure for the construction of PCR CDNA libraries from small amounts of plant tissue. Plant Mol. Biol. Rep. 9, 131–138.

Lee S B and Taylor J W 1990 Isolation of DNA from fungal mycelia and single spores. *In* PCR Protocols: A Guide to Methods and Applications. Eds. M A Innis, D H Gelfand, J J Sninsky and T J White. pp 282–287. Academic Press, New York.

Long S R 1989 Rhizobium–legume nodulation: life together in the underground. Cell 56, 203–214.

Lynn D G and Chang M 1990 Phenolic signals in cohabitation: implications for plant development. Annu. Rev. Plant Physiol. Plant Mol. Biol. 41, 497–526.

Margulis L 1981 Symbiosis in Cell Evolution. Freeman and Co., San Francisco. 419 p.

Marmeisse R, Gay G, Debaud J C and Casselton L A 1992 Genetic transformation of the symbiotic basidiomycete fungus *Hebeloma cylindrosporum*. Curr. Genet. 22, 41–45.

Millner P D 1989 A minireview of biotechnology as applied to vesicular-arbuscular mycorrhizae. Dev. Ind. Microbiol. 29, 149–157.

Morton J B and Benny G L 1990 Revised classification of arbuscular mycorrhizal fungi (Zygomycetes): A new order, Glomales, two new suborders, Glominae and Gigasporinae, and two new families, Acaulosporaceae and Gigasporaceae, with an emendation of Glomaceae. Mycotaxon. 37, 471–491.

Mosse B 1988 Some studies relating to 'independent' growth of vesicular-arbuscular endophytes. Can. J. Bot. 66, 2533–2540.

Mosse B and Hepper C M 1975 Vesicular-arbuscular mycorrhizal infections in root organ culture. Physiol. Plant Pathol. 5, 215–233.

Mugnier J and Mosse B 1987 Vesicular-arbuscular mycorrhizal infection in transformed root-inducing T-DNA roots grown axenically. Phytopathology 77, 1045–1050.

Mullis K B and Faloona F A 1987 Specific synthesis of DNA in vitro via a polymerase catalysed chain reaction. Methods Enzymol. 155, 335–350.

Nair M G, Safir G R and Siqueira J O 1991 Isolation and identification of vesicular-arbuscular mycorrhiza stimulatory compounds from clover (*Trifolium repens*) root. Appl. Environ. Microbiol. 57, 434–439.

Norris J R, Read D J and Varma A K 1992 Methods in Microbiology: Techniques for the Study of Mycorrhiza. Academic Press, London, U K. 450 p.

Pirozynski K A 1981 Interactions between fungi and plants through the ages. Can. J. Bot. 59, 1824–1827.

Pirozynski K A and Malloch D W 1975 The origin of land plants: a matter of mycotrophism. Biosystems 6, 153–164.

Sambrook J, Fritsch E F and Maniatis T 1989 Molecular Cloning: A laboratory manual. 2nd ed. Cold Spring Harbor Laboratory, New York.

Simon L, Lalonde M and Bruns T D 1992a Specific amplification of 18S fungal ribosomal genes from vesicular-arbuscular endomycorrhizal fungi colonizing roots. Appl. Environ. Microbiol. 58, 291–295.

Simon L, Lévesque R C and Lalonde M 1992b Rapid quantification by PCR of endomycorrhizal fungi colonizing roots. PCR Methods Applications 2, 76–80.

Siqueira J O, Nair M G, Hammerschmidt and Safir G R 1991a Significance of phenolic compounds in plant–soil–microbial systems. Crit. Rev. Plant Sci. 10, 63–121.

Siqueira J O, Safir G R and Nair M G 1991b Stimulation of vesicular-arbuscular mycorrhizal formation and growth of white clover by flavonoid compound. New Phytol. 118, 87–93.

Stachel S E, Messens E, Van Montagu M and P. Zambryski 1985 Identification of the signal molecules produced by wounded plant cells that activate T-DNA transfer in *Agrobacterium tumefaciens*. Nature 318, 624–629.

Strullu G D and Romand C 1986 Méthode d'obtention d'endomycorrhizes à vésicules et arbuscules en conditions axéniques. C.R. Acad. Sc. Paris 303, 245–250.

Tepfer D and Yacoub A 1986 Modèle de rhizosphère utile notamment pour l'étude des parasites des plantes. French patent 25757559, European patent 190161.

Timberlake W E and Marshall M A 1989 Genetic engineering of filamentous fungi. Science 244, 1313–1317.

Tsai S M and Phillips D A 1991 Flavanoids released naturally from alfalfa promote development of symbiotic *Glomus* spores in vitro. Appl. Environ. Microbiol. 57, 1485–1488.

Welsh J, Liu J P and Efstratiadis A 1990 Cloning of PCR-amplified total cDNA: construction of a mouse oocyte cDNA library. Genet. Anal. Techn. Appl. 7, 5–17.

White T J, Arnheim N and Erlich HA 1989 The Polymerase Chain Reaction. Trends Genet. 5, 185–189.

White T J, Bruns T D, Lee S B and Taylor J W 1990. Analysis of phylogenetic relationships by amplification and direct sequencing of ribosomal RNA genes. *In* PCR Protocols: A Guide to Methods and Applications. Eds. M A Innis, D H Gelfand, J J Sninsky and T J White. pp 315–322. Academic Press, New York.

Williams P G 1984 Obligate parasitism and axenic culture. *In* The Cereal Rust Eds. W R Bushnell and A P Roelfs. pp 399–430. Academic Press, New York, USA.

Williams P G 1990 Disinfecting vesicular-arbuscular mycorrhizas. Mycol. Res. 94, 995–997.

Williams P G 1992 Axenic culture of arbuscular mycorrhizal fungi. *In* Methods in Microbiology: Techniques for the Study of Mycorrhiza. Eds. J R Norris, D J Read and A K Varma. pp 203–220. Academic Press, London.

Wyss P, Mellor R B and Wiemken A 1990 Vesicular-arbuscular mycorrhizas of wild-type soybean and non-nodulating mutants with *Glomus mosseae* contain symbiosis-specific polypeptides (mycorrhizins), immunologically cross-reactive with nodulins. Planta 182, 22–26.

Reprinted from *Plant and Soil*, Volume 159, No. 1 (1994), pp. 171–178.

Host genotype and the formation and function of VA mycorrhizae

J.H. GRAHAM and D.M. EISSENSTAT
Citrus Research and Education Centre, University of Florida, IFAS, 700 Experiment Station Road, Lake Alfred, FL 33850, USA

Key words: carbon cost, carbon expenditure, citrus genotypes, host–fungus compatibility, mycorrhizal dependency, root colonization rate

Abstract

VA mycorrhizae, the most ancient type of mycorrhizal symbiosis, are present in the most phytogenetically advanced groups. Few plants have evolved mechanisms to completely prevent infection by VAM fungi. Yet, plant species that are less dependent on VA mycorrhizae for nutrient acquisition (e.g., grasses) generally have less root colonization in the field than more dependent species (e.g., *Citrus*). Among closely related Citrus genotypes, there is a greater tendency for less dependent species to limit the rate but not the extent of colonization, even in high-P soils. We hypothesize that colonization represents a significant carbon cost that may be regulated by the host genotype. Carbon expenditure on the fungus at high P may result in mycorrhizal-induced growth depression. The potential value of breeding plants for greater susceptibility to colonization will depend on the cost/benefit of VA mycorrhizae for the specific crop, soil and environmental conditions. Although the genetics and physiology of host control over VAM colonization are barely known, recently discovered mycorrhizal colonization mutants (myc$^-$) of pea offer great promise for the study of host–fungus compatibility.

Introduction

VA mycorrhizae are the most common of plant–fungal infections. Perhaps as much as 80% of all plant species form VA mycorrhizae with a comparatively small number of zygomycetous species in the order Glomales (Gianinazzi, 1991). The early origin of the zygomycetous symbiosis is supported by the fossil record and patterns of mycorrhizal habit in present-day plants (Trappe, 1987). The VA mycorrhizal association is ancient, yet has been maintained in even the most advanced plants. Complete resistance to infection by mycorrhizal fungi has evolved in few plant orders. However, phytogenetically advanced groups with fine roots that are less dependent on VA mycorrhizae for nutrient acquisition (e.g., grasses) generally have lower levels of root colonization in the field (Crush, 1973; Sparling and Tinker, 1978) than mycorrhizal dependent plants with coarse roots (e.g., *Citrus*) (Graham et al., 1991).

Given the lack of host specificity of VA mycorrhizal fungi, the genes for host–fungus compatibility would presumably be homologous among plant and fungus genotypes. The genes involve plant/fungus recognition, fungal infection/plant reaction, and the ultimate morphology and function of the mycorrhiza in carbon/nutrient exchange. The genetics of host–fungus compatibility are just now coming under study (Duc et al., 1989; Gianinazzi, 1991); thus, the precise physiological mechanisms underlying the different processes leading to mycorrhizal function are barely understood.

The purpose of this review is to appraise the costs of mycorrhizal colonization and the potential for crop benefit from plant breeding for

Florida Agricultural Experiment Station Journal Series No. R-02765.

A.D. Robson, L.K. Abbott & N. Malajczuk (eds.), Management of mycorrhizas in agriculture, horticulture and forestry, 179–185.

increased fungal colonization in agricultural situations. The prospects for study of the genetics of mycorrhizal formation and function will also be briefly discussed.

Functional relationship between P uptake and carbon expenditure

VA mycorrhizal colonization is most notably affected by P nutrition of the host (Stribley et al., 1980). Plants growing in low-P environments and deficient in tissue P are more readily colonized by mycorrhizal fungi than plants of high-P status. Host carbon must cross the plasma membrane at the host–fungus interface before becoming available to the fungus. Functionally speaking, phospholipid content of root cells regulates the symbiosis via membrane leakage of carbon to support the activities of the fungus (Graham et al., 1981). However, passive leakage alone may not completely explain the means of regulation since even when membrane permeability is reduced at high root-P content, sufficient carbon for fungal growth in the roots is available for colonization in high-P environments although at much lower rates (Schwab et al., 1991).

In low-P soils, rate of colonization is often correlated with growth responses (Abbott and Robson, 1984). However, the increase in leaf area and carbon assimilation may feed back carbon to the fungus even though tissue P is increased and membrane permeability reduced. In plant species with a greater growth response, more carbon will feed back from the shoot to the fungus than in less responsive plant species. Thus, a correlation between mycorrhizal dependency (VAM plant dry weight / non-mycorrhizal plant dry weight) and colonization at low P may represent a phenotypic response of the fungus to greater carbon availability. The correlation does not reveal the inherent ability of a host genotype to regulate carbon expenditure for the symbiosis (Graham et al., 1991).

Host genotype relationship

In addition to various environmental factors influencing colonization, host genotype is another important factor controlling the rate and extent of VAM formation. In high-P soil, colonization varies substantially among plant families, genera and even among closely related genotypes of the same species (Graham et al., 1991; Krishna et al., 1985; Manske, 1989; Toth et al., 1990). Where there is no growth benefit in high-P soils, colonization of plants cannot be promoted by carbon feedback. We hypothesize (Graham et al., 1991) that plants may be more fit (i.e., able to produce offspring) if the rate of carbon expenditure on the mycorrhizal fungus is limited during periods of little or no P benefit. When young roots of citrus genotypes were examined in an orchard high in P fertility (> 160 µg/g Mehlich I extractable), species of low mycorrhizal dependency had lower rates of VAM colonization than those of high mycorrhizal dependency (Graham et al., 1991). The rate rather than the extent of colonization tended to be limited in these genotypes. Presumably, the plant's expenditure of forming mycorrhizae was prolonged so that carbon could be expended for other competing needs that were more likely to provide a more immediate return on the investment, such as leaf area for carbon assimilation. Among citrus genotypes, this observation suggests that rate of colonization can represent a significant carbon cost and that there has been some evolutionary selection for this trait. The cost vs benefit of the host plant by restricting carbon expenditure on mycorrhizal formation would be a function of the amount of carbon saved and the likelihood that heavily colonized roots would be needed for P uptake in the near future. During their evolutionary history, plant genotypes that are highly mycorrhizal dependent would not exhibit reduced rates of colonization because they rarely grew in environments where they were not benefited by mycorrhizae.

In order for plants to be under selection pressure to limit the rate of carbon expenditure for mycorrhizal fungal growth during periods of no P benefit (i.e., high P fertility), carbon expenditure must be significant. In citrus species, which have substantial colonization even in high-P soil, growth is depressed by the mycorrhizal fungus. For instance, incidence of colonization of Volkamer lemon by *Glomus intraradices* Schenck and Smith is about 60% at 52 days after inoculation in high-P soil (Peng et al., 1993). This level of colonization is associated with a 37% higher rate of root plus soil respiration and a 10 to 20% lower specific daily carbon gain compared to non-mycorrhizal plants. Differences in soil/root respiration are linked to 19% higher root dry weight and 10% higher root growth rate. There is also a two-fold increase in total-fatty acid content of roots due to the presence of 16:0 and 16:1 fatty acids in the intraradical vesicles of *G. intraradices*. Quantities of fungal lipids in citrus roots are correlated with root construction cost, i.e., the amount of simple carbohydrate (e.g., glucose) required to build all the organic molecules present in the root (Williams et al., 1987). The cost of mycorrhizal formation at high P includes a 50% greater root allocation in mycorrhizal plants, 11% expenditure for building of lipid rich roots and the remaining 39% for maintenance of fungal tissue in the root and growth and maintenance of extramatrical hyphae. Therefore, the growth depression of plants at high P is attributed to carbon expenditure on construction of mycorrhizal roots and their maintenance (Peng et al., 1993). Similar mycorrhizal-induced growth reductions in high-P soils have been demonstrated for ryegrass, ferns, forage legumes, apple, tobacco and onions (Buwalda and Goh, 1982; Cooper, 1975; Crush, 1976; Miller et al., 1985; Modjo and Hendrix, 1986; Mosse, 1973). Thus, the potential among plant species for mycorrhizal fungi to cause growth depressions is quite widespread.

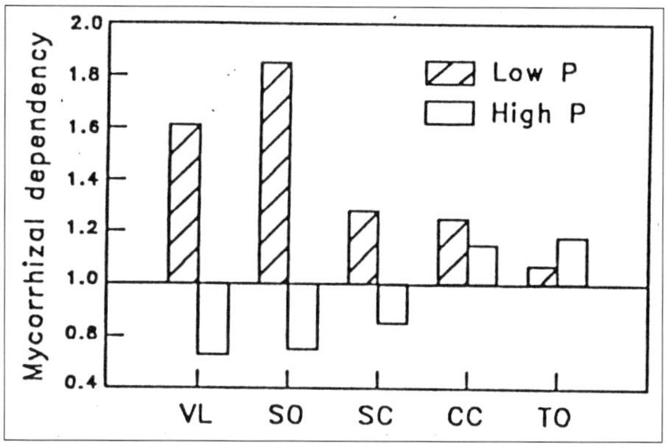

Fig. 1. Mycorrhizal dependency (ratio of VAM plant dry wt and nonmycorrhizal plant dry wt) for seedlings of five citrus genotypes grown at low- and high-P supply. VL = Volkamer (*Citrus volkameriana* Tan. and Pasq.), SO = Sour orange (*C. aurantium* L.), TO = trifoliate orange (*Poncirus trifoliata* (L.) Raf.), SC = Swingle citrumelo (*C. paradisi* Macf. X *P. trifoliata*), and CC = Carrizo citrange (*C. sinensis* (L.) Osb. X *P. trifoliata*).

Evaluation of carbon expenditure by citrus genotypes

We sought to link mycorrhizal dependency at low P with the tendency of the plant genotype to limit colonization and, therefore, carbon expenditure at high P.

To evaluate genotypic variation in carbon expenditure associated with mycorrhizal dependency, five citrus genotypes (Fig. 1) were grown at low- and high-P supply with and without *G. intraradices*. In the first experiment, total biomass was assessed after 155 days of growth in 125 cm^3 containers and mycorrhizal dependency of each cultivar expressed as the ratio of total dry weight of the mycorrhizal plant to that of the non-mycorrhizal plant (n = 5). Thus, values of less than one at high-P supply indicate growth depression. In the second experiment, the five genotypes were grown in 1,100 cm^3 pots for 125 days. At harvest, colonization was assessed as incidence of vesicles, arbuscules and hyphae in 20 root segments per plant (Graham et al., 1991) and intensity of vesicle formation was rated on a scale of 0–5 (0 = no vesicles present) (n = 5). Dried roots were ground and the fatty-acids extracted with methanol, saponified and derivatized to methyl esters. The Microbial Identification System (MIS) (Sasser, 1990) was used to quantify fatty-acid content of roots which was used as an index of fungal cost of mycorrhizal formation. Concentrations of individual fatty acids were expressed as relative amounts based on their integrated areas of the fatty-acid profile. Root construction costs were determined using the heat of combustion method (Williams et al., 1987).

In the first experiment, growth enhancement by *G. intraradices* was greater for the more

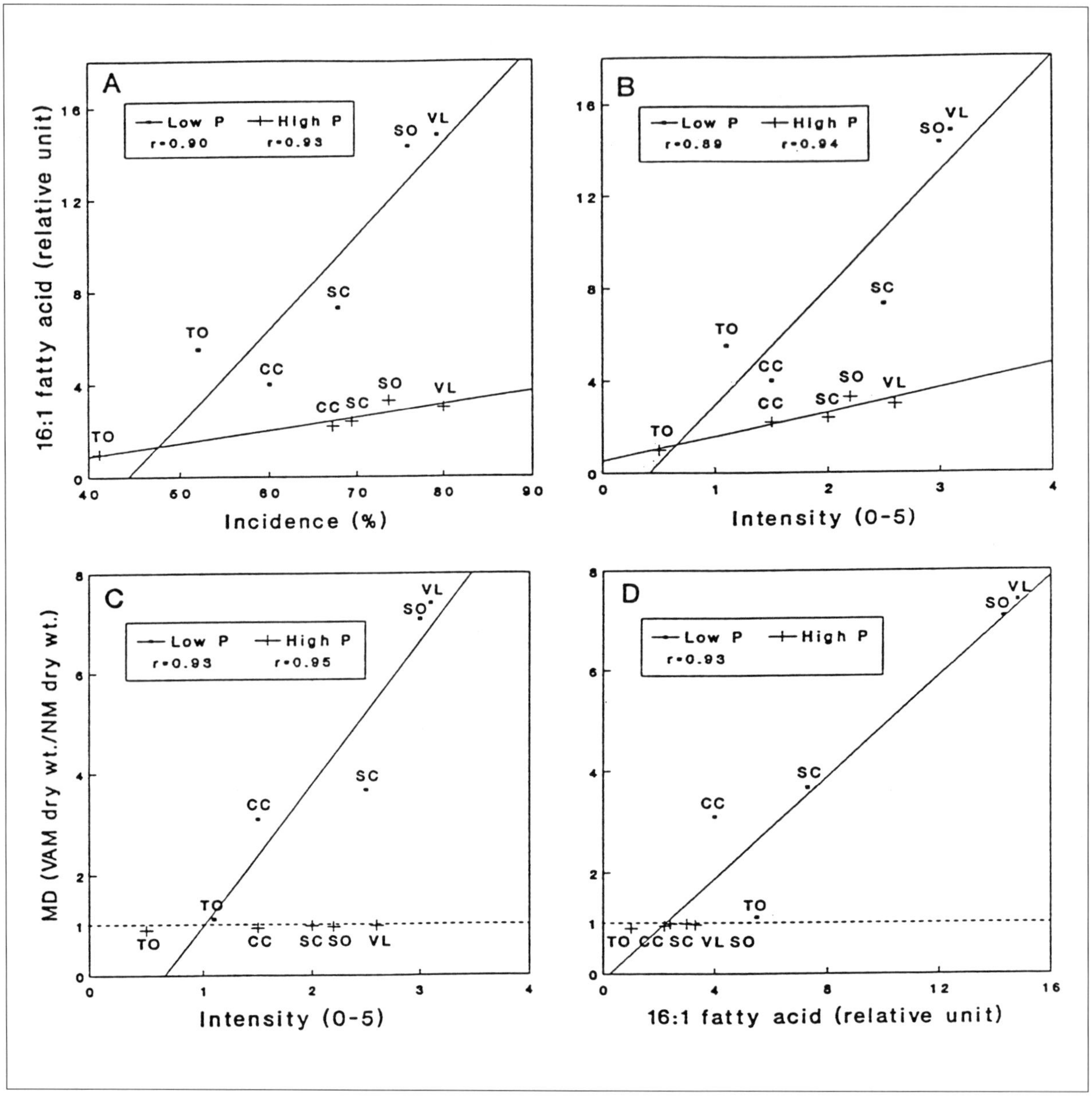

Fig. 2. Relationship among 16:1 fatty-acid content of roots, colonization by *Glomus intraradices* and mycorrhizal dependency of seedlings of five citrus genotypes growing at low- and high-P supply. See Fig. 1 for identification of genotypes. Correlation coefficients are significant at $P \leq 0.05$.

dependent genotypes, Volkamer lemon and sour orange, at low-P soil supply (Fig. 1). At high-P supply, growth of the dependent genotypes by *G. intraradices* was depressed, whereas growth of the least dependent genotypes was increased slightly. Thus, greater mycorrhizal dependency at low P appeared to be linked with the levels of growth depression at high P.

In the second experiment, incidence of colonization and intensity of vesicle formation by *G. intraradices* for the five citrus genotypes were highly correlated with the content of 16:1

fatty acid (Fig. 2a, b). This fatty acid was associated principally with infected roots and was in low quantities in non-mycorrhizal citrus roots. Furthermore, mycorrhizal dependency of the citrus genotypes was correlated with intensity of vesicle development at both low- and high-P supply (Fig. 2c) and was related to 16:1 fatty-acid content at low P (Fig. 2d). At high P, fatty-acid content of mycorrhizal roots did not vary substantially among the five citrus genotypes (Fig. 2d). This small variation in fatty-acid content suggests that there were only small differences in fungal carbon expenditure among citrus genotypes at high P in this experiment. This is consistent with the narrow range of mycorrhizal increases in root construction costs (-4 to 2%; data not shown) and of growth depression (2 to 6%; Fig. 2c) found among the genotypes. Consequently, growth depression at high P was not linked to mycorrhizal dependency at low P as in the first experiment.

In the second experiment, the failure to establish a relationship between mycorrhizal dependency and carbon expenditure on the fungus at high P may have occurred for several reasons. The range in colonization among citrus genotypes at high P was not great enough to detect clear differences in expenditure and a relationship to growth depression. The citrus genotypes may have varied in their root allocation in response to extrinsic factors other than the mycorrhizal fungus. The cost of resisting colonization by the different citrus genotypes may vary independently from the root allocation responses. In this study, plant growth, colonization levels and fatty-acids in roots were only determined at one point in the colonization process. These comparisons need to be made at different stages of root colonization and plant development.

Host genotype interactions and plant breeding

Several researchers have advocated the use of plant breeding to increase host 'susceptibility' to mycorrhizal fungi and, therefore, enhance the efficiency of mycorrhizae (Krishna et al., 1985; Lackie et al., 1988; Manske, 1989; Toth et al., 1984). This assumes that rapid and high levels of colonization are beneficial but ignores the potential costs particularly in high-P agricultural soils. Under P-limiting conditions, greater colonization within a group of plant genotypes has been correlated with increased tissue P concentrations and mycorrhizal dependency (Krishna et al., 1985; Manske, 1989). However, in some cases, a relationship between extent of mycorrhizal root length, P uptake, and greater growth response has not been demonstrated (Azcon and Ocampo, 1981; Crush and Caradus, 1980). A more complex interaction between colonization and plant growth response may exist than one related only to P or carbon economy of the genotype. Toth et al. (1990) recently reported that inbred lines of corn selected for a range of resistance to a variety of fungal leaf pathogens had lower mycorrhizal colonization, slower maturation and larger root systems.

Whether plant breeding programs that select for resistance to fungal pathogens are influencing mycorrhizal dependency of the host may partly depend on whether the relationship between cost and benefit of VA mycorrhizae is affected. If rapidly colonizing indigenous fungi cause growth depressions in high-P soils, as has been reported for field grown tobacco (Modjo and Hendrix, 1986), then selection of traits that reduce the rate or extent of colonization may be beneficial. By contrast, if the plant genotype is dependent on mycorrhizae for nutrient acquisition, then a delay in colonization may have adverse effects on yield. Cost vs benefit cannot be addressed unless the crop genotype, fungal genotype/populations, soil fertility and the environmental conditions are specified. For example, growth depression of wheat cultivars in a high-P soil occurred in a year of low light intensity in the greenhouse but not in a year of high light intensity (Manske, 1989). Likewise, we have demonstrated growth depression of young citrus trees in high-P soil resulting from colonization by indigenous mycorrhizal fungi in one orchard location but not in an adjacent field under similar soil conditions in another year.

The interaction of host and different fungus genotypes may also influence the cost/benefit of enhanced colonization. Fungi, like *G. intraradices*, that depress host growth, may be more opportunistic in their ability to colonize and consume carbon in plants susceptible to colonization than fungi that are less opportunistic and/or more efficient in their carbon utilization. Johnson et al. (1992) have hypothesized that yield decline with continuous corn and soybean monoculture is in part due to proliferation of less beneficial mycorrhizal fungi. Ultimately, the equation of cost and benefit will depend on the soil nutrient supply since this will have the most direct bearing on dependency of the host under a given cultural condition (Fig. 1). Thus, the merits of breeding for greater mycorrhizal formation will depend largely on the objectives of the crop management program (e.g., level of phosphate fertilizer inputs).

Genetics of host–fungus compatibility

The study of colonization in non-mycorrhizal plant species has provided some insight into the stages at which VAM fungus–host incompatibility may occur, but the basis for the incompatibility remains obscure (Glenn et al., 1988). Recently, chemically induced mutants of pea (*Pisum sativum*) were discovered that are characterized by aborted infections (Duc et al., 1989). Analysis of F1 crosses of the 'myc$^-$' mutants and the wild type indicate the involvement of at least three genes (Gianinazzi, 1991). The myc$^-$ character is constantly associated with the nod$^-$ genotype controlling nodulation by *Rhizobium leguminosarum*. This is intriguing because it suggests that there is a genetic linkage and presumably tight control of the two carbon consuming and potentially competing symbioses. The genetics of myc$^-$ in peas may be unique to legumes. Nevertheless, isogenic lines, except for the myc$^-$, nod$^-$ genes, offer unprecedented possibilities for study of the genes and gene products which control different stages of the VAM colonization process.

Acknowledgments

We gratefully acknowledge collaborations with S. Peng, K. Williams and C. Hodge in the studies presented here and the technical assistance of D. Drouillard and N. Timmer. This research was supported in part by the National Science Foundation (BSR-9111824).

References

Abbott L K and Robson A D 1984 The effect of mycorrhizae on plant growth. *In* VA Mycorrhiza. Eds. C L Powell and D J Bagyaraj. pp 113–130. CRC Press, Boca Raton, Florida.

Azcon R and Ocampo J A 1981 Factors affecting the vesicular-arbuscular infection and mycorrhizal dependency of thirteen wheat cultivars. New Phytol. 87, 677–685.

Buwalda J G and Goh G M 1982 Host–fungus competition for carbon as a cause of growth depressions in vesicular-arbuscular ryegrass. Soil Biol. Biochem. 14, 103–106.

Cooper K M 1975 Growth responses to the formation of endotrophic mycorrhizas in *Solanum leptospernum* and New Zealand ferns. *In* Endomycorrhizas. Eds. F E Sanders, B Mosse and P B Tinker. pp 391–407. Academic Press, London.

Crush J R 1973 Significance of endomycorrhizas in tussock grassland in Otago, New Zealand. N. Z. J. Bot. 11, 645–660.

Crush J R 1976 Endomycorrhizas and legume growth in some soils of the Mackenzie Basin, Canterbury, New Zealand. N. Z. J. Agric. Res. 19, 473–476.

Crush J R and Caradus J R 1980 Effect of mycorrhizas on growth of some white clovers. N. Z. J. Agric. Res. 23, 233–237.

Duc G, Trouvelot A, Gianinazzi-Pearson V and Gianinazzi S 1989 First report of non-mycorrhizal plant mutants (Myc$^-$) obtained in pea (*Pisum sativum*) and fababean (*Vicia faba* L.). Plant Sci. 60, 215–222.

Gianinazzi S 1991 Vesicular-arbuscular (endo-) mycorrhizas; cellular, biochemical and genetic aspects. Agric. Ecos. Environ. 35, 105–119.

Glenn M G, Chew F S and Williams P H 1988 Influence of glucosinolate content of *Brassica* (Cruciferae) roots on growth of vesicular-arbuscular mycorrhizal fungi. New Phytol. 110, 217–225.

Graham J H, Leonard R T and Menge J A 1981 Membrane-mediated decrease in root exudation responsible for phosphorus inhibition of vesicular-arbuscular mycorrhiza formation. Plant Physiol. 68, 548–552.

Graham J H, Eissenstat D M and Drouillard D L 1991 On the relationship between a plant's mycorrhizal dependency and rate of vesicular-arbuscular mycorrhizal colonization. Functional Ecol. 5, 773–779.

Johnson N C, Copeland P J, Crookston R K and Pfleger F L 1992 Mycorrhizae: Possible explanation for yield decline with continuous corn and soybean. Agron. J. 84, 387–390.

Krishna K R, Shetty K G, Dart P J and Andrews D J 1985 Genotype dependent variation in mycorrhizal colonization and response to inoculation of pearl millet. Plant Soil 86, 113–125.

Lackie S M, Bowley S R and Peterson R L 1988 Comparison of colonization among half-sib families of *Medicago sativa* L. by *Glomus versiforme* (Daniels and Trappe) Berch. New Phytol. 108, 477–482.

Manske G G B 1989 Genetical analysis of the efficiency of VA mycorrhiza with spring wheat. Agric. Ecos. Environ. 29, 273–280.

Miller D D, Domoto D A and Walker C 1985 Colonization and efficacy of different endomycorrhizal fungi with apple seedlings at two phosphorus levels. New Phytol. 100, 393–402.

Modjo H S and Hendrix J W 1986 The mycorrhizal fungus, *Glomus macrocarpum* as a cause of tobacco stunt disease. Phytopathology 76, 668–691.

Mosse B 1973 Plant growth responses to vesicular-arbuscular-mycorrhizae. IV. In soil given additional phosphate. New Phytol. 72, 127–136.

Peng S, Eissenstat D M, Graham J H and Williams K 1993 Growth depression in mycorrhizal citrus at high phosphorus supply: Analysis of carbon costs. Plant Physiol. (*in press*).

Sasser M 1990 Identification of bacteria through fatty acid analysis. *In* Methods in Phytobacteriology. Eds. Z Klement, K Rudolf and D Sands. pp 199–204. Akademiai Kiado, Budapest.

Schwab S M, Menge J A and Tinker P B 1991 Regulation of nutrient transfer between host and fungus in vesicular-arbuscular mycorrhizas. New Phytol. 117, 387–398.

Sparling G P and Tinker P B 1978 Mycorrhizal infection in Pennine grassland. I. Levels of infection in the field. J. Appl. Ecol. 15, 943–950.

Stribley D P, Tinker P B and Snellgrove R C 1980 Effect of vesicular-arbuscular mycorrhizal fungi on the relations of plant growth, internal phosphorus concentration and soil phosphate analysis. Soil Sci. 31, 655–672.

Toth R, Page T and Castleberry R 1984 Differences in mycorrhizal colonization of maize selections for high and low ear leaf phosphorus. Crop Sci. 24, 994–996.

Toth R, Toth D, Starke D and Smith D R 1990 Vesicular-arbuscular colonization in *Zea mays* affected by breeding for resistance to fungal pathogens. Can. J. Bot. 68, 1039–1044.

Trappe J M 1987 Phytogenetic and ecologic aspects of mycotrophy in angiosperms from an evolutionary standpoint. *In* Ecophysiology of VA Mycorrhizal Plants. Ed. G R Safir. pp 5–25. CRC Press. Boca Raton, Florida.

Williams K, Percival F, Merino J and Mooney H A 1987 Estimation of tissue construction cost from heat of combustion and organic nitrogen content. Plant Cell Env. 10, 725–734.

Reprinted from *Plant and Soil*, Volume 159, No. 1 (1994), pp. 179–185.

What is the potential for management of vesicular-arbuscular mycorrhizae in horticulture?

DORIS C.N. CHANG
Department of Horticulture, National Taiwan University, Taipei, Taiwan, Republic of China (l0617)

Key words: biological control, fruits, microorganisms, micropropagation, stress, transplanting, vegetables

Abstract

The effects of vesicular-arbuscular mycorrhizal (VAM) fungi on fruit, vegetable and ornamental crops include enhanced seedling growth, a saving in phosphate requirement, increased survival rate and growth of micropropagated plantlets, earlier flowering and fruiting of plants, increased resistance to water stress but not to acid rain, increased flower production and vase life of cut flowers, higher enzyme activities, greater development of xylem and vessel tissues and greater biological control of root diseases such as fungi, bacteria and nematodes.

The most promising areas for practical use of VAM in horticulture is for nursery or plug-growing seedlings and transplanting micropropagated plantlets. Inoculation could be done during seed sowing or transplanting seedlings and micropropagated plantlets out of test tubes. There are also potential applications for plants grown in fumigated soil, in soilless media and in arid or infertile areas. Most considerations should be directed towards certain perennials and high value crops to decrease phosphate required and to increased resistance of crops to root diseases. The combination of crop–VAM fungi should be tested before each large scale application.

The need to screen VAM fungi for ability to colonize plant roots in high phosphate soil or other media is extremely important in horticulture.

Introduction

Effects of vesicular-arbuscular mycorrhizal (VAM) fungi on the growth and development of horticultural crops have been studied by many investigators and more than 250 papers were published in the past ten years. Miller et al. (1986) presented an excellent review of vesicular-arbuscular mycorrhizae (VAM) in vegetable crops. Nemec (1987) reviewed VAM in horticultural systems. Other reviews dealt with potential use of mycorrhizae in control of soil-borne diseases (Caron, 1989; Liu and Chang, 1986; and Tellow et al., 1987), growth effects of VAM fungi on horticultural crops (Chang, 1992), interaction of VAM with beneficial soil microorganisms (Okon and Hadar, 1987) and the interaction of VAM and *Rhizobium* (Linderman and Paulitz, 1990; Patterson et al., 1990). Among horticultural hosts for VAM, fruit crops have been more studied than vegetable and ornamental crops.

The practical use of VAM fungi with horticultural crops is currently limited by a lack of supply of VAM fungi inocula on a commercial scale. Most VAM fungi inocula is produced in pot cultures although production in aeroponic culture appears promising (Hung and Sylvia, 1988).

Horticulturists are interested in the ability of VAM to increase the uptake of phosphorus and other nutrients and to increase resistance to stress

A.D. Robson, L.K. Abbott & N. Malajczuk (eds.), Management of mycorrhizas in agriculture, horticulture and forestry, 187–190.

and soil-borne diseases. Therefore, promising areas and limitations for the use of VAM in horticulture are presented.

Effects of VAM

Effects of mycorrhiza on fruit, vegetables and ornamental crops include enhanced seedling growth (Chang, 1990a), reduced phosphate requirement (Chang, 1992) and increased survival rate and growth of micropropagated plantlets (Chang, 1992; Gianinazzi-Pearson, 1989; Lin, 1986; Ponton et al., 1990). For micropropagated strawberry plantlets, VAM increased the resistance to parasitic mould infection (Chang, 1990b). Mycorrhizal plants flowered and fruited by at least a week before non-mycorrhizal plants (Backhaus, 1983; Chang, 1990b; Cheng, 1989; Cheng et al., 1991; Wen, 1991). Gerbera colonized by *Glomus etunicatum* flowered 16 days earlier than the non-mycorrhizal control (Wen, 1991) and zinnia inoculated with *Glomus mosseae* flowered 24 days before the control (Cheng, 1989).

Gerbera plants colonized by *G. mosseae* produced more flowers which lasted three days longer than non-mycorrhizal plants in the vase (Wen, 1991). The increased vase life of flowers from mycorrhizal plants may be due to the greater development of water conducting tissues in mycorrhizal plants than in non-mycorrhizal plants (Chang, 1992; Wen, 1991).

Mycorrhizal plants (bell pepper, citrus, papaya and rose) may also be more resistant to water stress (Auge and Stodola, 1990; Johnson and Hummel, 1985; Waterer and Coltman, 1989). There was no correlation between the extent of mycorrhizal colonization and the level of tolerance to acid rain for four plant species (Heijne et al., 1990). Finally, mycorrhizas may increase resistance to root diseases (Giovannetti, 1990; Liu and Chang, 1986; Sampangi and Bagyaraj, 1989).

Limitations of current VA mycorrhizal technology

There are several reasons why VAM are not more widely used in horticulture today. First there is an inability to culture VAM fungi axenically and to produce pathogen-free inoculum on a large scale. This has led to an absence of VAM inoculum on a commercial scale in most countries and to VAM fungi being more expensive than NPK fertilizers. Second, the use of high rates of fertilizers, particularly phosphate when growing horticultural crops may result in a low colonization rate by VAM fungi. Third, there is a lack of information on the successful use of VAM in horticultural production.

Current use of mycorrhizal fungi in horticulture

Case Studies

1. The use of VAM fungi infected muskmelon transplants in Taiwan. About 350 hectares and more than half a million muskmelons are grown using *Glomus clarum* infected transplants which were cultivated in jiffy pots containing sand medium in Taiwan. The survival rate of the mycorrhizal muskmelon in field soil was 99% compared to 91% for non-mycorrhizal seedlings (Cheng et al., 1991). Vine length for 6 and 7 week-old mycorrhizal muskmelon plants were 80 cm and 127 cm, while the vine lengths of the non-mycorrhizal plants at these two ages were only 57 cm and 104 cm respectively. Using mycorrhizal transplant seedlings in addition to high cultivation technology produced an excellent quality of muskmelons. If 9,000 seedlings/ha were planted and each plant was allowed to bear 2 fruits, on average, growing mycorrhizal muskmelon could gain 5,600 more US dollars/ha than the control. Cheng et al. (1991) showed the successful use of self-root system of mycorrhizal muskmelon to produce a second crop of muskmelon. For example, 80 days after the harvest of the first crop, an extra 22,900 kg of muskmelon could be produced and even more profits than the first crop could be made by the after-season production of muskmelon. In Taiwan, this is the largest scale of VAM application in horticultural production.

2. Micropropagated asparagus, gerbera and Harunoka strawberry. All the micropropagated asparagus, gerbera and Harunoka strawberry plantlets were inoculated with only one *Glomus* sp. during transplanting out of test tubes, and showed promising results (Lin, 1986; Wen, 1991; Chang, 1990b).

Inoculation of micropropagated Tai-nan selection No. 3 asparagus with a *Glomus* sp. increased survival rate by 20% during transplanting and the growth of mycorrhizal plantlets in pots was greatly increased compared to the non-mycorrhizal control (Lin, 1986). Vegetative and reproductive growth of micropropagated gerbera plantlets in pots were increased by inoculation with *Glomus* sp. Earlier flower production by 7–16 days was found in 'Tennensee' gerbera inoculated with *Glomus etunicatum*, and more flowers (5–7 flowers per plant) were produced in 30/25 (day/night C°) than the control. *Glomus* colonized gerbera plants produced 5–11% more flowers (from September to March) in field soil. Mycorrhizal 'Regina' gerbera flower could last 3 more days in 8–HQS solution and 'Tennensee' could even last 4–5 more days than the control (Wen, 1991).

Transplanting survival rate of Harunoka strawberry plantlets was increased by 27–36% by inoculation with a *Glomus* sp. Runner production was increased 55–73% by inoculation with *Glomus etunicatum* or *Glomus fasciculatum*. About 10% of the mycorrhizal strawberry plants showed 1–4 weeks earlier flowering and fruiting than non-mycorrhizal plants. Mycorrhizal Harunoka strawberry plants in field soil produced 8–28% more fruits and greater fresh weight (Chang, 1990b).

Menge (1983), Menge and Johnson (1978) and Nemec (1987) reported the use of VAM in the USA. In Brazil, mycorrhizal citrus seedlings and strawberry runners are available for commercial use (Lin, personal communication). Ericoid mycorrhizas stimulate fruit yield of blueberry (Powell and Bates, 1981).

Suggested Application of VAM in Horticulture

The suggested applications are:

1. Nursery or container (plug)-cultivated seedlings; for example, beans, citrus, green pepper, lily, marigold, rose, tomato seedlings (Chang, 1992).
2. Micropropagated plantlets and common transplants; for example, asparagus, citrus, gerbera, raspberry, strawberry.
3. Mycorrhizal dependent crops grown in fumigated soil; for example citrus, gerbera, lily and strawberry.
4. Cultivated crops grown in soilless medium.
5. Crops grown in arid or infertile areas; for example, onion, papaya and pineapple.
6. Perennials or high value crops; for example, apple, citrus, coffee, cocoa, grapevine, peach, raspberry and tea.

The first two applications are the most promising areas for practical use of VAM in horticulture. The combination of crop–VAM fungi should be tested before each new line of crop is inoculated with VAM fungi on a large scale.

Future Research Priorities

It is very important to screen VAM fungi for their ability to colonize plant roots in high phosphate containing soil or media. The physiology of spore germination should be thoroughly studied. Responses of crop seedling or micropropagated plantlets, from transplanting survival rate to shelf-life of mycorrhizal and non-mycorrhizal products should be studied and applied. The kinds of crops which might benefit from mycorrhizal inoculation should be identified. Gene transfer of the beneficial traits from mycorrhizal fungi to crops by biotechnology could be conducted.

Conclusion

Until recently, the effects of VAM on horticultural crops were not widely known by horticulturists. Thus special efforts should be made to increase their knowledge of mycorrhizas. As fertilizers were less expensive, high fertilizers were applied in most horticultural production systems. Thus, it is extremely important to breed or screen VAM fungi which can colonize crop roots at high levels of phosphate supply and for these fungi to be made available to the public. In the near future, it is very possible that nutrient film techniques or aeroponic cultures could produce vegetables or flowers on the above ground portion and at the same time, those plants could be used for mass production of VAM fungi spores and hyphae in the roots as inoculum.

References

Auge R M and Stodola A J W 1990 An apparent increase in symplastic in water contributes to greater turgor in mycorrhizal roots of droughted *Rosa* plants. New Phytol. 115, 285–295.

Backhaus G F 1983 Influence of vesicular-arbuscular mycorrhiza on the generative development of *Heliotropium* and *Fuchsia*. Gartenbauwissenschaft 48, 197–201.

Caron M 1989 Potential use of mycorrhizae in control of soil-borne diseases. Can. J. Plant Pathol. 11, 177–179.

Chang D C H 1990a Plug production of Asparagus seedlings and the use of vesicular-arbuscular mycorrhizae. J. Chin. Soc. Hortic. Sci. 36, 149–155.

Chang D C H 1990b Responses of micropropagated Harunoka strawberry plantlets to three *Glomus* spp. vesicular-arbuscular mycorrhizal fungi. J. Chin. Soc. Hortic. Sci. 36, 265–273.

Chang D C N 1992 Studies and prospect of horticultural vesicular-arbuscular mycorrhizae in Taiwan. Sci. Agric. 40, 45–52.

Cheng S F 1989 Single-spore culture of vesicular-arbuscular mycorrhizal (VAM) fungi and their effects on three flower crops. M.S. Thesis, Department of Horticulture, National Taiwan Univ. 101 p.

Cheng Y H, Tu C C, Cheng A S and Uren S T 1991 The use of endomycorrhiza on the cultivation of muskmelon. Taiwan Agric. 27, 53–56.

Gianinazzi-Pearson V 1989 Vesicular-arbuscular endomycorrhizae as determinants for plant growth and survival. *In* Interrelationships between Microorganisms and Plant in Soil. Proc. of an Internationall. Symposium, Liblice, Czechoslovakia, June, 22–27, 1987. Eds. V Vancura and F Kunc. pp 69–76. Elsevier S Sci. Pub. Co. Inc. Amsterdam.

Giovannetti M 1990 Mycorrhizas as agents of biological control of agricultural plant pathogens. Inf. Fitopatol. 40, 17–20.

Heijne B, Heil G W and Dam D van 1990 Relations between acid rain and vesicular-arbuscular mycorrhiza. Agric. Ecos. Environ. 29, 187–192.

Hung L L L and Sylvia D M 1988 Production of vesicular arbuscular mycorrhizal fungus inoculum in aeroponic culture. Appl. Environ. Microbiol. 54, 353–357.

Johnson C R and Hummel R L 1985 Influence of mycorrhizae and drought stress on growth of *Poncirus* X *Citrus* seedlings. Hortic. Sci. 20, 754–755.

Lin C H 1986 Effect of three *Glomus* endomycorrhizal fungi on the growth of micropropagated banana and asparagus plantlets. M.S Thesis, Department of Horticulture, Taiwan University of Taipei, Taiwan ROC. 117 p.

Linderman R G and Paulitz T C 1990 Mycorrhizal–rhizobacterial interactions. *In* Biological control of soil-borne plant pathogens. Ed. D Hornby. pp 261–283. CAB International, Wallingford, Oxon, UK.

Liu M L and Chang D C N 1986 Interactions between endomycoorhizae and plant pathogens. J. Chin. Soc. Hortic. Sci. 32, 205–213.

Menge J A and Johnson E L V 1978 Commercial production of mycorrhizal inoculum may benefit citrus growers. Citrograph 63, 139–143.

Menge J A 1983 Utilization of vesicular-arbuscular mycorrhizal fungi in agriculture. Can. J. Bot. 61, 1015–1024.

Miller J C Jr., Rajzpakse S and Garber R K 1986 Vesicular-arbuscular mycorrhizae in vegetable crops. Hortic. Sci. 21, 974–984.

Nemec S 1987 VA mycorrhizae in horticultural system. *In* Ecophysiology of VA Mycorrhizal Plants. Ed. Safir S R. pp 193–221. CRC Press Inc., Boca Raton, Florida.

Okon Y and Hadar Y 1987 Microbial inoculants as crop-yield enhancers. Crit. Rev. Biotech. 6, 61–85.

Patterson N A, Chet I, and Kapulnik Y 1990 Effect of mycorrhizal inoculation on nodule initiation, activity and contribution to legume productivity. Symbiosis 8, 9–20.

Ponton F, Piché Y, Parent S and Caron M 1990 Use of vesicular-arbuscular mycorrhizae in Boston fern production. II. Evaluation of four inocula. Hortic. Sci. 25, 416–419.

Powell C and Bates P M 1981 Ericoid mycorrhizas stimulate fruit yield of blueberry. Hortic. Sci. 16, 655–656.

Sampangi R K, Bagyaraj D J 1989 Root diseases and mycorrhizae—a review. J. Phytol. Res. 2, 1–6.

Tello J C, Vares F, Notario A and Lacasa A 1987 Mycorrhizae: a potential against plant diseases. Inf. Tec. Econ. Agraria 18, 40–64.

Waterer D R and Coltman R R 1989 Response of mycorrhizal bell peppers to inoculation timing, phosphorus, and water stress. Hortic. Sci. 24, 688–690.

Wen C L 1991 Effect of temperature and *Glomus* spp. on the growth and cut flower quality of micropropagated *Gerbera jamesonii*. M.S. thesis, Department of Horticulture, National Taiwan University. 158 p.

What is the potential for management of mycorrhizas in agriculture?

J.P. THOMPSON
Queensland Wheat Research Institute, PO Box 2282, Toowoomba, Qld 4350, Australia

Key words: crops, fallow, inoculum potential, mycorrhizal dependency, rotation, tillage

Abstract

The potential for profitably managing VAM in agriculture can be achieved through a combination of approaches. The levels of inoculum of VAM fungi in the soil and colonization of crop roots can be kept high by cropping with hosts, selectively using fungicides and fertilizers, and avoiding excessive clean fallow, cropping with non-hosts, topsoil removal, soil disturbance, heating and water-logging. Potential crop profitability can be optimized by taking VAM into consideration by choosing crop species and varieties on their mycorrhizal dependency to match the VAM inoculum level of the soil, by using P and Zn fertilizers judiciously, by rotating VAM-building crops to avoid pathogen buildup and by managing other constraints to crop growth especially water and nitrate. All the above are management options that are currently available.

There is a large future potential to breed crop varieties that can better exploit the VAM symbiosis. Two directions could readily be followed to produce high yielding varieties that are either mycorrhizal dependent and suitable for sowing into fields with high densities of VAM propagules, or mycorrhizal independent for sowing into fields with low densities. If and when wide-scale inoculation of agricultural crops with VAM fungi will become economic cannot be foretold as further successful research is needed in the area of mass production of inoculum. In the meantime our knowledge of potentially responsive situations for inoculation is increasing and the extent of adoption would ultimately depend on the benefit/cost ratios for different agricultural systems.

Introduction

Management of VAM in agriculture is required where there are obvious benefits to be derived, either economic or environmental. The three most important factors determining the magnitude of benefits from improved VAM management are (a) mycorrhizal dependency of the host (b) nutrient status of the soil and (c) inoculum potential of the VAM fungi present.

The potential to manage VAM in agriculture will undoubtedly expand with time as research and development proceed. Currently, effective management of VAM is feasible via agronomic practices. In the near future, practical plant breeding will be a feasible approach to develop crop varieties with desirable response patterns to VAM.

Mycorrhizal dependency

Considerations of mycorrhizal dependency (Bayliss, 1975; Plenchette et al., 1983) are most important in developing effective crop management strategies. For example, if we know that a particular crop species derives little benefit from VAM or that conversely another crop species is highly dependent on VAM for nutrition and growth then we can manage each accordingly.

Clearly, crops that are non-hosts of VAM (Table 1) have no mycorrhizal dependency. Although some plant families comprise a high proportion of non-mycorrhizal species and others few, it is unwise to assume the host status of a particular species from its family membership

A.D. Robson, L.K. Abbott & N. Malajczuk (eds.), Management of mycorrhizas in agriculture, horticulture and forestry, 191–200.

because of the clear exceptions (Newman and Reddell, 1987).

Although most agricultural crop species are hosts of VAM not all benefit equally. The relative field mycorrhizal dependency (RFMD) index proposed by Plenchette et al. (1983) expresses the difference in dry mass between mycorrhizal and non-mycorrhizal treatments as a percentage of the mass of the mycorrhizal treatment. Mycorrhizal and non-mycorrhizal treatments were obtained by growing plants in natural and fumigated field soil respectively. The approach is quite useful in ranking crop hosts within a single experiment but the absolute values of RFMD will depend on the soil nutrient status. Plenchette et al. suggested that RFMD be appended with a superscript to indicate the extractable P level of the soil to aid comparisons across experiments. One difficulty in this suggestion is the wide range of extraction methods for 'available' phosphorus that are in use (Ojala et al., 1983).

Soil nutrient status

In most cases beneficial effects of mycorrhizas are associated with increased phosphate uptake (Abbott and Robson, 1984). Therefore, it is unlikely that management of mycorrhizas will provide benefits if phosphate supply is not limiting the growth of non-mycorrhizal plants. Abbot and Robson (1991) have emphasized the need to assess the value of VAM to a host across a full range of soil P levels by determining the response curves for both mycorrhizal and non-mycorrhizal plants.

In some soils, the response of some crop species to VAM is expressed as a zinc as well as a phosphate benefit (Wellings and Thompson, 1993). In a vertisol, non-mycorrhizal linseed (*Linum usitatissimum* L.) grew poorly unless high rates of both P and Zn were applied (Fig. 1). In contrast, mycorrhizal linseed required virtually no applied P and one half the rate of zinc as did non-mycorrhizal linseed. Thus in some circumstances the benefits of VAM need to be measured in additional terms to that of phosphate alone.

Inoculum potential of VAM fungi

Agricultural systems are disturbed biological systems in which several practices may be detrimental to VAM fungi. Among practices that

Table 1. Agriculturally important crop species that are non-hosts of vesicular-arbuscular mycorrhizal (VAM) fungi

Plant family	% non-mycorrhizal species[a]	Non-host crop species	References
Cruciferae	87	rapeseed (*Brassica napus* L. var *napus*)	Ocampo et al., 1980
		canola (*Brassica campestris* L.)	Glenn et al., 1988
		mustard (*Sinapsis alba* L.)	Fitter and Nichols, 1988
		kale (*Brassica oleracea* L.)	Black and Tinker, 1979
		chinese cabbage (*Brassica chinensis*)	Yost and Fox, 1979
		swede (*Brassica campestris* L.)	Ocampo et al., 1980
Chenopodiaceae	61	sugarbeet (*Beta vulgaris* L. var. *rapa*)	Ocampo et al., 1980
Amaranthaceae	–	grain amaranth (*Amaranthus viridis* L.)	Khan, 1974
		chinese spinach (*Amaranthus bicolor*)	C. Murdoch, *pers. comm.*
Polygonaceae	37	buckwheat (*Fagopyrum esculentum*)	Mosse and Thompson, 1984
Leguminosae	4	lupins (*Lupinus angustifolius, L. cosentinii, L. luteus, L. albus*)	Morley and Mosse, 1976; Trinick, 1977; Thompson and Wildermuth, 1989

[a]From Newman and Reddell (1987)

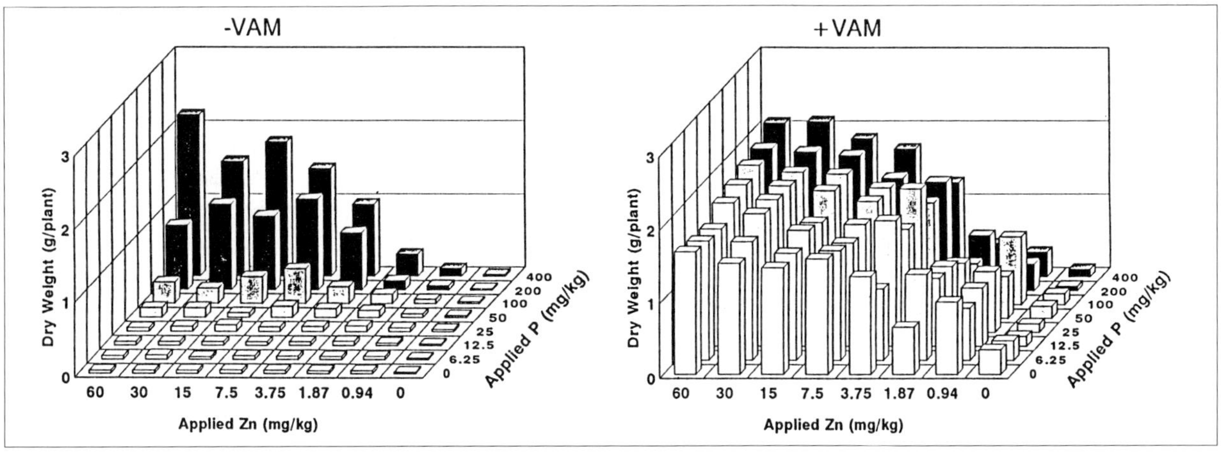

Fig. 1. Responses of mycorrhizal and non-mycorrhizal linseed in a Vertisol to jointly varying phosphate and zinc fertilizer rates (Wellings and Thompson, 1991).

can influence VAM inoculum potential are clean fallowing, tillage, topsoil removal, fires, waterlogging and the use of chemicals like pesticides and fertilizers.

Fallowing

Fallowing, whereby land is left for an extended period without a crop, is an integral part of many agricultural systems. In some systems, weeds are allowed to grow and fallows are grazed by livestock. In other dry-land agricultural systems, fallows are used to accumulate soil water and nitrate and so are kept weed-free. Usually, crops grow better the longer the fallow. However, clean fallowing results in a decline in VAM inoculum and in VAM colonization of the next crop to be sown (Salt, 1977; Black and Tinker, 1979; Thompson, 1987). In the vertisols of the north-east Australian grain belt a problem known as long-fallow disorder results from depletion of VAM inoculum by fallowing (Thompson, 1987). The problem affects a range of crop species with different severities depending on the mycorrhizal dependency of the crop. A feature of this area is the low use of P fertilizers and it is now clear that normal crops are benefiting greatly from VAM. The problem is noticed when a farmer sows a crop across land of two fallow lengths and after 3–12 weeks the crop on the longer fallow is comparatively stunted with symptoms of P and or Zn deficiency. At this stage the crop on the longer fallow has a lower percentage of roots colonized by VAM fungi (Thompson, 1987, 1991).

In replicated field experiments, mycorrhizal-dependent crops like linseed and maize were sown after treatments consisting of various periods of fallow from 2 to 21 months (Thompson, 1991). The longer the fallow, the lower was the number of intact VAM spores, and the poorer the root colonization with VAM and the growth of the crop without P and Zn fertilizers.

Crop rotation

If VAM inoculum declines during fallow then the converse is equally true that VAM inoculum builds up during cropping. However, not all crops are equal in this regard. Crop species that are non-hosts of VAM are equivalent to fallow in their effects on VAM populations (Black and Tinker, 1979; Ocampo et al., 1980; Thompson, 1991).

Host species differ in terms of their residual value of VAM inoculum for a subsequent crop. This probably depends not only on the percent colonization of the root system with VAM but also on the total mass of mycorrhizal root produced (Thompson and Wildermuth, 1989). Some plant species may have other properties that make them superior sources of VAM spores (Struble and Skipper, 1988).

Selection of crop species to grow preceding a highly mycorrhizal-dependent crop species is therefore an important management factor. By growing hosts that leave a high density of VAM propagules in the soil, natural VAM populations can be effectively managed to achieve results that might be considered the objective of artificial inoculation. However, it must be recognized that the soil contains a complex microbial population including root pathogens whose populations may also be responsive to the previous crop. On the Darling Downs of Queensland the root-lesion nematode *Pratylenchus thornei* is widespread. Like VAM fungi, root-lesion nematodes obligately obtain nutrition from the cortex of plant roots but cause pathological changes, stunting and loss of yield in some crops. In one experiment (Thompson unpublished), a vertisol site that contained root-lesion nematode was clean-fallowed for 5.5 years before a range of winter and summer crops was grown. Plots were then split for linseed (a strongly mycorrhizal-dependent crop but a non-host of root-lesion nematode) and wheat (a weakly mycorrhizal-dependent crop and an intolerant host of root-lesion nematode). There were low numbers of VAM spores after fallow and the non-mycorrhizal host rapeseed (Fig. 2). Numbers after the winter crops and six months fallow were next lowest with the exception of the winter-legume chickpea. The long-season summer-legume pigeonpea produced the most spores.

There was no relation between the effect of hosts on VAM inoculum and number of root-lesion nematodes. Wheat cultivars generally left the most nematodes.

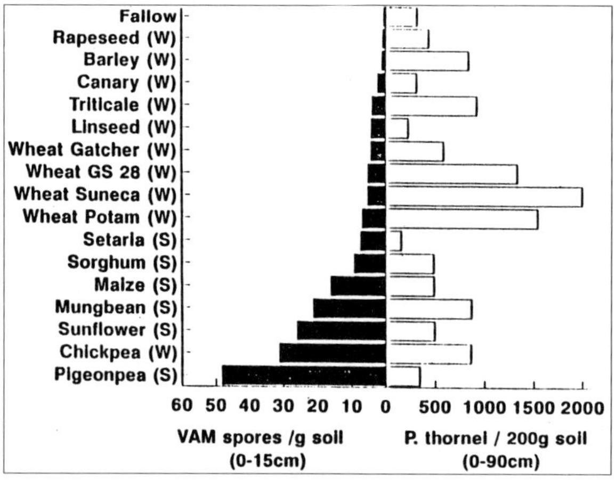

Fig. 2. Residual populations of VAM spores and root-lesion nematodes (*Pratylenchus thornei*) after the growth of various crops on a long-fallowed vertisol, where W and S signify winter and summer crops respectively. Values are the means of three replicates (J.P. Thompson, unpublished data).

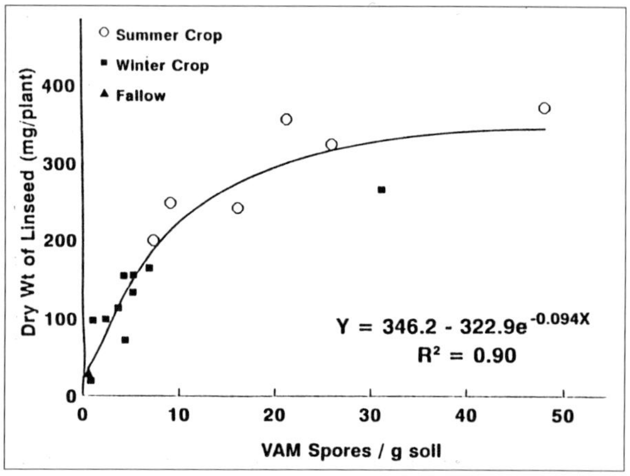

Fig. 3. The growth of linseed in response to VAM spores residual after the growth of various winter and summer crops. Each point is the mean of three replicate field plots (J.P. Thompson, unpublished data).

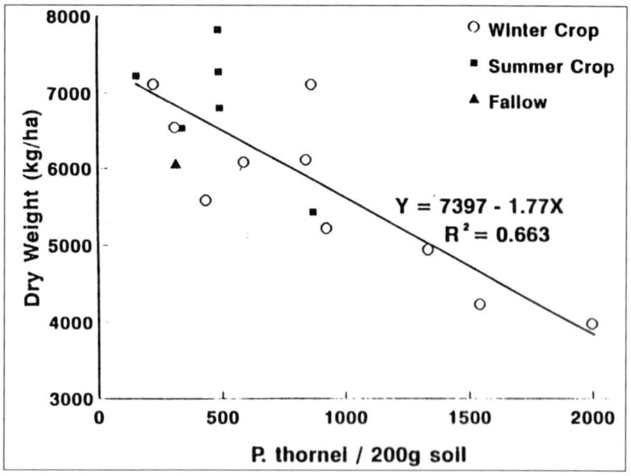

Fig. 4. The effect of root-lesion nematodes (*Pratylenchus thornei*) residual from various pre-crop treatments on the dry matter production of wheat. Each point is the mean of three replicate field plots (J.P. Thompson, unpublished data).

When linseed was grown after the pre-crop treatments there was an asymptotic relationship between the dry weight of the linseed plants and the residual spore populations (Fig. 3). The growth of highly mycorrhizal-dependent crops like linseed can be increased by pre-cropping with crops that leave a high level of VAM inoculum.

In contrast to linseed, wheat cv. Gatcher was most influenced by the root-lesion nematodes that were residual from the previous crops (Fig. 4). For wheat, this was a more important factor than variation in VAM inoculum density. Nonetheless, variation in VAM was still an important factor influencing the growth of the wheat. When multiple regression was applied to the data, 95% of the variation in wheat dry matter production was explained by linear and quadratic terms for VAM spore numbers, root-lesion nematodes and extractable phosphate and zinc.

This example serves to emphasize that in crop rotation experiments or in farm practice, variation in VAM populations can also be accompanied by variation in populations of plant pathogens. It is therefore important to recognize the common pathogens and to know their host ranges. The wide host range of VAM compared with most pathogens enables rotations to be devised that maintain VAM at high levels for mycorrhizal-dependent crops but pathogens at low levels for susceptible crops.

Flooding rice may lead to lower inoculum potential of VAM. There was no increase in the number of propagules during the growth of flooded rice crops compared to an eightfold increase after non-flooded rice (Ilag et al., 1987).

The key factors for managing crop rotations to obtain maximum benefits from VAM can be summarized as follows. If the VAM inoculum in the soil is low after such practices as clean fallowing, growth of a non-host crop or rice under water-logged conditions, then sow a crop of low mycorrhizal dependency. If a crop of high dependency is to be grown for other reasons like disease control then a high P fertilizer rate, and possibly Zn fertilizer should be used. In contrast, if the VAM inoculum level is high, it is feasible to sow the most profitable crops of high mycorrhizal dependency. Finally, if soil-borne fungal or nematode pathogens are present then crop rotations should be designed to keep VAM populations high but to avoid sequencing host crops of the pathogen in quick succession.

Top soil removal

Most studies show that the density of VAM inoculum decreases down the soil profile (e.g. Habte, 1989). Therefore, removal of the VAM rich topsoil through soil erosion, land planing or the construction of contour banks or bunds can result in reduction of VAM inoculum and in poor VAM colonization of the next crop. The problem can be magnified because the subsoil is often poorer in available phosphate for the less mycorrhizal roots to access. In addition, the subsoil may be poorer in other nutrients which limit the crops ability to respond should the VAM and P status be rectified. Experience after land planing of vertisols in Queensland indicates that initially poor crop growth improves with successive crops to become normal after about three years. During this period, growing crops that are hosts of VAM but that have low mycorrhizal dependency can help build VAM populations without undue crop loss.

Tillage

Severe soil disturbance can reduce the infectivity of VAM fungi in agricultural systems. Excessive secondary tillage and traffic increased soil bulk density, and decreased root growth, VAM colonization and top growth of dry edible bean (*Phaseolus vulgaris* L.) (Mulligan et al., 1985).

After 7 years of continuous maize, untilled soil contained more residues of phosphate fertilizers and a greater root length with slightly higher VAM colonization than did a mouldboard-ploughed soil (Anderson et al., 1987).

Several studies into the effects of soil disturbance on VAM and P nutrition by M.H. Miller and co-workers in Canada have shown the following. Disrupting the macrostructure of long-term, zero-tilled soils reduced VAM colonization and P and Zn uptake of the mycorrhizal hosts maize and wheat but not of the non-mycorrhizal hosts spinach (*Spinacea oleracea* L.) or rapeseed (*Brassica napus* L.)

(Evans and Miller, 1988). Injection of the fungicide benomyl (Evans and Miller, 1988) decreased VAM colonization of maize and P concentration of the maize shoots in undisturbed soil to values comparable to disturbed soil. The effects of disturbance appear to be due to killing the VAM hyphae in the soil (Evans and Miller, 1990; McGonigle et al., 1990).

When maize was grown through successive 3-week cycles in an initially disturbed soil and then either left undisturbed or disturbed again between cycles, VAM colonization and P uptake were markedly greater in the former than in the latter (Fairchild and Miller, 1988). Soil pasteurization which killed VAM fungi prevented the improved P uptake from the undisturbed soil in a similar experiment (McGonigle and Miller, 1992).

In studies in Queensland (Peck et al., 1992), bioassays with linseed in intact cores taken from various tillage treatments at the end of six-month fallow periods have shown only modest differences in VAM colonization. With zero-tillage there was greater VAM colonization at 5–10 cm depth than at either 0–5 or 10–15 cm. Increasing frequency of tillage, possibly through mixing the layers, tended to decrease the per cent colonization at 5–10 cm and increase it at 0–5 and 10–15 cm depths. Earlier, it was shown that long-fallow disorder as assessed by symptoms of sequentially sown linseed could be induced earlier in the fallow by aggressive tillage with a rotary hoe (J.K. Leslie, pers. comm.).

Overall it would seem that decreased frequency and aggressiveness of tillage will be preferred for best management of VAM. With zero tillage systems, this may best be combined with some fertilizer use because of the accumulation of P at the soil surface.

Fire and heating

Colonization of wheat plants with VAM fungi in soil previously heated with aerated steam for 30 minutes was severely destroyed at 50°C and virtually eliminated at 60°C (Thompson, 1990). Similar soil temperatures (50–60°C) resulting from burning forest litter significantly reduced VAM colonization in Sudan grass bioassay plants (Klopatek et al., 1988). Decrease in VAM colonization was significantly correlated with soil temperature with hardly any colonization in two dry soils that reached 80–90°C.

In an agricultural context, clearing and burning of eucalypt woodland in the Northern Territory of Australia has resulted in VAM depletion and poor growth of maize where the logs have been burnt in windrows (Thiagalingam et al., 1988). No doubt, similar effects occur with shifting cultivation following forest burns which could result in poor growth of VAM-dependent crops on high P-fixing soils such as the latosolic soils that occur in many tropical areas.

Pesticides

There have been several reviews on the effects of pesticides on VAM fungi (e.g. Smith, 1978; Menge, 1982; Johnson and Pfleger, 1992). Generally, the broad-spectrum biocidal fumigants like methyl bromide and methyl isothiocyanate are very toxic to VAM fungi under ideal conditions. Because methyl bromide is denser than air it can kill VAM propagules deep in the soil profile (Menge, 1982). Although such products are little used for commercial agriculture they are excellent research tools.

Johnson and Pfleger (1992) indicated two classes of fungicides that are toxic to VAM fungi. These are the substituted aromatic hydrocarbons comprising botran, chlorothalonil, lanstan and quintozene or PCNB, and the benzimidazoles comprising benomyl, carbendazim, thiabendazole, thiophanate, and thiophanatemethyl. In some agricultural systems there is regular application of such systemic fungicides to cereals to control foliar pathogens.

Generally, insecticides and nematicides at recommended rates do not inhibit VAM fungi. The latter may even promote VAM colonization through control of fungivorous nematodes. Likewise, herbicides generally do not directly inhibit VAM fungi although they may reduce VAM inoculum through eliminating potential hosts.

Another aspect of VAM/herbicide interactions is the possibility that VAM may improve uptake of residual herbicides. Mycorrhizal soybeans suffered more damage than non-mycorrhizal soybeans in soil with low concentrations of residual atrazine (0.01 mg kg^{-1} soil) (Busse and Ellis, 1987). Hyphae of *Glomus* spp. were shown to directly remove ^{14}C-residues of atrazine from soil and transfer them to maize (Nelson and Khan, 1992). Possibly the density of VAM inoculum in soil is an important factor determining the sensitivity of mycorrhizal crop and weed species to residual atrazine.

Fertilizers

Much research has shown that phosphatic fertilizers can reduce VAM colonization. However, additions of P fertilizers to soils very low in phosphate can increase per cent colonization of the root system possibly through a direct effect on the VAM fungus (Bolan et al., 1984). At increasing rates of P fertilizer the effect is to decrease per cent colonization but to increase length of VAM colonized root because root growth is stimulated by the P additions. At still higher rates of P fertilizer, per cent colonization is so severely depressed that total length of mycorrhizal root is also decreased by the P fertilizer. In an experiment to analyse these effects, Amijee et al. (1989) found that the inhibiting effect of high rates of P on VAM colonization appeared to result from a reduction in the formation of secondary entry points.

Hayman (1982) reviewed experiments that showed high rates of nitrogen fertilizers also inhibited VAM fungi. However, Thompson (1986) using multiple regression found that colonization of maize and wheat roots with *Glomus* spp. was positively correlated with nitrogen concentration of roots and negatively related to P concentration. Colonization of lettuce roots was also found to increase with higher N and lower P concentrations in the roots (Hepper, 1983). Somewhat contrasting results were obtained by Sylvia and Neal (1990) who found that P decreased colonization of the roots of onion by *Glomus etunicatum* under N-sufficient but not under N-limited conditions.

Many of these fertilizer experiments have been conducted in pots with extremely high rates. More information is required from field experiments at commercial rates of fertilizer application. Since the major benefit that crops receive from VAM fungi is increased efficiency of P extraction, the reduction in VAM colonization by P fertilizer is unlikely to be expressed in decreased yield unless there is some unsatisfied benefit from VAM fungi, e.g. Zn uptake, that is diminished by the P fertilizer. For best use of P resources for crop production and environmental care P fertilizer rates that do not suppress VAM fungi should be used.

Liming

Soil pH is a major factor controlling the distribution of different species and strains of VAM fungi. For example, *Acaulospora laevis* and *Glomus* spp. in Western Australia were restricted to soils of low and high pH respectively (Porter et al., 1987). That the VAM fungal composition can change with agricultural practices that change soil pH was clearly demonstrated by Wang et al. (1985). In a long-term experiment in England, liming changed the mycorrhizal colonization of spring oats from exclusively fine endophyte (*Glomus tenuis*) at pH 4.5 to exclusively coarse endophytes (seven other *Glomus* spp., and *Acaulospora* spp.) at pH 7.5 (Fig. 5). If pH modification were to select VAM fungal strains of lesser or greater symbiotic effectiveness than pre-existing strains, then differences in crop performance could be

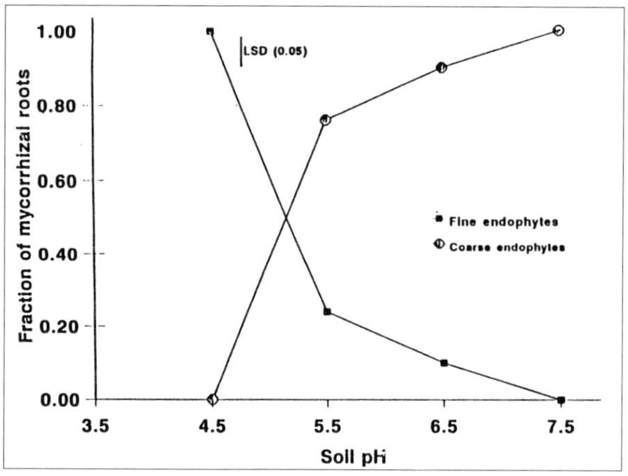

Fig. 5. Fraction of total length of mycorrhizal root of spring oats colonised by fine and coarse VAM fungi in response to a range of soil pH established by long-term liming (Wang et al., 1985).

expected. Such a possibility needs to be considered where soil pH modification is a component of an agricultural system.

Plant breeding

The potential value of managing VAM by specifically breeding appropriate crop varieties is probably quite large and realizable in the next decade. Genetic variation among plant genotypes in resistance to fungal pathogens has long been exploited effectively by plant breeding to protect agricultural crops. Much is known about the genes that control resistance in hosts and virulence in pathogens. There are now a sufficient number of studies to demonstrate the genetic control that host genotype has over the VAM symbiosis in terms of level of fungal colonization and mycorrhizal dependency of the host. These studies should catalyse plant breeders and mycorrhiza researchers to work toward producing varieties for specific situations that have an advantage because of their genetic control over the VAM symbiosis. Crops for which genetic variation in VAM colonization or dependency have been demonstrated are; wheat (Bertheau et al., 1980; Azcon and Ocampo, 1981; Manske, 1990 and Stoppler et al., 1990), pearl millet (Krishna et al., 1985), sorghum (Raju et al., 1990), maize (Toth et al., 1990), cowpea (Mercy et al., 1990), chickpea (Daft, 1991) and peanut (Daft, 1991).

Schwab (1987) considered approaches in breeding mycorrhizal plants for enhanced mineral uptake. Breeding crops for less extensive root systems with high rates of root exudation (to stimulate VAM fungi) combined with inoculation with effective strains of VAM fungi was considered to provide for a high ratio of P uptake per photosynthate cost. Even though it is currently uneconomical to artificially inoculate with VAM fungi in most agricultural systems, Schwab's suggestion would be relevant for intensively cropped systems with high densities of natural VAM inoculum. For other agricultural systems with low levels of natural VAM inoculum it may be preferable to select high yielding cultivars of low mycorrhizal dependency.

The value of VAM to various genotypes can also be assessed in terms of P benefits. Raju et al. (1990) found that mycorrhizal sorghum plants were less P efficient than non-mycorrhizal plants, where P efficiency is calculated as dry matter produced per unit P absorbed. However, P efficiency here may merely reflect the deficiency stress level which can be relieved by VAM or P fertilizer resulting in increased tissue P concentration. The value vs cost of VAM was greatest when soil P was lowest and decreased with increasing P application. A further aspect to bear in mind is the finding of Toth et al. (1990) that maize inbred lines with high overall resistance to a number of fungal diseases also had low colonization with VAM fungi. Further research is required with other crop species to see if this correlation is general.

Inoculation

Although there may be some specific agricultural circumstances where inoculation with VAM fungi is warranted, current costs of inoculum production are too great for general use in agriculture. Inoculation could be economic for transplanted crops like finger millet (Govinda Rao et al., 1983) or extremely mycorrhizal dependent crops, like cassava (Howeler et al., 1982). Inoculation is more feasible when targeted to responsive situations. The likelihood of worthwhile responses to inoculation will be greatest with hosts of high mycorrhizal dependency on sites of low P status and inadequate natural VAM densities. It is also possible that some sites which contain poorly effective strains of VAM fungi would be responsive to inoculation with more effective strains (Abbott and Robson, 1982). One scientific breakthrough on how VAM inoculum could be mass-produced cheaply could dramatically change our perspective of inoculation in agriculture.

References

Abbott L K and Robson A D 1982 The role of vesicular arbuscular mycorrhizal fungi in agriculture and the selection of fungi for inoculation. Aust. J. Agric. Res. 33, 389–408.

Abbott L K and Robson A D 1984 The effect of mycorrhiza on plant growth. *In* VA Mycorrhizae. Eds. C D Powell and D J Bagyaraj. pp 113–130. CRC Press, Boca Raton, Florida.

Abbott L K and Robson A D 1991 Field management of VA mycorrhizal fungi *In* The Rhizosphere and Plant Growth. Eds. D L Keister and P B Cregan. pp 355–362. Kluwer Academic Publishers, Dordrecht, The Netherlands.

Amijee F, Tinker P B and Stribley D P 1989 The development of endomycorrhizal root systems. VII. A detailed study of effects of soil phosphorus on colonization. New Phytol. III, 435–446.

Anderson E L, Miller P D and Kunishi H M 1987 Maize root length density and mycorrhizal infection as influenced by tillage and soil phosphorus. J. Plant Nutr. 10, 1349–1356.

Azcon R and Ocampo J A 1981 Factors affecting the vesicular-arbuscular infection and mycorrhizal dependency of thirteen wheat cultivars. New Phytol. 87, 677–685.

Bayliss G T S 1975 The magnolioid mycorrhiza and mycotrophy in root systems derived from it. *In* Endomycorrhizas. Eds. F E Sander, B Mosse and P B Tinker. pp 373–389. Academic Press, London and New York.

Bertheau Y, Gianinazzi-Pearson V and Gianinazzi S 1980 Dèveloppement et expression de l'association endomycorhizienne chez le blè. I. Mise en èvidence d'un effet variètal. Ann. Ameliorante Planta 30, 67–78.

Black R and Tinker P B 1979 The development of endomycorrhizal root systems. II. Effect of agronomic factors and soil conditions on the development of vesicular-arbuscular mycorrhizal infection in barley and on endophyte spore density. New Phytol. 83, 401–413.

Bolan N S, Robson A D and Barrow N J 1984 Increasing phosphorus supply can increase the infection of plant roots by vesicular-arbuscular mycorrhizal fungi. Soil Biol. Biochem. 16, 419–420.

Busse M E and Ellis J R 1987 VAM effect on atrazine uptake by soybeans. *In* Mycorrhizae in the Next Decade: Practical Applications and Research Priorities, Proc. 7th NACOM. Eds. D M Sylvia, L L Hung and J H Graham. p 242, Institute of Food and Agricultural Sciences, University of Florida, Gainesville, Florida.

Daft M J 1991 Influences of genotypes, rock phosphate and plant densities on mycorrhizal development and the growth responses of five different crops. Agric. Ecos. Environ. 35, 151–169.

Evans D G and Miller M H 1988 Vesicular-arbuscular mycorrhizas and the soil-disturbance-induced reduction of nutrient absorption in maize. I. Causal relations. New Phytol. 110, 67–74.

Evans D G and Miller M H 1990 The role of the external mycelial network in the effect of soil disturbance upon vesicular-arbuscular mycorrhizae colonization of maize. New Phytol. 114, 65–71.

Fairchild G L and Miller M H 1988 Vesicular-arbuscular mycorrhizas and the soil-disturbance-induced reduction of nutrient absorption in maize. II Development of the effect. New Phytol. 110, 75–84.

Fitter A H and Nichols R 1988 The use of benomyl to control infection by vesicular-arbuscular mycorrhizal fungi. New Phytol. 110, 201–206.

Glenn M G, Chew F S and Williams P H 1988 Influence of glucosinolate content of *Brassica* (Cruciferae) roots on growth of vesicular-arbuscular mycorrhizal fungi. New Phytol. 110, 217–225.

Govinda Rao Y S, Bagyaraj D J and Rai P V 1983 Selection of an efficient VA mycorrhizal fungus for finger millet. Zentralbl Mikrobiol. 138, 415–419.

Habte M 1989 Impact of simulated erosion on the abundance and activity of indigenous vesicular-arbuscular mycorrhizal endophytes in an Oxisol. Biol. Fert. Soils 7, 164–167.

Hayman D S 1982 Influence of soils and fertility on activity and survival of vesicular-arbuscular mycorrhizal fungi. Phytopathology 72, 1119–1125.

Hepper C M 1983 The effect of nitrate and phosphate on the vesicular-arbuscular mycorrhizal infection of lettuce. New Phytol. 92, 389–399.

Howeler R H, Cadavid L F and Burckhardt E 1982 Response of cassava to VA mycorrhizal inoculation and phosphorus application in greenhouse and field experiments. Plant Soil 69, 327–339.

Ilag L I, Rosales A M, Elazegui F A and Mew T W 1987 Changes in the population of infective endomycorrhizal fungi in a rice-based cropping system. Plant Soil 103, 67–73.

Johnson N C and Pfleger F L 1992 VA mycorrhizae and cultural stresses *In* Mycorrhizae in Sustainable Agriculture. Ed. G J Bethlenfalvay. Wisconsis, Madison, USA.

Khan A G 1974 The occurrence of mycorrhizas in halophytes, hydrophytes and xero-phytes, and of *Endogone* spores in adjacent soils. J. Gen. Microbiol. 81, 7–14.

Klopatek C C, Debano L F and Klopatek J M 1988 Effects of simulated fire on vesicular-arbuscular mycorrhizae in pinyon-juniper woodland soil. Plant Soil 109, 245–249.

Krishna K R, Shelty K E, Dart P J and Andrews D J 1985 Genotype dependent variation in mycorrhizal colonization and response to inoculation of pearl millet. Plant Soil 86, 113–125.

Manske G G B 1990 Genetical analysis of the efficiency of VA mycorrhiza with spring wheat. Agric. Ecos. Environ. 29, 273–280.

McGonigle T P and Miller M H 1992 Effect of disturbance of unpasteurized, pasteurized and of pasteurized and inoculated soil on P absorption by maize. Abstracts, International Symposium on Management of Mycorrhizas in Agriculture, Horticulture and Forestry, University of Western Australia, Perth. p 94.

McGonigle T P, Evans D G and Miller M H 1990 Effect of degree of soil disturbance on mycorrhizal colonization and phosphorus absorption by maize in growth chamber and field experiments. New Phytol. 116, 629–636.

Menge J A 1982 Effect of soil fumigants and fungicides on vesicular-arbuscular fungi. Phytopathology 72, 1125–1132.

Mercy M A, Shivashankar G and Bagyaraj D J 1990 Mycorrhizal colonization in cowpea is host dependent and heritable. Plant Soil 121, 292–294.

Morley C D and Mosse B 1976 Abnormal vesicular-arbuscular mycorrhizal infections in white clover induced by lupin. Trans. Br. Mycol. Soc. 67, 510–513.

Mosse B and Thompson J P 1984 Vesicular-arbuscular endomycorrhizal inoculum production. I. Exploratory experiments with beans (*Phaseolus vulgaris*) in nutrient flow culture. Can. J. Bot. 62, 1523–1530.

Mulligan M F, Smucker A J M and Safir G F 1985 Tillage modifications of dry edible bean root colonization by VAM fungi. Agron. J. 77, 140–144.

Nelson S D and Khan S U 1992 Uptake of atrazine by hyphae of *Glomus* vesicular-arbuscular mycorrhizae and root systems of maize (*Zea mays* L.). Weed Sci. 40, 161–170.

Newman E I and Reddell P 1987 The distribution of mycorrhizas among families of vascular plants. New Phytol. 106, 745–751.

Ocampo J A, Martin J and Hayman D S 1980 Influence of plant interactions on vesicular-arbuscular mycorrhizal infections. I. Host and non-host plants grown together. New Phytol. 84, 27–35.

Ojala J C, Jarrell W M, Menge J A and Johnson E L V 1983 Comparison of soil phosphorus extractants as predictors of mycorrhizal dependency. Soil Sci. Soc. Am. J. 47, 958–962.

Peck D M, Thompson J P and Haak M I 1992 Effects of conservation tillage practices on vesicular-arbuscular mycorrhizae build-up after a long fallow. Abstracts, International Symposium on Management of Mycorrhizas in Agriculture, Horticulture and Forest, University of Western Australia, Perth. p 130.

Plenchette C, Fortin J A and Furlan V 1983 Growth responses of several plant species to mycorrhizae in a soil of moderate P-fertility. I. Mycorrhizal dependency under field conditions. Plant Soil 70, 199–209.

Porter W M, Robson A D and Abbott L K 1987 Field survey of the distribution of vesicular-arbuscular mycorrhizal fungi in relation to soil pH. J. Appl. Ecol. 24, 659–662.

Raju P S, Clarke R B, Ellis J R, Duncan R R and Marranville J W 1990 Benefit and cost analysis and phosphorus efficiency of VA mycorrhizal fungi colonizations with sorghum (*Sorghum bicolor*) genotypes grown at varied phosphorus levels. Plant Soil 124, 199–204.

Salt G A 1977 A survey of fungi in cereal roots in Rothamsted, Woburn and Saxmundham, 1970–75. Rothamsted Report for 1976, Part 2. pp 153–168.

Schwab S M 1987 Considerations of vesicular-arbuscular mycorrhiza physiology in breeding for enhanced mineral uptake by plants. *In* Genetic Aspects of Plant Mineral Nutrition. Eds. H W Gabelman and B C Loughman. pp 603–615. Martinus Nijhoff, Dordrecht.

Smith T F 1978 Some effects of crop protection chemicals on the distribution and abundance of vesicular-arbuscular endomycorrhizas. J. Aust. Inst. Agric. Sci. 44, 82–88.

Stoppler H, Kolsch E and Vogtmann H 1990 Vesicular-arbuscular mycorrhiza in varieties of winter wheat in a low external input system. Biol. Agric. Hort. 7, 191–199.

Struble J E and Skipper H D 1988 Vesicular-arbuscular mycorrhizal fungal spore production as influenced by plant species. Plant Soil 109, 277–280.

Sylvia D M and Neal L H 1990 Nitrogen affects the phosphorus response of VA mycorrhiza. New Phytol. 115, 303–310.

Thiagalingam K, Thompson J P and Watson P 1988 Preliminary investigation of the presence of vesicular-arbuscular mycorrhizae (VAM) in a heavy and a light (Oolloo) Tippera soil. Northern Territory Department of Industries and Development, Seasonal Report 1987–1988. 7 p.

Thompson J P 1986 Soilless culture of vesicular-arbuscular mycorrhizae of cereals: Effects of nutrient concentration and nitrogen source. Can. J. Bot. 64, 2282–2294.

Thompson J P 1987 Decline of vesicular-arbuscular mycorrhizae in long fallow disorder of field crops and its expression in phosphorus deficiency of sunflower. Aust. J. Agric. Res. 38, 847–867.

Thompson J P 1990 Soil sterilization methods to show VA-mycorrhizae aid P and Zn nutrition of wheat in vertisols. Soil Biol. Biochem. 22, 229–240.

Thompson J P 1991 Improving the mycorrhizal condition of the soil through cultural practices and effects on growth and phosphorus uptake by plants. *In* Phosphorus Nutrition of Grain Legumes in the Semi-Arid Tropics. Eds. G Johansen, K K Lee and K L Sahrawat. pp 117–137. ICRISAT, Patancheru, A P, India.

Thompson J P and Wildermuth G B 1989 Colonization of crop and pasture species with vesicular-arbuscular mycorrhizal fungi and a negative correlation with root infection by *Bipolaris sorokiniana*. Can. J. Bot. 67, 687–693.

Toth R, Toth D, Starke D and Smith D R 1990 Vesicular-arbuscular mycorrhizal colonization in *Zea mays* affected by breeding for resistance to fungal pathogens. Can. J. Bot. 68, 1039–1044.

Trinick M J 1977 Vesicular-arbuscular infection and soil phosphorus utilization in *Lupinus* spp. New Phytol. 78, 297–304.

Wang G M, Stribley D P, Tinker P B and Walker C 1985 Soil pH and vesicular-arbuscular mycorrhizas. *In* Ecological Interactions in Soil. Ed. A F Fitter, pp 219–224, Blackwell Scientific Publications, Oxford.

Wellings N P and Thompson J P 1993 Effects of VAM and P fertilizer rate on Zn fertilizer requirement of linseed. *In* Proceedings Second Asian Conference on Mycorrhizae, Chiang Mai, Thailand (*in press*).

Yost R S and Fox R L 1979 Contribution of mycorrhizae to P nutrition of crops growing on an oxisol. Agron. J. 71, 903–908.

The potential for management of ectomycorrhiza in forestry

T.S. GROVE and N. MALAJCZUK
CSIRO, Division of Forestry, Private Bag, P.O. Wembley, WA 6014, Australia

Key words: conifers, *Eucaluptus*, host–fungus specificity, inoculation, plantation

Abstract

Ectomycorrhizal fungi are the predominant mycobionts of commercially important forest species. There appears to be a large scope to manipulate these fungal symbionts to improve the establishment and early growth of trees. Management of ectomycorrhiza in forestry is favoured by the diverse range of inoculant fungi which can be readily cultured and handled, and by cultural practices in forest nurseries. However, forest managers must take account of the variable nature of host-fungus specificity and the broad range of cultural/environmental situations into which fungal symbionts are introduced. Inoculation procedures can be introduced relatively easily into nurseries, although changes to the plant rooting medium, fungicide use and nutrient application may be required to favour the development of ectomycorrhiza. In the field, increased tree growth due to inoculation with ectomycorrhizal fungi has been reported for a broad range of host species and for diverse habitats worldwide. However, in most forest regions we lack the information required to explain the variable nature of field responses and to identify responsive sites. The selection of fungal isolates for inoculation, and the management of these introduced fungi in the field, is hindered by a poor knowledge of the effects these fungi have on plant growth in relation to other silvicultural practices and to biotic and abiotic factors of the site.

Introduction

In contrast to the long history of improvements in agricultural crops, the relatively recent expansion of plantation forestry worldwide, combined with the long rotations required for tree crops, suggests that substantial gains in productivity can be made through improved forestry practices including inoculation with ectomycorrhizal fungi. The function of ectomycorrhiza in absorption of nutrients and growth of forest trees has been extensively researched and documented. There is substantial evidence from field studies that demonstrates the potential for improving tree survival and forest production by inoculating seedlings with specific ectomycorrhizal fungi. However, inoculation has yet to be widely adopted as part of standard silvicultural management. Significant advances in the management of ectomycorrhiza are required if we are to fully exploit the tree-fungus association for the increased productivity of forests.

In this paper we focus on the management of ectomycorrhiza for increasing the survival and growth of commercially important plantation trees. Although benefits of ectomycorrhiza to tree growth have been demonstrated, the potential to manage the association depends on the extent to which responses to inoculation can be expected in the field and on the intrinsic features of forest trees and forest practices which provides the scope for management. These aspects will therefore be considered before discussing management of ectomycorrhiza in forest nurseries and on forest sites, and indicating research directions which may lead to increased effectiveness and utilization of fungal inoculation practices.

A.D. Robson, L.K. Abbott & N. Malajczuk (eds.), Management of mycorrhizas in agriculture, horticulture and forestry, 201–210.

Responses to inoculation

Striking examples of the requirement for ectomycorrhizal fungi by forest trees were the failures of exotic pines introduced into new regions early this century (Kessell, 1927; Mikola, 1973). These early problems were overcome by introducing soil or organic matter from the species natural environment containing compatible ectomycorrhizal fungi. However, there are inherent risks with this method because of the possibility of introducing pathogenic fungi, fungi that are poisonous to man, or noxious weeds (Mikola, 1973). Furthermore, although the early problems with exotic conifers were remedied, there was little attempt to further improve growth through selecting more effective fungi from the diverse fungal flora associated with trees in their natural habitats.

The ability of a broad range of fungal species to colonize roots and enhance the growth of plantation trees in nurseries and in the field have been examined in numerous studies. Past research has concentrated on conifers, partly because of their economic importance in forest plantations but also because of their demonstrated dependence on ectomycorrhizal fungi. Over the past 15 years, for example, significant responses to inoculation have been reported for many nursery and field experiments in the southern United States (see Marx and Cordell, 1988) and in France (see Le Tacon et al., 1988; Grove and Le Tacon, 1993). Other studies are now indicating that responses to inoculation can be obtained in the field for a broad range of commercially important tree species and for diverse habitats throughout the world. For example, in recent studies the inoculation of seedlings of fast-growing eucalypts with specific fungi increased early tree growth (Malajczuk, 1987; Garbaye et al., 1988; De La Cruz et al., 1988; Grove et al., 1991a).

Responses to inoculation with ectomycorrhizal fungi are not confined to eroded and degraded lands and have been recorded at sites where propagules of ectomycorrhizal fungi are likely to be abundant in soils. To date, studies of the effects of inoculation on early tree growth have involved more than 500 specific fungal–host tree associations and nearly half of those have shown a positive effect of inoculation on growth (Castellano, personal communication). This may not represent a true picture, however, as it is likely that a disproportionate number of studies with positive effects may have been reported. On the other hand, at forest sites where there was no effect or a negative effect of inoculation, it is also possible that inoculation with an alternative fungal species more suited to the environment and host species may have produced a positive effect. An example of this is the low frequency of responses of conifers to inoculation with *Pisolithus tinctorius* in south-western Oregon where *Rhizopogon* spp. are better suited to the soils and environment and are more effective in promoting early growth (Castellano and Trappe, 1991).

Factors influencing management

There are potentially greater opportunities for management of ectomycorrhiza in plantation forestry than for symbiotic associations in other cropping systems. This is associated with the ability to handle ectomycorrhizal fungi in culture and with the characteristics of forest tree growth and forest practices which favour the use of inoculation. Isolates of ectomycorrhizal fungi can often be readily obtained from fruiting bodies and pure cultures of the fungi can be built up as a source of inoculum. Alternatively, for fungi which fruit profusely, basidiocarps or spores can be used for inoculation.

The potential for management of ectomycorrhiza is related to the significant degree of host–fungus specificity (Molina et al., 1992). Some fungal species are restricted to single or to a few host tree species. Others have a broad host range, although individual strains are generally most effective in colonizing roots and stimulating growth of tree species with which they were originally associated (Malajczuk et al., 1990). The specificity associated with ectomycorrhiza is probably one factor explaining the large biodiversity of ectomycorrhizal fungi which occurs within and between natural forest habitats. 'Early' and 'late' succession fungi which can dominate at different stages of stand development (Dighton and Mason, 1985; Gardner and Malajczuk, 1988) also contribute to this large diversity of ectomycorrhizal fungi. Even where soils contain infective propagules of ectomycorrhizal fungi, the large diversity of fungi and variation in host–fungus specificity provide the potential to improve growth by introduction of an inoculant fungus, particularly with exotic tree species.

A characteristic of forest stand development is that nutrient requirement for growth is greatest in the initial period of plantation establishment and this corresponds with the time when inoculant fungi most strongly colonize roots. Where ectomycorrhiza increase nutrient uptake and storage in young trees, they may help to extend tree growth into periods when uptake of nutrients and water by feeder roots is restricted by soil moisture availability.

A further feature of plantation forestry practice is that seedlings are grown in nursery beds or containers and are not exposed to indigenous fungi until outplanting. Thus, inoculant fungi established on roots in the nursery and selected on the basis of their effectiveness in improving seedling survival and growth in the field initially have a competitive advantage over natural fungi resident in field soils. Inoculation in nurseries also avoids delays in the colonization of roots by indigenous fungi which, in turn, may improve seedling survival and growth in regions where the period for establishment is restricted by the onset of summer drought or cold winters.

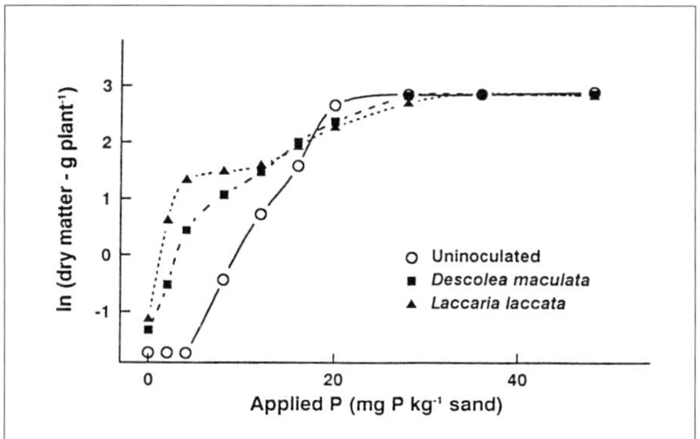

Fig 1. Dry matter production of 12 week-old uninoculated and inoculated *Eucalyptus diversicolor* seedlings in relation to levels of applied P. Seedlings were grown in steam-sterilized yellow sand in a temperature controlled glasshouse (after Bougher et al., 1990).

One question likely to determine the potential to manage ectomycorrhiza and to improve tree growth by inoculation is whether the supply of phosphate in soils is limiting to plant growth (Fig. 1; Grove et al., 1991a; Abbott and Robson, 1991). Forest plantations are often established on degraded lands or on soils of relatively low fertility not suitable for agriculture or horticulture, and these are situations which are likely to be most favourable for managing ectomycorrhiza.

Management in forest nurseries

Fungal inoculation is becoming more widely adopted as the efficiency of nursery practices increase and hygiene is improved (e.g. fumigation or pasteurization of container media). Species of ectomycorrhizal fungi such as *Thelephora terrestris* that commonly produce basidiocarps in or around forest nurseries often colonize the roots of seedlings. Although possibly less effective than inoculant fungi at increasing growth (e.g. Villeneuve et al., 1991), these fungi may still provide some benefit where inoculation practices are not in place. The incidence of such 'contaminant' fungi will be reduced by the use of containerized systems, by inoculation with selected fungi and by improvements in nursery hygiene.

The primary objective of inoculation in nurseries is to produce healthy seedlings suitable for planting into the field that have roots colonized with a selected fungus. Growth responses to inoculation in forest nurseries are generally of secondary importance although benefits can be obtained for slow growing conifers where the time required before seedlings are planted into the field can be significantly reduced (Le Tacon and Bouchard, 1986). Increased growth due to inoculation appear to be more common in bare-rooted seedlings than in containerized seedlings, possibly because ectomycorrhiza can increase P uptake more from the soils of nursery beds than from organic based media used in containerized nurseries (Grove and Le Tacon, 1993).

There is considerable flexibility in selecting the inoculant fungus, the form of inoculum, the method of applying inoculum, and the cultural conditions which are best suited to ectomycorrhiza formation and to seedling growth. The choices depend to some extent on the level of sophistication in particular nurseries. However, some form of inoculation is feasible for nurseries ranging from small scale labour-intensive nurseries to the large mechanized operations which have developed in some countries.

Form and application of inoculum

Forms of inoculum and developments in inoculation techniques have been extensively reviewed in other papers (e.g. Marx et al., 1982; Tommerup et al., 1987), and are not considered in detail here. Apart from natural soil and litter, a wide range of inoculum forms have been used and include spores, crushed sporophores and mycelium. Inoculation with spores has been widely practiced, as spores are relatively easily applied by mixing into nursery soils prior to planting or watering onto nursery beds and seedling containers. Spore pellets and spore coating of seeds has also been used in order to increase ectomycorrhiza formation (Theodorou and Bowen, 1973; Lamb and Richards, 1974; De La Cruz et al., 1988; Marx et al., 1989a). Major drawbacks to the use of spore inoculum are the restrictions to fungal species that produce abundant sporocarps, the poor germination or low viability of spores of some species and the large spore numbers often required for colonization of roots (Fries, 1984; Cordell et al., 1987). Advantages of mycelial forms of inoculum are that they are often more effective than spores in colonizing roots and they allow large-scale production of single strains of fungi (Cordell et al., 1987; Marx et al., 1989b). Mycelia are commonly contained in peat–vermiculite mixes, liquid media, or alginate beads. There is increasing interest in the use of mycelium encapsulated in alginate beads, although there are difficulties in beading some fungal species. Beaded forms of inoculum have the advantages of high effectiveness per unit volume of inoculum, the ability to store the inoculum for several months, protection of the living mycelium against adverse conditions until roots become colonized and suitability for mechanized planting systems such as fluid drills (Hardy et al., 1991; Kuek et al., 1992).

Plant growth conditions

Inoculant fungi are selected primarily for their effectiveness in colonizing and persisting on roots and increasing the growth of trees in the field and not in the forest nursery environment (Grove and Le Tacon, 1993; Dodd and Thomson, Chapter 15). When inoculation procedures are first introduced into forest nurseries, cultural practices must often be modified to produce conditions which are optimal for the development of ectomycorrhiza. As nursery systems are generally devised for greatest efficiency (e.g. minimization of space, labour costs and material costs), and to provide uniform seedlings suited to field planting techniques, conditions for seedling growth often differ markedly between the nursery and field (Table 1).

In forest nurseries factors such as the composition, pH, moisture content and temperature of the rooting media, fumigation and pasteurization practices, and the use of biocides and fungicides can be readily manipulated to provide conditions which are optimal to both root colonization and seedling production (Marx et al., 1982; Cordell et al., 1991). For large-scale nursery operations there is generally less scope to control soil temperatures than to modify other environmental factors affecting the development of ectomycorrhiza. Well aerated container media or soils are required to ensure excessive soil moisture does not inhibit the formation of ectomycorrhiza (Bougher and Malajczuk, 1990) and some components of these media (e.g. sawdust, bark) may have fungistatic properties. Following fumigation or pasteurization of the media, dual inoculation with bacteria isolated from sporophores or ectomycorrhiza can stimulate ectomycorrhiza formation (Garbaye and Bowen, 1989; Duponnois and Garbaye, 1991), although this will add to the complexity and cost of ectomycorrhizal inoculation practices.

High rates of fertilizer, particularly those

Table 1. Common differences between containerized nursery seedlings and field trees in cultural and management factors which can affect ectomycorrhiza formation and growth responses to inoculation

	Nursery	Field
Climatic variables		
—Temperature	Often higher than field with greater fluctuations	Moderated by soil depth and by vegetation and organic matter cover
—Soil Moisture	Controlled; adequately aerated media	Dependent on seasonal rainfall and soil texture
Growing medium	'Open' aerated organic based mixes—low P sorption capacity and heavily leached	Wide range of mineral soils which may be strongly P-sorbing and vary in pH
Nutrient application	Supply is often regulated through liquid feed systems to maximize infection	Less well controlled with fertilizer input at the time of planting
Environmental stress (e.g. salt, drought)	Minimized or absent	Site dependent
Pest/disease control	Controlled through use of biocides	Minimal control
Phytotoxic factors	Mostly absent	Allelopathic effects can arise from previous vegetation or from weed species
Microbial interactions	Minimal in pasteurized media—starter bacteria may be introduced	Stimulation or depression by microorganisms. Competition from other ectomycorrhizal fungi.

containing N and P, suppress the development of ectomycorrhiza on containerized seedlings (Beckjord et al., 1985; Shaw et al., 1987). Glasshouse studies have shown marked interactions between effects of inoculation and soil P supply on the development of ectomycorrhiza (Bougher et al., 1990; Jones et al., 1990). Growth responses to inoculation are greatest where P supply is sufficient for ectomycorrhizal seedlings to achieve about half of their maximum rate of growth (Fig. 1), and where fungal colonization is greatest. High concentrations of soil N also reduce colonization, with NO_3^--N often having a stronger inhibitory effect than NH_4^+-N (Theodorou and Bowen, 1969; Wallander and Nylund, 1991). Nitrogen concentrations that maximize the development of ectomycorrhiza have been defined for specific nursery conditions (Beckjord et al., 1985; Gagnon et al., 1988). The effects of P and N supply on colonization by ectomycorrhizal fungi are often linked to their effects on the concentration of soluble carbohydrates in roots, although complex interactions between a number of regulatory mechanisms are likely (Nylund, 1988; Grove and Le Tacon, 1993).

There is a lack of quantitative information on the extent that interactions among nutritional and non-nutritional factors affect root colonization and growth responses to inoculation. Nutrient studies have only examined effects of single nutrients, combined fertilizers containing fixed ratios of N and P (Ruehle and Wells, 1984) or interactions among the effects of N,P and K at single rates of addition (Menge et al., 1977; Newton and Pigott, 1991). P supply, N supply and fungal isolate each have a marked effect on root colonization of *Eucalyptus diversicolor* grown under controlled glasshouse conditions (Fig. 2), and there is a marked interaction between the effects of all three factors on seedling growth (Grove et al., 1991b). These results suggest that inconsistent findings in earlier studies of effects of the supply of single nutrients may be explained by interactions with other nutrients and emphasize that interactions

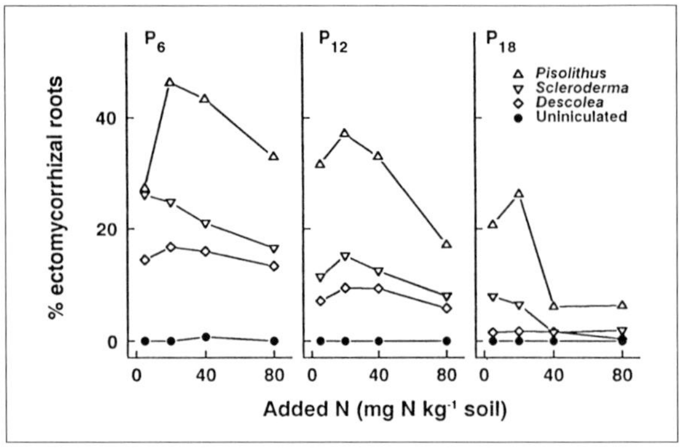

Fig 2. Effects of fungal inoculation and applied N and P on the percentage of ectomycorrhizal fine roots of *Eucalyptus diversicolor* seedlings growing in yellow sand in the glasshouse (Grove, Dell and Malajczuk, unpublished results).

must be more fully understood in devising strategies for the production of ectomycorrhizal seedlings in forest nurseries.

Management on forest sites

The characteristics of sites on which forest plantations are established vary widely. Sites include areas of degraded land, recently clear-felled forest and agricultural land with long histories of cultivation and fertilizer application. Consequently, there will be a large variation in the likelihood of responses to inoculation and the choice of appropriate management strategies to enhance effects of ectomycorrhizal fungi. Key factors in assessing the scope for management of ectomycorrhiza to increase tree growth are: the identification of sites where responses to inoculation are likely; determination of management procedures to maximize benefits from ectomycorrhiza fungi; and economic considerations which may have a bearing on management decisions.

Identifying responsive sites

For specific planting sites it is generally difficult to predict whether inoculation will increase tree growth. An ability to identify sites where trees will respond to inoculation would help to maximize benefits from inoculation programs, ensure fungal screening trials are located in suitable areas, and may facilitate management of naturally occurring fungi.

There are two factors that are of most significance in determining whether tree growth can be increased by inoculation on specific sites. The first of these is that responses to inoculation are most likely on sites where the soil supply of nutrients, particularly P, is the primary factor which limits growth. Under controlled glasshouse conditions, the improvement in growth is greatest where P supply to roots is moderately to severely limiting and there is no response to inoculation where P supply is adequate (Fig. 1; Bougher et al., 1990; Jones et al., 1990). It is expected that there will be a similar interaction between inoculation and P supply in the field, although this aspect has received little attention. An unexpected result was obtained in experiments on clear-felled eucalypt forest sites in south-western Australia. Growth was improved by inoculation on two of three sites and the responses occurred across a wide range of P rates (Fig 3; Grove et al., 1991a; Grove, Malajczuk, Burgess, unpublished report). The lack of interaction between effects of inoculation and soil P supply suggests that factors other than enhanced P uptake also contribute to the growth response to inoculation and points to the need for further field experiments.

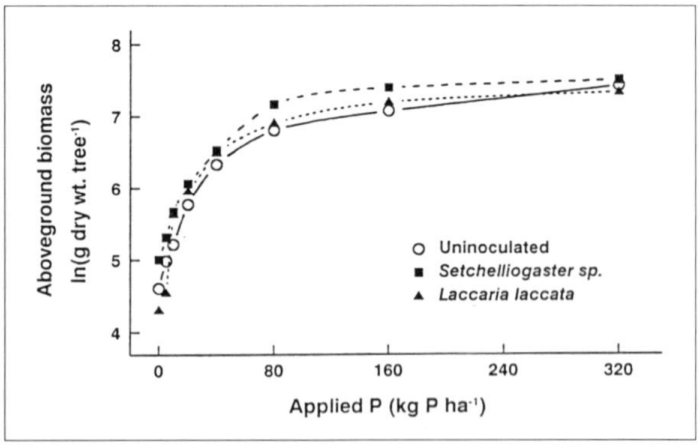

Fig 3. Dry matter production of uninoculated and inoculated *Eucalyptus globulus* one year after planting in relation to levels of applied P on a recently clearfelled eucalypt forest site in south-western Australia (after Grove et al., 1991a).

The second factor is the abundance of propagules of indigenous fungi and their effectiveness in colonizing roots. The influence of indigenous fungi will vary markedly between sites and depends particularly on previous land use. Responses to inoculation are likely to be greatest where there are low populations of indigenous fungi, or where the indigenous fungi are incompatible with the host tree.

Soils on clear-cut forest sites generally contain abundant propagules of a diverse range of fungal species associated with plants in the original forest, although propagules of a few species may dominate (Perry et al., 1982; Dahlberg and Stenström, 1991). Perry et al. (1987) suggested that major influences on the formation of ectomycorrhiza by these indigenous fungi during reforestation are the temporal changes in the types and numbers of propagules, the regeneration of non-crop host plants and the extent of fungal diversity. High background numbers of indigenous fungi are generally expected to reduce the likelihood of responses to inoculation. However in studies with Douglas fir in the USA (Castellano et al., 1985) and France (Le Tacon et al., 1988; Villeneuve et al., 1991), and with eucalypts in south-western Australia (Grove et al., 1991a) inoculation of seedlings with selected fungi significantly increased tree growth on recently clear-cut sites. In some cases increased growth was obtained with the same tree species that had previously occupied the site, suggesting that the indigenous fungi may be dominated by 'late-succession' species which are less effective than inoculant fungi from the nursery.

On agricultural lands there is likely to be a lesser diversity and abundance of indigenous ectomycorrhizal fungi, and propagule numbers in the soil will depend on the quantity and survival of wind-blown spores originating from nearby vegetation (Grove and Le Tacon, 1993). Fungal species will probably be restricted to those associated with this vegetation and to those producing basidiocarps from which spores are readily disseminated. There is a lack of information on the diversity, abundance and spatial distribution of propagules or of their effectiveness in colonizing roots on agricultural lands and other areas devoid of trees for long periods. Positive responses to inoculation have been obtained in France on agricultural soils, although there is generally a wide variation among sites (Le Tacon et al., 1988). Our studies on pasture lands in south-western Australia are indicating that responses of *Eucalyptus globulus* to inoculation may be smaller than those on forest sites. This may be due to high concentrations of available nutrients in soils (Chu-Chou and Grace, 1987) or to strong competition from indigenous fungi (Dodd and Thomson, Chapter 15).

Measurements of the abundance, diversity and infectivity of indigenous ectomycorrhizal fungi in soils using various bioassay techniques (Perry et al., 1982; Parke et al., 1984; McAfee and Fortin, 1986; Warcup, 1991) can be useful in determining whether indigenous fungi reduce responses to inoculation. Although these techniques will be valuable in gaining a better understanding of the infectivity of indigenous fungi and how they interact with introduced fungi, it is unlikely that they would be used on a broad scale in the management of fungal symbionts. On the other hand, site classification systems which can be used as a guide to fertilizer requirements (e.g. Turvey, 1980) may be useful in identifying sites where soil P supply is poor and a positive response to inoculation is most likely.

Silvicultural practices

The largest effects of inoculation on the host are expected at an early stage of tree growth when there is a high nutrient requirement in developing tissues, a high dependence on the external nutrient supply for growth and when introduced fungi are most likely to colonize the major portion of fine roots (Grove and Le Tacon, 1993). It is at this initial stage that silvicultural practices will exert their largest effects on ectomycorrhiza.

During the establishment phase of plantation forests, site preparation, transplant shock, and fertilizer practices may all influence the effectiveness of inoculation in increasing growth. On forest sites, the disturbance to surface soils caused by clearfelling and site preparation generally reduces numbers of indigenous fungal propagules and the formation of ectomycorrhiza (Harvey et al., 1980; Perry et al., 1982; Parke et al., 1984). Spore numbers and viability are likely to be less reduced than the numbers of vegetative propagules which will decline with delays in reforestation (Perry et al., 1987). Burning of logging residues can markedly alter the species composition of fungi which first colonize seedlings (Warcup, 1991) as well as significantly reducing inoculum in the soil (Harvey et al., 1980; Parke et al., 1984). Knowledge of the effects of disturbance has been used to suggest management strategies which can minimize the loss of infective propagules on forest sites (Harvey et al., 1980; Perry et al., 1987).

Greater attention is now being directed to gaining a greater understanding of the effects of site, site preparation methods and planting practices on the numbers and diversity of ectomycorrhiza (e.g. Dahlberg and Stenström, 1991). Emphasis may be given in future to retaining indigenous fungal populations in regrowth forest where a high priority is given to maintaining ecosystem biodiversity, and in production forestry where the available inoculant fungi are not suited to particular sites or where effective inoculation practices are not in place. Alternatively, in intensively managed plantation forests it may be more desirable to develop practices which give introduced fungi a competitive advantage over indigenous fungi.

Physiological stresses and physical damage to roots when nursery seedlings are outplanted may have marked effects on the ability of inoculant fungi to colonize new roots and affect early tree growth. Methods to minimize deleterious effects of transplanting on ectomycorrhiza established in the nursery have been described by Marx and Hatchell (1986) and Cordell et al. (1991).

Fertilizer practices can be expected to have large effects on tree–fungus associations and to affect tree growth responses to inoculation. Although effects of soil P and N supply on root colonization and host growth have been quantified for seedlings growing under glasshouse and nursery conditions (e.g. Figs. 1, 2), there have been few comparable studies on field sites. High rates of P and N fertilizers suppress ectomycorrhiza development in the field (Menge et al., 1977; Newton and Pigott, 1991) and high concentrations of soil N can also reduce the number and relative abundance of different ectomycorrhiza types (Alexander and Fairley, 1983).

Despite extensive field experimentation examining effects of specific ectomycorrhizal fungi on the survival and growth of forest trees, there is a dearth of information on the effects of fertilizer type, rate, form, placement and timing on the development of ectomycorrhiza and on growth responses to inoculation. As many factors can be varied, there is probably scope to select fertilizer practices which are optimal for tree growth but which are also favourable for the colonization of roots by introduced fungi. The persistence of, and continuing benefits from, the fungi may be of greatest importance in soils where the effects of applied fertilizers are shortlived.

Conclusions

The improved growth of forest trees after inoculation with specific fungi has been demonstrated on a wide range of forest sites. These responses, and features of tree–fungal associations, forest environments and forestry practices suggest that benefits can be obtained on many forest sites through the management of ectomycorrhiza. Many options are available to establish ectomycorrhiza seedlings in forest nurseries by inoculation with specific fungi. Inoculation practices of some form can be used in most forest nurseries, irrespective of the scale of operation and level of sophistication. Successful inoculation may require some changes to the growing media, and to biocide and fertilizer application.

There is a much greater knowledge of how to manage ectomycorrhiza in forest nurseries than in the field. The effectiveness of inoculant fungi in colonizing roots and increasing the early growth of trees on forest sites is likely to depend mostly on their ability to persist and compete with indigenous fungi and on whether soil P supply is limiting tree growth. The ability to manage ectomycorrhiza in the field depends on improving our ecological understanding of associations of tree roots with indigenous and introduced fungi on forest sites and on increasing our knowledge of the effectiveness of inoculant fungi in relation to site preparation and tree establishment practices. It may then be possible to modify forestry practices to increase the benefits from ectomycorrhizal fungi while still achieving the other primary objectives of forest managers.

References

Abbott L K and Robson A D 1991 Field management of VA mycorrhizal fungi. *In* The Rhizosphere and Plant Growth. Eds. D L Keister and P B Cregan. pp 355–362. Kluwer Academic Publishers, Netherlands.

Alexander I J and Fairley R I 1983 Effects of N fertilization on populations of fine roots and mycorrhizas in spruce humus. Plant Soil 71, 49–53.

Beckjord P R, Melhuish J H Jr. and McIntosh M S 1985 Effects of nitrogen and phosphorus fertilization on growth and formation of ectomycorrhizae of *Quercus alba* and *Q. rubra* seedlings by *Pisolithus tinctorius* and *Scleroderma auranteum*. Can. J. Bot. 63, 1677–1680.

Bougher N L and Malajczuk N 1990 Effects of high soil moisture on formation of ectomycorrhizas and growth of karri (*Eucalyptus diversicolor*) seedlings inoculated with *Descolea maculata*, *Pisolithus tinctorius* and *Laccaria laccata*. New Phytol. 114, 87–91.

Bougher N L, Grove T S and Malajczuk N 1990 Growth and phosphorus acquisition of karri (*Eucalyptus diversicolor* F. Muell.) seedlings inoculated with ectomycorrhizal fungi in relation to phosphorus supply. New Phytol. 114, 77–85.

Castellano M A and Trappe J M 1991 *Pisolithus tinctorius* fails to improve plantation performance of inoculated conifers in southwestern Oregon. New For. 5, 349–358.

Castellano M A, Trappe J M and Molina R 1985 Inoculation of container-grown Douglas-fir seedlings with basidiospores of *Rhizopogon vinicolor* and *R. colossus*: effects of fertility and spore application rate. Can. J. For. Res. 15, 10–13.

Chu-Chou M and Grace L J 1987 Mycorrhizal fungi of *Pinus radiata* planted on farmland in New Zealand. N. Z. J. For. Sci. 17, 76–82.

Cordell C E, Marx D H, Maul S B and Owen J H 1987 Production and utilization of ectomycorrhizal fungal inoculum in the eastern United States. *In* Mycorrhizae in the Next Decade: Practical Applications and Research Priorities, Proc. 7th NACOM. Eds. D M Sylvia L L Hung and J H Graham. pp 287–289. Institute of Food and Agricultural Science, University of Florida, Gainesville, Florida.

Cordell C E, Marx D H and Caldwell C 1991 Operational application of specific ectomycorrhizal fungi in mineland reclamation. *In* Proc. National Meeting of the American Society for Surface Mining and Reclamation Durango p 8.

Dahlberg A and Stenström E 1991 Dynamic changes in nursery and indigenous mycorrhiza of *Pinus sylvestris* seedlings planted out in forest and clearcuts. Plant Soil 136, 73–86.

De La Cruz R E, Bartolome HT and Aggangan N S 1988 Pilot testing of mycorrhizal tablets for pines and *Eucalyptus* in the Philippines. *In* Proc. UNESCO Regional Workshop on Development and Production of Mycorrhizal Inoculants. Biotech, UPLB College Lagunn, Philippines.

Dighton J and Mason P A 1985 Mycorrhizal dynamics during forest tree development. *In* Developmental biology of higher fungi. Eds. D Moore L A Casselton D A Wood and J C Frankland. pp 117–139, Cambridge University Press, Cambridge.

Duponnois R and Garbaye J 1991 Effect of dual inoculation of Douglas fir with the ectomycorrhizal fungus *Laccaria laccata* and mycorrhization helper bacteria (MHB) in two bare-root forest nurseries. Plant Soil 138, 169–176.

Fries N 1984 Spore germination in the higher basidiomycetes. Proc. Indian Acad. Sci. 93, 205–222.

Gagnon J, Langlois C G and Fortin J A 1988 Growth and ectomycorrhiza formation of containerized black spruce seedlings as affected by nitrogen fertilization, inoculum type, and symbiont. Can. J. For. Ecol. 18, 922–929.

Garbaye J and Bowen G D 1989 Stimulation of ectomycorrhizal infection of *Pinus radiata* by some microorganisms associated with the mantle of ectomycorrhizas. New Phytol. 112, 383–388.

Garbaye J, Delwaulle J C and Diangana D 1988 Growth response of Eucalypts in the Congo to ectomycorrhizal inoculation. For. Ecol. Manag. 24, 151–157.

Gardner J H and Malajczuk N 1988 Recolonization of rehabilitated bauxite mine sites in Western Australia by mycorrhizal fungi. For. Ecol. Manag. 24, 27–42.

Grove T S and Le Tacon F 1993 Mycorrhiza in plantation forestry. *In* Mycorrhiza Synthesis. Advances in Plant Pathology Vol 9. Ed. I C Tommerup. pp 191–227 Academic Press, London.

Grove T S, Malajczuk N, Burgess T, Thomson B D and Hardy G 1991a Growth responses of plantation eucalypts to inoculation with selected ectomycorrhizal fungi. *In* The Role of Eucalypts. Proc. IUFRO Symposium Intensive Forestry, Durban, South Africa. pp 86–93.

Grove T S, Malajczuk N and Dell B 1991b Effects of applied nitrogen and phosphorus on ectomycorrhizal formation, plant growth and nutrient uptake in *Eucalyptus diversicolor* seedlings. *In* Productivity in Perspective. Proc. Third Aust. For. Soils and Nutri. Conf., Melbourne. Ed. P J Ryan. p 209. For. Comm. NSW, Sydney.

Hardy G E St J, Thomson B D, Malajczuk N and Grove T S 1991 Commercial inoculation of eucalypts with ectomycorrhizal fungi. *In* Mycorrhizas in Ecosystems—Structure and Function, Abstracts 3rd ECM, Sheffield.

Harvey A E, Jurgensen M F and Larsen M J 1980 Clearcut harvesting and ectomycorrhizae: survival of activity on residual roots and influence on a bordering forest stand in western Montana. Can. J. For. Res. 10, 300–303.

Jones M D, Durall D M and Tinker P B 1990 Phosphorus relationships and production of extramatrical hyphae by two types of willow ectomycorrhizas at different soil phosphorus levels. New Phytol. 115, 259–267.

Kessell S L 1927 Soil organisms. The dependence of certain pine species on a biological soil factor. Emp. For. J. 6, 70–74.

Kuek C, Tommerup I C and Malajczuk N 1992 Hydrogel bead inocula for the production of ectomycorrhizal eucalypts for plantations. Mycol. Res. 96, 273–277.

Lamb R J and Richards B N 1974 Inoculation of pines with mycorrhizal fungi in natural soils. Effects of density and time of application of inoculum and phosphorus amendment on seedlings yield. Soil Biol. Biochem. 6, 173–177.

Le Tacon F and Bouchard D 1986 Effects of different ectomycorrhizal fungi on growth of larch, Douglas fir, Scots pine and Norway spruce seedlings in fumigated nursery soil. Acta Ocological/Ocologia Applicata 7, 389–402.

Le Tacon F, Garbaye J, Bouchard D, Chevalier G, Olivier J M, Guimberteau J, Poitou N and Frochot, N 1988 Field results from ectomycorrhizal inoculation in France. *In* Canadian Workshop on Mycorrhizae in Forestry. Eds. M Lalonde and Y Piché. pp 51–74. Universite Lavel, Ste-Foy.

McAfee B J and Fortin J A 1986 Comparative effects of the soil microflora on ectomycorrhizal inoculation of conifer seedlings. New Phytol. 108, 108–443.

Malajczuk N 1987 Ecology and management of ectomycorrhizal fungi in regenerating forest ecosystems in Australia. *In* Mycorrhizae in the Next Decade—Practical Applications and Research Priorities. Eds. D M Sylvia, L L Hung and J H Graham. pp 118–120. University of Florida, Gainesville.

Malajczuk N, Lapeyrie F and Garbaye J 1990 Infectivity of pine and eucalypt isolates of *Pisolithus tinctorius* on roots of *Eucalyptus urophylla* in vitro I. Mycorrhiza formation in model systems. New Phytol. 114, 627–631.

Marx D H and Cordell C E 1988 Specific ectomycorrhizae improve reforestation and reclamation in eastern United States. *In* Canadian Workshop on Mycorrhizae in Forestry. Eds. M Lalonde and Y Piché. pp 75–85. Universite Lavel, Ste-Foy.

Marx D H and Hatchell G E 1986 Root stripping of ectomycorrhizae decreases field performance of loblolly and longleaf pine seedlings. Southern J. Appl. For. 10, 173–179.

Marx D H, Ruehle J L, Kenney D S, Cordell C E, Riffle J W, Molina R J, Pawuk W H, Navratil S, Tinus R W and Goodwin O C 1982 Commercial vegetative inoculum of *Pisolithus tinctorius* and inoculation techniques for development of ectomycorrhizae on container-grown tree seedlings. For. Sci. 28, 373–400.

Marx D H, Cordell C E, Maul S B and Ruehle J L 1989a Ectomycorrhizal development on pine by *Pisolithus tinctorius* in bare-root and container seedling nurseries II. Efficacy of various vegetative and spore inocula. New For. 3, 57–66.

Marx D H, Cordell C E, Maul S B and Ruehle J L 1989b Ectomycorrhizal development on pine by *Pisolithus tinctorius* in bare-root and container seedling nurseries I. Efficacy of various vegetative inoculum formulations. New For. 3, 45–56.

Menge J A, Grand L F and Haines L W 1977 The effect of fertilization on growth and mycorrhizae numbers in 11-year-old loblolly pine plantations. For. Sci. 23, 37–44.

Mikola P 1973 Application of mycorrhizal symbiosis in forestry practice. *In* Ectomycorrhizae—Their Ecology and Physiology. Eds. G C Marks and T T Kozlowski. pp 383–411. Academic Press, New York.

Molina R, Massicotte H and Trappe J M 1992 Specificity phenomena in mycorrhizal symbiosis: Community-ecological consequences and practical implications. *In* Mycorrhizal functioning: an integrative plant–fungal process. Ed. M F Allen. pp 357–423. Chapman and Hall, New York.

Newton A C and Pigott C D 1991 Mineral nutrition and mycorrhizal infection of seedling oak and birch II. The effect of fertilizers on growth, nutrient uptake and ectomycorrhizal infection. New Phytol. 117, 45–52.

Nylund J-E 1988 The regulation of mycorrhiza formation-carbohydrate and hormone theories reviewed. Scand. J. For. Res. 3, 465–479.

Parke J L, Linderman R G and Trappe J M 1984 Inoculum potential of ectomycorrhizal fungi in forest soil from southwest Oregon and Northern California. For. Sci. 30, 300–304.

Perry D A, Meyer M M, Egeland D, Rose S L and Pilz D 1982 Seedling growth and mycorrhizal formation in clearcut and adjacent undisturbed soils in Montana: a greenhouse bioassay. For. Ecol. Manag. 4, 261–273.

Perry D A, Molina R and Amaranthus M P 1987 Mycorrhizae, mycorhizospheres, and reforestation: current knowledge and research needs. Can. J. For. Res. 17, 929–940.

Ruehle J L and Wells C G 1984 Development of *Pisolithus tinctorius* ectomycorrhizae on container-grown pine seedlings as affected by fertility. For. Sci. 30, 1010–1016.

Shaw C G III, Jackson R M and Thomas G W 1987 Fertilizer levels and fungal strain influence the development of ectomycorrhizae on Sitka spruce seedlings. New For. 3, 215–223.

Theodorou C and Bowen G D 1969 The influence of pH and nitrate on mycorrhizal associations of *Pinus radiata* D. Don. Aust. J. Bot. 17, 59–67.

Theodorou C and Bowen G D 1973 Inoculation of seeds and soil with basidiospores of mycorrhizal fungi. Soil Biol. Biochem. 5, 765–771.

Tommerup I C, Kuek C and Malajczuk N 1987 Ectomycorrhizal inoculum production and utilization in Australia. *In* Mycorrhizae in the Next Decade: Practical Applications and Research Priorities, Proc. 7th NACOM. Eds. D M Sylvia, L L Hung and J H Graham. pp 293–295. Institute of Food and Agricultural Science, University of Florida, Gainesville, Florida.

Turvey N D 1980 A forest soil survey: The application of soil survey information to forestry operations. Aust. For. 43, 172–177.

Villeneuve N, Le Tacon F and Bouchard D 1991 Survival of inoculated *Laccaria bicolor* in competition with native ectomycorrhizal fungi and effects on the growth of outplanted Douglas-fir seedlings. Plant Soil 135, 95–107.

Wallander H and Nylund J-E 1991 Effects of excess nitrogen on carbohydrate concentration and mycorrhizal development of *Pinus sylvestris* L. seedlings. New Phytol 119, 405–411.

Warcup J H 1991 The fungi forming mycorrhizas on eucalypt seedlings in regeneration coupes in Tasmania. Mycol Res 95, 329–332.

Management of mycorrhizas in revegetation

D.A. JASPER
Soil Science and Plant Nutrition, School of Agriculture, The University of Western Australia, Nedlands, WA 6009, Australia

Key words: disturbed soils, effectiveness, infectivity, inoculation, mycorrhizal fungi, revegetation, soil management

Abstract

Revegetation of disturbed soils should aim to re-establish a stable, ecosystem with fully-functioning nutrient-cycling processes. Mycorrhizal fungi are likely to be important in helping achieve that aim by increasing nutrient uptake by plants and contributing to the restoration of soil structure. Yet the populations of mycorrhizal fungi are often reduced in soils to be revegetated, due in particular to the absence of host plants, soil disturbance and storage of soils in stockpiles.

Mycorrhizal fungi are likely to be most beneficial in revegetated ecosystems where the proportion of plants capable of forming mycorrhizas is high and nutrient deficiencies are an important limitation to plant growth. These fungi can make an important contribution to the success of revegetation through the following mechanisms:

i) enhancing establishment and growth of plants, by increasing nutrient uptake,
ii) maintaining diversity, by boosting the ability of host plants to compete for resources
iii) contributing to efficient recycling of nutrients and thus to long-term stability, and
iv) stabilizing the soil.

The keystone in the protocol to manage mycorrhizal fungi for optimum revegetation is to determine the infectivity and effectiveness of the mycorrhizal fungi that may be present in the soil to be revegetated. Further, the likely rate of increase in the infectivity of effective fungi would be crucial in determining the probable benefit from inoculation.

Where the timing and nature of ecosystem disturbance is predictable, then management of the disturbance may be possible, to maximize the survival of mycorrhizal fungi. Management strategies should focus on ameliorating the disturbance effects and on managing revegetation to optimize the recovery of the populations of mycorrhizal fungi.

Inoculation may be required if the populations of infective and effective mycorrhizal fungi are not adequate. Inoculation techniques that have been developed are most applicable to intensive plant production where precise mechanical placement of seed and inoculum is possible. Revegetation of disturbed soils is likely to differ from intensive plant industries in several important areas, each impacting on the selection of an inoculation technique.

If substantial changes are required to revegetation practices to allow inoculation of mycorrhizal fungi, then any increases in costs must first be clearly demonstrated to be justified in terms of increased survival, growth, and diversity of revegetation.

Introduction

Revegetation of disturbed soils should aim to re-establish a stable, ecosystem with fully-functioning nutrient-cycling processes. Mycorrhizal fungi are likely to be important in helping achieve that aim by increasing nutrient uptake by plants and contributing to the

A.D. Robson, L.K. Abbott & N. Malajczuk (eds.), Management of mycorrhizas in agriculture, horticulture and forestry, 211–219.

restoration of soil structure (Miller and Jastrow, 1991). Yet the populations of mycorrhizal fungi are often reduced in soils to be revegetated, due in particular to the absence of host plants, soil disturbance and storage of soils in stockpiles (Jasper et al., 1987).

Re-creating functioning natural ecosystems in disturbed soils or in near-sterile materials, such as those resulting from mining, is the most challenging of revegetation tasks. This is due to the wide diversity of highly-adapted plant species in most natural ecosystems, which in turn depend on an active and diverse population of organisms, including mycorrhizal fungi, for the efficient functioning of nutrient-cycling processes in the soil. Typically, in excess of 80% of plant species in natural ecosystems form mycorrhizal associations (Brundrett, 1991). This review will discuss the likely importance of mycorrhizal fungi in re-establishing natural ecosystems, and focus on strategies to maximize the populations of infective and effective fungi in the soil to be revegetated. A protocol for the management of mycorrhizal fungi in revegetation will be presented, and the need for further research will be highlighted.

The protocol (Fig. 1), which is an extension of that suggested by Abbott and Robson (1985) has two major steps:

i) to determine the importance of mycorrhizal fungi to the success of revegetation, and
ii) to assess the infectivity and effectiveness of mycorrhizal fungi in the soils to be revegetated.

Each assessment or course of action requires a level of scientific knowledge which in many cases has not been achieved to date. The adequacy of existing knowledge and the need for further research will be considered under the general headings from the protocol.

The importance of mycorrhizal fungi to the success of revegetation

Mycorrhizal fungi are likely to be most beneficial in revegetated ecosystems where the proportion of plants capable of forming mycorrhizas is high and nutrient deficiencies are an important limitation to plant growth. Mycorrhizal fungi can make an important contribution to the success of revegetation through the following mechanisms as discussed below:

i) enhancing establishment and growth of mycorrhizal plants, by increasing nutrient uptake,
ii) maintaining diversity, by boosting the ability of mycorrhizal plants to compete for resources,
iii) contributing to efficient recycling of nutrients and thus to long-term stability, and
iv) stabilizing the soil.

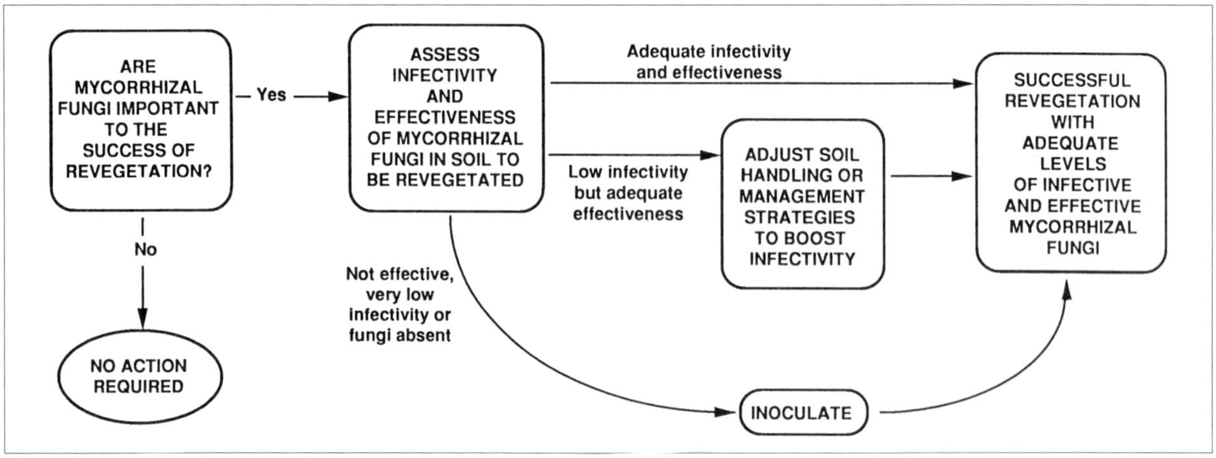

Fig. 1. A schematic representation of a protocol for managing mycorrhizal fungi in soils to be revegetated.

i) Enhancing establishment and growth by increasing nutrient uptake

A wide range of plants from native ecosystems have now been reported to respond to inoculation with mycorrhizal fungi at nutrient levels typical of those found in field soils (Brundrett, 1991). In general the most responsive species are trees, woody shrubs and herbs with most grasses gaining little or no benefit from inoculation (Brundrett, 1991). Some plant species are almost totally dependent on mycorrhizal associations for nutrient uptake (e.g. Warcup and McGee, 1983). Improved growth and survival of such species will help establish a diverse ecosystem, an important criteria for revegetation success.

Fitter (1985) has highlighted the lack of clear demonstrations of the benefit of mycorrhizas in field conditions, suggesting possible explanations as ineffective strains, inter-plant connections, grazing by soil fauna and short-lived roots. In disturbed soil to be revegetated, grazing by soil fauna and inter-plant connections will initially be low. Further if inoculation becomes feasible, then only effective fungi will be used. Therefore, the presence of infective, effective fungi is most likely to be beneficial in disturbed soil. A current limitation to quantifying this benefit in inoculation experiments is that the lack of techniques for inoculating large areas prevents establishing plots of sufficient size to accommodate the large variability in plant growth on disturbed soils.

At many sites the major limitations to successful revegetation may be climatic or soil factors which override any potential response to increased nutrient uptake by mycorrhizas. For example, in a study by Miller (1987) the establishment of mycorrhizal hosts in Wyoming was limited not by low mycorrhizal infectivity, but by the lack of micro-habitats with adequate soil moisture. Mycorrhizal fungi may alleviate to some extent the effect of limitations such as low soil moisture (Nelsen, 1987) or soil chemical toxicities, by enhancing nutritional status. However, the benefits may be small if such non-nutritional growth limitations are severe.

To predict the potential for mycorrhizal fungi to alleviate limitations to plant growth, the mycorrhizal associations of plants used in revegetation must be known. Frequently this knowledge is limited, and may be restricted to generalizations from studies of related species in other environments. Understanding the mycorrhizal associations of plants in particular ecosystems and the biology of the fungi that form them, is an important area that requires further research.

ii) Maintaining diversity by boosting the ability of mycorrhizal plants to compete for resources

A diversity of plant species in an ecosystem depends on access of those plants to sufficient nutrients for continued growth and reproduction. Non-mycorrhizal plant species have various adaptations to enhance nutrient uptake, such as cluster roots or many long root hairs (Pate, this volume). Plants which are normally mycorrhizal will be at a competitive disadvantage in revegetation if phosphorus is limiting and if there are few effective mycorrhizal fungi in the soil. Therefore, the presence of effective mycorrhizal fungi is likely to be essential if a diverse ecosystem is to be re-established (Janos, 1980).

The ability of mycorrhizal fungi to influence the outcome of competition between plants has been demonstrated in pot experiments (e.g. Hall, 1978; Perry et al., 1989) and in microcosms (Grime et al., 1987). However, field demonstrations are difficult, particularly in establishing comparable areas of soil which differ only in mycorrhizal infectivity. Allen and Allen (1986) used infective topsoil as an inoculum in a field experiment but recorded no increase over one season in percentage cover of the total plant community as a result of increased mycorrhiza formation. The use of soils with different disturbance and storage histories to establish different levels of mycorrhizal infectivity makes diversity comparisons difficult because of possible differences in other important soil characteristics (Miller, 1987). Gange et al. (1990) using a soil-applied fungicide were able to demonstrate that reducing the infectivity of mycorrhizal fungi reduced cover abundance and species diversity of a grass/forb mixture in the first season of growth. Re-created ecosystems must be stable and maintain a diversity of species indefinitely. Therefore, studies of benefits of mycorrhizas in revegetation are required which extend over long time periods, until a stable plant community is established.

iii) Nutrient cycling

The long-term stability of the revegetation ecosystem depends on an adequate nutrient pool being maintained without continued fertilizer applications. Nutrient availability may be high at the time of establishment of revegetation due to mineralization in the disturbed soils and to fertilizer applications, but may quickly become less available, for example by phosphorus adsorption in high-fixing soils (Barrow, 1975) or leaching in sandy soils (Diggle and Bell, 1984). Therefore, management strategies that ensure that effective mycorrhizal fungi are present in these soils will enhance the efficient capture and recycling of nutrients.

iv) Enhancing soil structure

Physical disturbance of soil such as that occurring during mining is likely to severely damage the soil structure. Restoration of a stable soil structure in revegetation is important both in resisting erosion and because it is an important factor in soil productivity, influencing all physical, chemical and biological processes (Miller and Jastrow, 1991). Rapid stabilization of soil structure after disturbance is most likely to be achieved by extensive growth of roots of mycorrhizal plants because these roots and their associated hyphae can act to bind sand grains or soil micro-aggregates into macro-aggregates (Tisdall, 1991).

Assessing infectivity and effectiveness of resident mycorrhizal fungi

The keystone in the protocol to manage mycorrhizal fungi for optimum revegetation is to determine the infectivity and effectiveness of the mycorrhizal fungi that may be present in the soil to be revegetated

The infectivity of mycorrhizal fungi can be defined as the rate and extent of mycorrhiza formation (Abbott and Robson, 1982), and depends therefore on the number of propagules in the soil and the capacity of those propagules to produce infective hyphae which reach the root surface. Effectiveness of mycorrhizal fungi is a measure of their ability to benefit plant growth or otherwise increase fitness, either directly through increased uptake of nutrients or indirectly by improving soil conditions (Abbott et al., 1991).

Substantial basic research is required to allow identification of the fungi that are present in the soil to be revegetated and prediction of their infectivity, effectiveness and their likely rate of increase in infectivity (Abbott et al., 1991). The principles that contribute to fungi being infective and effective in soils need to be understood before they can be used to predict the behaviour of mycorrhizal fungi in each revegetation situation.

A basic requirement for evaluating the indigenous VA mycorrhizal fungi is to isolate them in pot cultures. The fungi can then be identified and important characteristics such as propagule type, rate of mycorrhiza formation, effectiveness in nutrient uptake, amount and distribution of hyphae and responses to changes in the soil environment can be described. This basic knowledge is still lacking for fungi in most ecosystems.

i) Assessing infectivity of mycorrhizal fungi in soils to be revegetated

Prediction of infectivity prior to plant establishment is essential if inoculation with effective mycorrhizal fungi is to be considered or to allow adjustments in fertilizer applications. Glasshouse bioassays of field soil using standard plant hosts and growing conditions (Deacon et al., 1983; Abbott and Robson, 1990) are the most useful indicators of infectivity, overcoming for VA mycorrhizal fungi, the limitations of predicting infectivity directly from numbers of fungal propagules in the soil (Abbott and Robson, 1985).

An infectivity test must be correlated with actual infectivity in the field if it is to be used for managing mycorrhizal fungi in revegetation. An important complement to measurements of the mean level of infectivity of soil to be revegetated is defining the spatial distribution of infective propagules in that soil (Jasper et al., 1992). Patchy distribution of propagules, even if highly infective, may still limit the success of revegetation.

ii) Effectiveness of the mycorrhizal fungi in each environment

Knowledge of the effectiveness of mycorrhizal fungi has important applications in two areas of their management. Firstly, where some or all of the indigenous fungi remain in disturbed soil, it must be established if those fungi are effective in the changed soil environment. The return of topsoil to areas to be revegetated may ensure little change in important chemical characteristics, but physical structure and levels of organic matter and surface litter will be different. These changes affect the persistence (Gardner and Malajczuk, 1988) and therefore presumably the effectiveness of individual species of both ectomycorrhizal and VA mycorrhizal fungi. Characterization of the effectiveness of indigenous fungi in the likely soil conditions in revegetation is an important area of further research.

The second important area in which the effectiveness of mycorrhizal fungi must be established is inoculation of fungi into soils or materials which have no indigenous mycorrhizal fungi. The inoculant fungi must be tolerant of conditions in these materials in order to colonize the host plants, explore the soil and increase nutrient uptake and then must produce a large number of propagules to ensure their persistence (Abbott et al., 1991).

iii) The rate of return of mycorrhizal fungi in revegetation

The mycorrhizal fungi in soil at the time of revegetation may be effective but present only at very low levels. The rate of increase in infectivity of those fungi would be crucial in determining the need for inoculation. Monitoring the increase in infectivity of each of the fungal species present is difficult because of the lack of simple techniques for identifying fungi (especially VA mycorrhizal fungi) in soils and roots. Quantifying spores of VA mycorrhizal fungi is not a reliable guide to the level of activity for all species (e.g. Friese and Allen, 1991). Further development of techniques such as using fluorescent antibodies (Friese and Allen, 1991) to directly identify fungi in soil and hyphae associated with roots, is important .

As an alternative, it may be possible to discriminate among colonizing fungi at the generic level or possibly at the species level by identifying the fungi according to the morphology of their hyphal structures in the root (Abbott, 1982), and thus correlate the activity of particular fungi or genera with changing soil conditions. This technique was used to record successional changes in infective VA mycorrhizal fungi at two Australian bauxite mines, using a glasshouse bioassay. The pattern of return of infective fungi with revegetation was similar at each site; an early build-up and continued major presence of one or more Glomus-type fungi, an early but slow return of a second fungus, with a third distinct fungus appearing at 4 to 5 years after revegetation and building-up quickly (D. Jasper, unpublished results). Dominant fungi in early revegetation may simply be those with a large number of propagules surviving disturbance (r-selected). However, such studies can indicate fungi suited to conditions at the time of sowing revegetation, when a diverse range of plants have the best opportunity to establish. Therefore, these studies provide a starting point for selecting inoculant fungi.

Soil management or handling strategies to boost infectivity

Where the timing and nature of disturbance of an ecosystem is predictable, then it may be possible to manage the disturbance to maximize the survival of mycorrhizal fungi. Management strategies should focus firstly on ameliorating the disturbance effects and secondly on managing the revegetation process to optimize the recovery of the populations of mycorrhizal fungi (Allen et al., 1992), and in so doing, optimizing the growth and diversity of plants in revegetation.

Propagules of VA mycorrhizal fungi such as spores and mycorrhizal roots are likely to be more tolerant of disturbance than hyphae (Jasper et al., 1989, 1991). Therefore survival of the fungi will be enhanced if the numbers of these robust propagules are initially high. After disturbance, strategies should aim to avoid the possibility of deleterious conditions developing during soil storage and to minimize the time that soil remains without growing plants.

An additional strategy is to manage revegetation to enhance subsequent invasion of the soils by the fungi. A key factor may be the proximity of undisturbed areas as sources of propagules. Further, in an arid site, the establishment of shrub islands has been shown to increase capture of wind-dispersed propagules (Allen, 1987).

i) Timing and nature of disturbance events

Disturbance of surface soil is unavoidable during logging or topsoil-salvage prior to mining, and in most cases there may be little opportunity to amend the methods used. However, optimizing the timing of the disturbance event may be beneficial. Disturbance of an ecosystem should be timed ideally to occur when there is a maximum number of propagules that are likely to survive both soil disturbance and the absence of host plants. Such timing therefore requires a good understanding of the biology of the fungi. In particular it is important to establish what are the important propagules of each fungus (Jasper et al., 1991) and what is the effect of disturbance on each propagule type (Jasper et al., 1989, 1991).

For ectomycorrhizal fungi, changes in the soil environment as a result of disturbance frequently favour the appearance of early-stage fungi which are not prevalent in mature ecosystems (Mason et al., 1983; Gardner and Malajczuk, 1988). Thus, the rate of mycorrhiza formation by these early-stage fungi may be an important factor contributing to successful revegetation (Amaranthus and Perry, 1987). Studies of the effect of clearcutting in forests in the northwest United States suggest that in general there is an abundance of infective early-stage fungi in disturbed soils (e.g. Schoeneberger and Perry, 1982).

It would be valuable to be able to predict the survival of soil disturbance by mycorrhizal fungi, by identifying correlations with vegetation-type and climate. In mesic environments, the number of robust propagules and thus infectivity after disturbance, is likely to remain high in productive ecosystems with a high proportion of mycorrhizal plants (Jasper et al., 1991). In a contrasting environment in arid areas of the United States, the number of VA mycorrhizal propagules, and their survival of disturbance, is also high in disturbed topsoils (Rives et al., 1980, Miller et al., 1985), perhaps reflecting an accumulation of propagules in this arid climate. Further work is required, before the tolerance of mycorrhizal fungi of disturbance in specific ecosystems can be predicted.

ii) Effect of stockpiling on the infectivity of propagules of mycorrhizal fungi

Storage of soils in stockpiles is widely practised but in most cases is deleterious to populations of mycorrhizal fungi (e.g. Abdul-Kareem and McRae, 1984). In an arid environment, the level of moisture in stored soils is likely to be the most important factor affecting the survival of propagules of mycorrhizal fungi (Miller et al., 1985). This is supported by studies in higher rainfall areas which indicate that the loss of infectivity of propagules is greatest at depth in stockpiles with high soil water potential (Harris et al., 1987), presumably due to the effects of anaerobiosis (Abdul-Kareem and McRae, 1984).

For optimum survival of mycorrhizal propagules, the depth of stockpiles should be restricted to the rooting depth of covering vegetation (Miller and Jastrow, 1991). This requirement is less critical in arid environments if the soil is stored dry (Miller et al., 1985), while in mesic environments ensuring a cover crop of deep-rooted mycorrhizal plants will be an important complementary strategy to minimize the accumulation of moisture in the soil.

This area of soil management requires further research to quantify the losses of infectivity in stockpiles in mesic environments. Ultimately a demonstration of the cost of these losses (if any) in terms of poor quality revegetation will be required if there is to be any deviation from the widely-practised technique of storing soil in deep stockpiles for long periods.

Inoculation

Inoculation will be necessary if the presence of infective and effective mycorrhizal fungi are not adequate to ensure the success of revegetation (Fig. 1). Inoculation is most likely to be required in materials which have never previously supported vegetation such as mining overburdens and wastes from mineral processing, and where there has been a long absence of mycorrhizal host plants, for example stockpiled soils. Inoculation may also be warranted in revegetation of soils which are less severely disturbed but in which the return of effective fungi is too slow to ensure establishment of a diverse range of plant species before canopy closure (Jasper et al., 1992).

Several inoculation techniques have been developed for VA mycorrhizal fungi (Hayman, 1987; Sylvia, this volume) and ectomycorrhizal fungi (see reviews in Sylvia et al., 1987; Kuek, this volume). In general, current techniques are most likely to be feasible in intensive plant production such as nurseries and horticulture, where precise mechanical placement of seed and inoculum is possible. Revegetation of disturbed soils, as currently practised in Australia, is likely to differ from these intensive plant industries in several important areas (Table 1), each of which will impact on selection of an optimum inoculation technique. Any inoculation procedure developed for revegetation will need to accommodate these differences.

Table 1. Characteristics of revegetation techniques in the Australian mining industry which will be important factors in selecting techniques for inoculation with VA mycorrhizal fungi

	Characteristic
1	Sites to be revegetated may not be accessible to machinery.
2	Areas to be revegetated may be large, with up to 400 ha revegetated annually at each mine site.
3	Establishment of vegetation is usually by broadcasting of seed at low rates and with limited transplanting of nursery seedlings.
4	Sites to be revegetated are often in areas of low or uncertain rainfall and great extremes of temperature.
5	Native plant species used in revegetation are frequently slow to germinate and relatively slow growing, especially in the soils of low fertility that are most often revegetated.

The most effective inoculation procedure will depend on the physical and environmental constraints of the site to be revegetated. The constraint of poor machinery access (Table 1) contributes to a large degree to the common practice of broadcasting seed. However, application of inoculum to the soil surface, directly or by seed coating, is unlikely to prove effective (Ferguson and Menge, 1986), especially in harsh environments (Friese and Allen, 1991). At some sites, but not all, it is possible that the current techniques of site preparation could be changed to allow mechanical placement of inocula in soil to give the optimum opportunity for mycorrhiza formation. This would of course require a clear demonstration of the benefit to be gained. It is unlikely however, that revegetation techniques could be changed to the extent that hand-planting of nursery-inoculated seedlings could be considered as an effective option for establishment of mycorrhizal fungi in large areas.

Two constraints on procedures for inoculation which can not be altered, are the large areas of land to be revegetated and the prevailing climates, in which low or erratic rainfall and high temperatures are typical. Large areas to be inoculated require that the inoculum must be of reasonable cost. The harsh environment at many sites demands that the propagules must be able to survive for long periods in soil and that they are well protected from high temperatures.

While rapid and extensive colonization may be a requirement of inoculation in intensive plant production, the optimum rate of colonization and the demand for nutrients is not known for native plants in revegetation, but could be expected to be less than for intensive agriculture. The relatively-slow germination of seeds of native species, followed in many cases by slow growth rates, may provide a framework in which adequate development of mycorrhizas from a localized inoculum is easier to achieve.

If substantial changes are required to revegetation practices to allow inoculation of mycorrhizal fungi, then any increases in costs must first be clearly demonstrated to be justified in terms of increased survival, growth, and diversity of revegetation.

Conclusions

While mycorrhizal fungi are likely to be important in successful revegetation, field scale demonstrations are few. However, increased nutrient uptake by mycorrhizal plants is well demonstrated under experimental conditions and is likely to be important in the field, and the fungal hyphae are important in restoring soil structure. Therefore, managing disturbance and revegetation processes to enhance the survival of mycorrhizal fungi may be justified.

In general some losses of infectivity of mycorrhizal fungi are unavoidable during ecosystem disturbance. The challenge is to minimize that loss. Restricting the time that soil spends without mycorrhizal plants and avoiding storage of soil in deep stockpiles are the most important principles in maximizing the infectivity of the fungi in the soil to be revegetated. Optimizing the survival of mycorrhizal fungi could provide a useful focus for revegetation strategies, giving wider benefits in enhancing the overall microbial activity of the soil.

Soil management strategies are currently the only practical option to enhance the infectivity of mycorrhizal fungi because of the lack of an inoculation technique for large areas. Development of an inoculation procedure would be useful, not only for revegetation of microbially sterile soils, but also to allow quantitative evaluation of the contribution of mycorrhizal fungi to the diversity and stability of revegetated ecosystems.

Acknowledgements

This research is supported by Alcoa of Australia Ltd., AMC Mineral Sands Ltd., Argyle Diamond Mines, Comalco Aluminium Ltd., Curragh Queensland Mining Ltd., Hamersley Iron Pty Ltd., Newlands Coal Ltd., Oaky Creek Coal Ltd., Tiwest Joint Venture, and Westralian Sands Ltd., through the Australian Mineral Industries Research Association. I thank Julie Davy, Karen Clarke and Simone Mercer for their technical assistance.

References

Abbott L K 1982 Comparative anatomy of vesicular-arbuscular mycorrhizas formed on subterranean clover. Aust. J. Bot. 30, 485–499.

Abbott L K and Robson A D 1982 Infectivity of vesicular arbuscular mycorrhizal fungi in agricultural soils. Aust. J. Agric. Res. 33, 1049–1059.

Abbott L K and Robson A D 1985 Assessing the potential for wide-scale VA mycorrhiza formation. *In* Proceedings of the 6th North Am. Conf. on Mycorrhizae. Ed. R Molina. pp 76–79. Forest Research Laboratory, Oregon State University, Corvallis, Oregon

Abbott L K and Robson A D 1990 Field management of VA mycorrhizal fungi. *In* The Rhizosphere and Plant Growth. Eds. D L Keister and P B Gregan. pp 355–362. Kluwer Academic Publishers, Dordrecht , Netherlands

Abbott L K, Robson A D and Gazey C 1991 Selection of inoculant VAM fungi. *In* Methods in Microbiology Vol 24. Eds. J R Norris, D J Read and A K Varma. pp 1–21. Academic Press, London.

Abdul-Kareem A W and McRae S G 1984 The effects on topsoil of long-term storage in stockpiles. Plant Soil 76, 357–363.

Allen E B and Allen M F 1986 Water relations of xeric grasses in the field: interactions of mycorrhizas and competition. New Phytol. 104, 559–571.

Allen M F 1987 Ecology of vesicular-arbuscular mycorrhizae in an arid ecosystem: use of natural processes promoting dispersal and establishment. *In* Mycorrhizae in the Next Decade: Practical Applications and Research Priorities. Proceedings of the 7th North Am. Conf. on Mycorrhizae. Eds. D M Sylvia, L L Hung and J H Graham. pp 133–135.University of Florida, Gainesville.

Allen M F, Jasper D A and Zak J 1992 Soil ecology. *In* Principles of Restoration Ecology: An Integrated Approach. Eds. E B Allen and Z Naveh. Springer-Verlag. (*in press*).

Amaranthus M P and Perry D A 1987 Effect of soil transfer on ectomycorrhiza formation and the survival and growth of conifer seedlings on old, non-reforested clearcuts. Can. J. For. Res. 17, 944–950.

Barrow J 1975 The response to phosphate of two annual pasture species. I. Effect of the soil's ability to adsorb phosphate on comparative phosphate requirement Aust. J. Agric. Res. 26, 137–143.

Brundrett M C 1991 Mycorrhizas in natural ecosystems. Adv. Ecol. Res. 21, 171–313

Deacon J W, Donaldson S J and Last F T 1983 Sequences and interactions of mycorrhizal fungi on birch. Plant Soil 71, 257–262.

Diggle A J and Bell L C 1984 Movement of applied phosphorus following the mining and revegetation of mineral sands on Australia's east coast. Aust. J. Soil Res. 22, 135–148.

Ferguson J J and Menge J A 1986 Response of citrus seedlings to various field inoculation methods with *Glomus deserticola* in fumigated nursery soils. J. Am. Soc. Hortic. Sci. 111, 288–292.

Fitter A H 1985 Functioning of vesicular-arbuscular mycorrhizas under field conditions. New Phytol. 99, 257–265.

Friese C F and Allen M F 1991 Tracking the fates of exotic and local VA mycorrhizal fungi : methods and patterns. Agric. Ecos. Environ. 34, 87–96.

Gange A C, Brown V K and Farmer L M 1990 A test of mycorrhizal benefit in an early successional plant community. New Phytol. 115, 85–91.

Gardner J H and Malajczuk N 1988 Recolonization of rehabilitated bauxite mine sites in Western Australia by mycorrhizal fungi. For. Ecol. Manag. 24, 27–42.

Grime J P, Mackey J M L, Hillier S M and Read D J 1987 Floristic diversity in a model system using experimental microcosms. Nature. 328, 420–422.

Hall I R 1978 Effects of endomycorrhizas on the competitive ability of white clover. N. Z. J. Agric. Res. 21, 509–515.

Harris J A, Hunter D, Birch P and Short K C 1987 Vesicular-arbuscular mycorrhizal populations in stored topsoil. Trans.Br. Mycol. Soc. 89, 600–603.

Hayman D S 1987 VA mycorrhizas in field crop systems. *In* Ecophysiology of VA mycorrhizal plants. Ed. G R Safir. pp 171–192.CRC Press, Boca Raton, Florida.

Janos D P 1980 Mycorrhizae influence tropical succession. Biotropica 12, 56–95.

Jasper D A, Robson A D and Abbott L K 1987 The effect of surface mining on the infectivity of vesicular-arbuscular mycorrhizal fungi. Aust. J. Bot. 35, 641–652.

Jasper D A, Abbott L K and Robson A D 1989 Hyphae of a vesicular-arbuscular mycorrhizal fungus maintain infectivity in dry soil except when the soil is disturbed. New Phytol. 112, 101–107.

Jasper D A, Abbott L K and Robson A D 1991 The effect of disturbance on vesicular-arbuscular mycorrhizal fungi in soils from different vegetation types New Phytol. 118, 471–476.

Jasper D A, Abbott L K and Robson A D 1992 Soil disturbance in native ecosystems—the decline and recovery of infectivity of VA mycorrhizal fungi. *In* Mycorrhizas in Ecosystems. Eds. D J Read., D H Lewis, A H Fitter and I J Alexander. pp 151–155. CAB International, Oxon, UK.

Mason P A, Wilson J, Last F T and Walker C 1983 The concept of succession in relation to the spread of sheathing mycorrhizal fungi on inoculated tree seedlings in unsterile soils. Plant Soil 71, 247–256.

Miller R M 1987 Mycorrhizae and succession. *In* Restoration Ecology: A Synthetic Approach to Ecological Research. Eds. W R Jordan, M E Gilpin and J D Aber. pp 205–219. Cambridge University Press, Cambridge.

Miller R M and Jastrow J D 1991 The application of VA mycorrhizae to ecosystem restoration and reclamation. *In* Mycorrhizal Functioning: An Integrative Plant–Fungal Process. Ed. M F Allen. Routledge, Chapman and Hall, New York.

Miller R M, Carnes B A and Moorman T 1985 Factors influencing survival of vesicular-arbuscular mycorrhiza propagules during topsoil storage. J. Appl. Ecol. 22, 259–266.

Nelsen C E 1987 The water relations of vesicular-arbuscular mycorrhizal systems. *In* Ecophysiology of VA Mycorrhizal Plants. Ed. G R Safir. CRC Press, Boca Raton, Florida

Perry D A, Margolis H, Choquette C, Molina R and Trappe J M 1989 Ectomycorrhizal mediation of competition between coniferous tree species. New Phytol. 112, 501–511.

Rives C S, Bajwa M I, Liberta A E and Miller R M 1980 Effects of topsoil storage during surface mining on the viability of VA mycorrhiza. Soil Sci. 129, 253–257.

Schoeneberger M M and Perry D A 1982 The effect of soil disturbance on growth and ectomycorrhizae of Douglas-fir and western hemlock seedlings: a greenhouse bioassay. Can. J. For. Res. 12, 343–353.

Sylvia D M, Hung L L and Graham J H 1987 Mycorrhizae in the next decade: Practical applications and research priorities. Proceedings of the 7th North Am. Conf. on Mycorrhizae. University of Florida, Gainesville.

Tisdall J M 1991 Fungal hyphae and structural stability of soil. Aust. J. Soil Res. 29, 729–743.

Warcup J H and McGee P A 1983 The mycorrhizal associations of some Australian Asteraceae. New Phytol. 95, 667–672.

Issues concerning the production and use of inocula of ectomycorrhizal fungi in increasing the economic productivity of plantations

CLEM KUEK
Department of Biological Sciences, Faculty of Business and Technology, The University of Western Sydney, Macarthur, P.O. Box 555, Campbelltown, NSW 2560, Australia

Key words: cost, economic value, ectomycorrhizal fungi, inoculum production

Abstract

Despite the clear evidence from small-scale experiments that ectomycorrhizal fungi can improve the growth rate of host plants, the use of inoculation to exploit this in plantation forestry is not widespread. The factors which tend to limit the use of inoculation as a management option are the lack of data from large-scale productivity trials to support the economic benefit achievable, and the lack of commercial producers of appropriate inocula. An economic analysis of the value of growth improvement in plantation forestry indicates that significant increases in revenue can be obtained. Such analyses can be used to persuade usage of the technology, argue the cost of inocula, and to encourage commercial interest in inoculum production. The linkage between field data and economic benefit dictates that only defined inocula containing known and field-tested fungi will be of significance in the commercial adoption of inoculation technology on a large-scale. Criteria for effective and practical inocula are presented. The relative merits of production of inocula via solid-substrate culture and submerged aerobic culture are considered and the superiority of the latter for higher quality inocula is indicated.

Introduction

There is now a wealth of information on the ecology and biology of ectomycorrhizal associations with plants (see Harley and Smith, 1983; Kropp and Langlois, 1990). Of particular interest to plantation forestry is the potential for improvement in biomass productivity achievable through exploiting the growth stimulating effects of ectomycorrhiza. These effects are real and measurable. It is now known that; (i) some ectomycorrhizal fungi are better than others in improving the growth of host plants (e.g. Dixon et al., 1984; Gagnon et al., 1987; Grove et al., 1991a), (ii) to maximize benefits of the fungi, plants should be inoculated with particular fungi at the time of seedling production, (iii) there are few technical barriers to the exercise of inoculation as an option in current production practices for seedlings, and (iv) unlike vesicular-arbuscular mycorrhizal fungi, ectomycorrhizal fungi are relatively easy to culture artificially.

Two factors determining whether ectomycorrhizal fungi will be used in plantations.

Despite the fact that under many conditions, the benefits of inoculation are clear and the option technically exercisable, use of ectomycorrhizal fungi in plantations is not as widespread as the situation might suggest. That inoculation technology is yet to be routinely applied in wood production shows that the evidence which is currently available is viewed conservatively. Two of the major factors determining whether ectomycorrhizal fungi will be used in plantations are the economic efficacy of inocula, and the

A.D. Robson, L.K. Abbott & N. Malajczuk (eds.), *Management of mycorrhizas in agriculture, horticulture and forestry*, 221–230.

availability of practical and efficacious inocula.

Economic efficacy of inoculation in growth improvement

Management options in forestry are often simply different ways of minimizing costs and maximizing revenue. The efficacy of any new option has to be decided against practices already in place and which may be relatively well understood.

Field relevance. For inoculation to be economically efficacious, it must produce improvements in growth rates. The question which is always asked first of ectomycorrhizal technology is whether the growth improvements readily demonstrated under controlled conditions are reproducible in the field. The answer requires broad-scale productivity trials under various field conditions and over reasonable trial periods. Evidence of the beneficial effects of ectomycorrhiza is easily found in the literature (e.g. Ruehle, 1982; Valdes, 1986; Marx and Cordell, 1988). However, not all these results can be used to justify routine inoculation to increase the economic productivity of plantations because the conditions used may not be representative (Jeffries and Dodd, 1991) or of sufficient scale under realistic conditions (Kropp and Langlois, 1990). For example, the choice of nutrient levels may dictate that control plants perform poorly. In many cases, the obligatory requirement for mycobionts such as in pines, means that inoculations often lead to improvements over uninoculated controls. The results from inoculations in rehabilitation forestry can be accentuated by the rehabilitation site no longer possessing an appropriate population of ectomycorrhizal fungi. Lastly, ecological variability of field sites will tend to make extrapolations from trial results difficult.

Inoculation may not be advantageous on some sites because of existing populations of symbiotic fungi and the need to develop a capability to predict the advisability of inoculation on particular sites has been identified (Kropp and Langlois, 1990). Field evidence will be more difficult to obtain on sites where a naturally occurring population of relevant ectomycorrhizal fungi already exists because uninoculated plants may still form a relationship with fungi existing on the site. In these situations, the rationale for inoculation is based on the use of fungi which are demonstrably superior to the fungi found at the site. This has an important consequence for the type of inoculation that can be practiced, i.e. specific cultures rather than undefined cultures must be used to obtain efficacy data. Data obtained using undefined inocula such as soil or basidiospores can only support the general proposition that hosts are better off with mycorrhizal fungi than without. Such evidence does not satisfy Koch's postulates in that a relationship between specific microorganisms (by genus, species and strain) and hosts cannot be verified. As discussed later, this has implications for the selection of inoculum form and production method.

In the production of mycorrhizal seedlings, the nutrient and watering regime in nurseries must be appropriately managed if mycorrhiza are to develop (Beckjord et al., 1980; Beckjord et al., 1984; Bougher et al., 1990). After outplanting, the inoculated plant may lose its inoculant fungus due to succession of other ectomycorrhizal fungi (Fleming, 1985; Gardner and Malajczuk, 1988), and the nutrient status of the soil will influence the effect of the mycorrhiza (Grove et al., 1991a; Grove et al., 1991b). Thus, it may not be surprising that quantitative field evidence of the use of inoculations in wood production has been slow to emerge. In the case of eucalypt plantations, there have been few instances of the use of ectomycorrhizal fungi (Grove et al., 1991a) and accordingly, there is little field evidence. However, a small-scale field trial in the Congo showed a volume increase of 30% at 50 months after outplanting of eucalypts inoculated with *Pisolithus tinctorius* (Garbaye et al., 1988). Similarly, in south-western Australia, growth increases of up to 80% were found 7 years after inoculation with *Amanita xanthocephala* (Malajczuk et al., 1993). Such positive albeit limited results show the promise of ectomycorrhizal technology. The demonstration of the benefits of ectomycorrhiza under plantation conditions is vital to the adoption of the technology and the current use of inoculation is limited by a lack of positive demonstration (Jeffries, 1987).

The corresponding low level of demand does not attract commercial production of inocula.

The cost of inoculation technology. If gains in biomass productivity can be shown to be attainable under field conditions, the second question that is asked by the potential end-users is about the cost of inoculation. Plantation economists or managers and indeed even mycorrhizal scientists (e.g. Cordell et al., 1987; Jeffries, 1987; Le Tacon et al., 1988) often pronounce that for inoculation technology to be adopted, it should be cheap. 'Cheap' is often defined with reference to the cost of seedling production. Indeed, an economic justification for inoculation has been made on the basis of improvement in seedling survival and thus reduction in overall costs associated with seedling establishment (Kropp and Langlois, 1990). This way of putting the cost of inoculation technology in perspective, while short-sighted, is understandable because to inoculate is to incur an immediate cost while the potential economic reward is years away. The method of costing which is likely to be most beneficial towards commercialization of inoculation technology is to relate cost to the economic value of increased biomass productivity. Thus, if a biological advantage can be conferred by ectomycorrhizal fungi, then the second question should be about its economic benefit, not the cost of using it.

The economic value of improved growth rate. Since inoculation as a management option will increase operational costs, the advisability of adopting ectomycorrhizal technology will be a measure of cost (inoculation) against benefit (increased productivity). The rationale for incurring the cost of inoculation is that the gain in improved biomass productivity will be greater than the cost of inoculation. The ratio of cost to benefit which will be acceptable, will depend on the overall economics of the plantation operation and the degree of certainty of growth improvement. The potential economic benefit is in the future because the monetary gains in improved biomass productivity are realized at the end of the rotation period of the plantation stand. The economic value of such a deferred return can be estimated using net present value analysis to correct for the time value of money.

Table 1. Major costings used in an analysis of the economic value of improvement in the growth rate of plantation *E. globulus*

Item	\$ ha^{-1}
Annual lease for land (6% of land value)	72
Site preparation and establishment	668
Additional weeding (year 2)	31
Additional fertilizer (year 5)	96
Annual maintenance (years 1–5)	38
Annual maintenance (years 5–10)	25

Representative costs (Australian dollars) for woodchip plantations in south-western Australia (Table 1) were used in an analysis completed in 1989 for Biosynthetica Pty. Ltd. (formerly Interbac Australasia Pty. Ltd.) by the Australian Agricultural Consulting and Management Company Pty. Ltd., both of Perth, Western Australia. The analysis was based on the planting of *Eucalyptus globulus* on 40 hectare, leased, ex-pasture sites. A discount rate of 5% was used to equate all future revenue to present day values. Estimation was made of the monetary effect of the increase in mean annual increment (MAI) of the biomass of plantation trees in the order of 10, 20 and 30% over a base MAI of 22 m^3 ha^{-1} year^{-1}. At the base MAI rate, it was assumed that the stands would be harvestable in ten years. In the analysis, the increase in MAI was translated to a shortening of rotation period and the effect on revenue was calculated on that basis. If the growth improvements are attributable to inoculated ectomycorrhizal fungi, then the analysis would in effect be one of the economic value of inoculation. It can also be used to compare the relative merits of the growth stimulating capabilities of specific ectomycorrhizal fungi.

Three selected scenarios from the analysis representing single, coppice and multiple rotation operations are presented in Figs. 1–3. As expected, all scenarios showed an increase in revenue resulting from increment in MAI. The type of plantation operation affects the size of the potential return with single rotations poorest and multiple rotations the best. Coppiced rotations would best reflect the average scenario in plantation operations in Western Australia. With such rotations, an increase in MAI of 30%

and a stumpage price of \$20 m^{-3}, would result in an extra \$952 ha^{-1} (net present value) return from the plantation at harvest. Therefore, in Western Australia, should ectomycorrhizal fungi be capable of improving MAI by 30% then significant gains in revenue can be realized. In terms of the commercialization of inoculation technology, the significance of such analyses goes beyond quantitative data to persuade usage of ectomycorrhizal fungi, to a revelation of the price which inocula may command by applying a cost to benefit ratio to the gain in revenue. For example, at an increased return of \$952 ha^{-1}, a reasonable cost to benefit ratio of 1:5 indicates an inoculum cost of \$190.40 ha^{-1}. At a tree density of 1000 ha^{-1}, this is a cost of 19 cents per seedling. The 1:5 ratio also means that with ectomycorrhiza boosting growth by 30%, the plantation can afford to spend 95.2 cents more per seedling for inoculation and be no worse off. Herein lies the potential for inoculum producers to argue the cost of inoculum. Experience with fermentation methods for the production of biomass indicate that the cost to produce a single inoculum dose would be well inside 19 cents. Thus, such analyses can also provide an economic rationale for the investment of research and development funds in inoculation technology for now these may be tallied against potential returns from sales of inocula as estimated from prices which may be commanded. The analyses can also be used by ectomycorrhizal scientists to gauge the potential value of different fungi or to set minimum targets in growth increment in their search for efficacious fungi. Using the same example above, if for various reasons, the price of inoculum had to be a minimum of 19 cents per dose and the end-user requires a cost to benefit ratio of greater than 1:5, then the growth increment which would have to be delivered has to be greater than 30%.

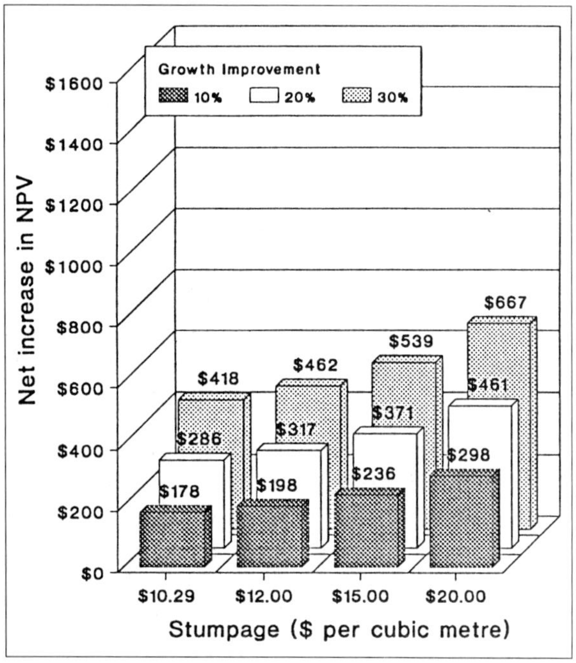

Fig. 1. The economic value of increased growth rates as indicated by the net increase in the net present value (NPV) of revenue at harvest in single-rotation eucalypt plantations in south-western Australia.

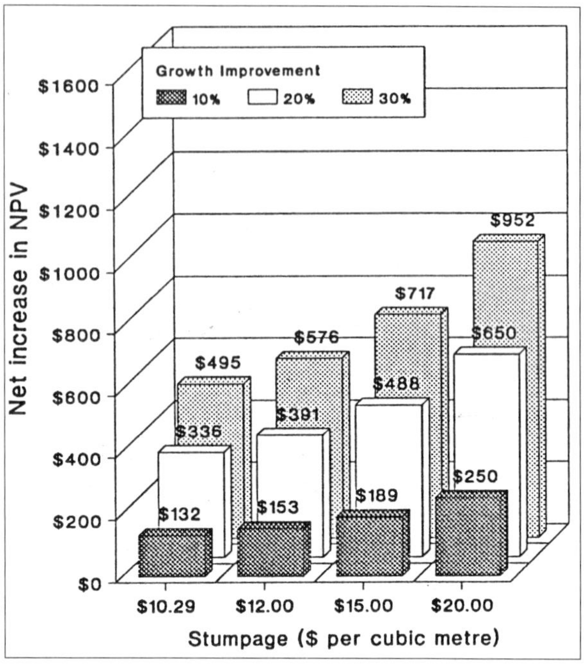

Fig. 2. The economic value of increased growth rates as indicated by the net increase in net present value (NPV) of revenue at harvest in coppiced eucalypt plantations in south-western Australia. The analysis was based on two coppice rotations with yields which were 110% and 90% of the initial rotation. Growth increment by ectomycorrhizal fungi was assigned to only the initial rotation.

Clearly for such economic analyses to be of any practical significance, firstly, they must be

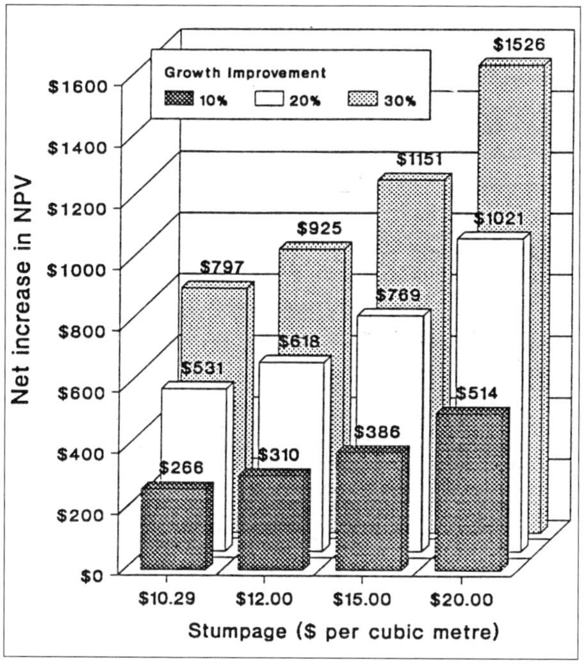

Fig. 3. The economic value of increased growth rates as indicated by the net increase in the net present value (NPV) of revenue at harvest in multiple-rotation eucalypt plantations in south-western Australia.

reliable. Given the influence of the environment on the ecology of the mycorrhizal association, the data must either be available for the specific environmental regimes or be of a fundamental nature, capable of intra- and extrapolation. In general, business sense dictates that the higher the risk and cost of an option, the greater is the return on investment which is expected. If field evidence is poor, then the potential economic gain must be high. This means that the price which may be commanded by inocula will be correspondingly low. This has a negative effect on encouraging the development of inoculation technology. Secondly, the data must be defined specifically by fungus (pure cultures used in trials), host tree, and environment for it is only this way that the results can be duplicated. The value of such data as argument for the use of ectomycorrhizal technology is diminished when the combination cannot be duplicated or knowingly not duplicated. The routine duplication of results obtained in field tests under operational plantation conditions will require that the specific fungi used be available as inoculum before the economic advantage of the ectomycorrhizal fungi can be exploited.

Availability of practical and effective inocula

Inocula. Ectomycorrhizal fungi sought as inocula will be those which field testing and economic analysis reveal as desirable. These must then be capable of being supplied as inocula on demand. Inocula whose use are justified on the basis of a verified beneficial relationship between specific fungi and hosts must in the first instance be capable of producing ectomycorrhiza of specific fungi on inoculated plants. This has implications on the type of inocula which are relevant for plantation forestry. The previous arguments make it clear that in economic terms, those inocula which are not strain specific cannot be justified and are therefore not considered here. Unspecific or undefined inocula include soil, forest litter, basidiospores and sporocarps. What then are the types and forms of inocula which are best able to deliver the promise of field trial results and which are practical and effective?

Criteria for practical and effective inocula. The effectiveness of an inoculum is associated with its characteristics as a source of propagules. Practicality is related to the ability to produce it satisfactorily, and to its ease of usage. In general, effectiveness and practicality are inversely related. Therefore, the ideal inoculum is a compromise between these two attributes. Fungal propagules (biomass) may be used directly as inoculum. However, for ease of handling or because of the culture process, inocula often consist of biomass and a carrier or particulate bulking material such as peat-vermiculite or cereal grains. Thus, inocula are available in different forms depending on the production process or other considerations such as the method of inoculation. Effective and practical inocula should satisfy the following criteria (Tommerup et al., 1987); (i) the inoculant microorganism must be efficacious, (ii) it should be produced axenically and have a physiological state which is consistent and appropriate for storage, inoculation and initiation of mycorrhiza, (iii) carrier or bulking material if used, should protect the biomass against physiological stress during production, and (iv) the inoculum should be of a form which will allow close control of growth conditions in large-scale production, facilitate handling, and be capable of being provided in repeatable dosages associated with consistent amounts of biomass. Most experimental and commercial inocula currently available are sufficiently below ideal as to pose a potential limitation on the routine practice of inoculation in plantation forestry.

Large-scale production of inocula. Inoculation can only be routinely practiced in plantation forestry if production of inocula can be scaled up to the size required. Foreshadowing large-scale usage of inocula, and having defined what constitutes effective and practical inocula, the production technology which is required will involve the use of pure isolates in axenic-mass culture under controlled conditions. As most ectomycorrhizal fungi do not form yeast phases or asexual spores in culture, the production of vegetative mycelium will provide the basis for manufacturing inocula of many ectomycorrhizal fungi on a large-scale. The two main techniques for large-scale production are solid-substrate and submerged aerobic culture.

Solid-substrate culture. This form of culture is where the substrate is largely undissolved in an aqueous phase. Such culture may be static such as those within bags and other containers, or stirred such as those within rotary drum reactors. Examples of inocula produced using solid-substrate culture is mycelium grown on a substrate of cereal grain or a mixture of peat and vermiculite (Marx and Kenney, 1982; Cordell et al., 1987; Le Tacon et al., 1988). The solid-substrate gives the inoculum physical form and no further formulation is usually required.

As there is little water in the system, the chemical environment of a fungus growing in solid-substrate culture consists of the film of water that may be present around substrate particles, and the zone within each substrate particle where diffusion of chemicals can occur. Regulation of the moisture level in the culture determines the amount of free water available but this can be difficult to achieve (Laukevics et al., 1984; Larroche and Gros, 1986). The immediate gaseous environment consists of the inter-particulate space not occupied by water. In static systems, the supply of oxygen and removal of carbon dioxide is determined entirely by gaseous diffusion, the degree of which is in turn determined by the size and continuity of the inter-particulate space. Thus, in solid-substrate culture, the diffusion of gases and nutrients is of overriding importance in determining the rate and amount of growth, and is a serious limitation. This contributes to long incubation periods e.g. 2 to 3 months in a 2 L container for fast growers like *Pisolithus tinctorius* and 8 to 10 months for slow growers like *Cenoccocum geophilum* (Marx and Kenney, 1982). The regulation of incubation temperature is another problem in static cultures because it is reliant upon the inefficient heat exchange within and without the culture. These considerations tend to limit the size of culture batches. The mixing of cultures that is achieved in stirred solid-state fermentors reduces the problems of diffusion and temperature control to some extent. However, mixing is a self-limiting solution because excessive mixing disrupts fungal growth (Silman, 1980; Moo-Young et al., 1983; Larroche and Gros, 1986).

The lack of water in solid-substrate culture makes the measurement of the physico-chemical environment such as pH, gaseous and nutrient status difficult. This means that fine control of the environment to optimize growth is not possible. Even if mensuration of the physico-chemical parameters is made easier, there remains the problem of control because of the inherent heterogeneity of the culture and the lack of a good medium (water) through which control may be exerted.

The characteristics of solid-substrate culture contribute to heterogeneity in the growth conditions among individual inoculum particles, and within and between production batches. There are difficulties in large-scale production especially in static culture. The process has a large requirement for space in terms of the volume of inocula produced. At a commercial scale, quality and axenicity have been found to be difficult to maintain (Kendrick and Berch, 1985; Le Tacon et al., 1988). The inocula produced often contain undesirable levels of residual nutrients which can adversely affect the success of inoculation (Marx, 1980). On the positive side, any ectomycorrhizal fungus which can be isolated and cultured on agar plates are likely to be culturable via solid-substrate culture. The main advantage of solid-substrate culture is that it is a relatively simple technology requiring relatively low capital investment.

Submerged aerobic culture. In this form of culture, microorganisms are grown within a liquid medium which is stirred and/or sparged with air in a bioreactor. Bioreactors are specialized vessels for the conduct of axenic microbial growth. Mixing of the culture is generally achieved by impellers. Control of pH, oxygen tension, degree of agitation and temperature of the culture broth is readily implemented using sensor probes and correcting actuators e.g. automatic addition of acid or base to maintain a set pH. Thus, the main advantages of submerged aerobic culture are the high level of homogeneity which may be achieved within an entire culture, and the ability to precisely control the physico-chemical conditions for growth. Such control enables the physiological state of the microorganism to be finely regulated. Incubation periods are shorter than solid-substrate culture and are typically measured in weeks (Boyle et al., 1987; Gagnon et al., 1988) or even days (Le Tacon et al., 1985). Scale-up of production is not a technical problem because large-scale production of fungi has long been practised in the fermentation industry, e.g. in the production of antibiotics. The superior features of submerged aerobic culture enable the production of inocula which are consistent as propagules, predictable, and reproducible in quality within a batch and among batches. Homogeneity in inoculum quality means that inoculation doses can be minimal as opposed to maximal in order to compensate for ineffective 'propagules' when inoculum quality is heterogeneous.

However, submerged aerobic culture may not be suitable for all ectomycorrhizal fungi. Some fungi are presently difficult to grow in submerged aerobic culture (Harvey et al., 1988) and the culture of slow-growers may not be economical. This difficulty is related in part to the requirement to fragment mycelia during culture. In submerged aerobic culture, batch volumes are increased sequentially where the contents of a batch is used as seed culture for the following batch which would typically be 8 to 9 times the volume of the seed culture. Since ectomycorrhizal fungi only form vegetative mycelium in submerged aerobic culture, and since a maximum number of growing points is optimal, the seed culture is fragmented, usually by homogenization prior to use in the subsequent batch. Fragmentation of mycelia is also performed post-culture (see below) to evenly distribute the fungus in a slurry and maximize the number of propagules. But, ectomycorrhizal fungi can differ in their ability to withstand fragmentation (Boyle et al., 1988) and it is most likely that such treatment at least reduces viability. More studies on aseptic fragmentation techniques for large volumes of mycelial suspensions would advance the technology. The increased handling of the cultures and translation of small initial volumes to what may ultimately be very large volumes in fermentation under growth-optimized conditions also means that the axenicity of fungal cultures must be absolute. This is a potential problem area not always apparent when the culture of ectomycorrhizal fungi is restricted only to small scale agar or other solid-substrate culture. Biomass from submerged aerobic culture may be used directly as an inoculum slurry with or without prior fragmentation (Boyle et al., 1987; Gagnon et al., 1988). It can also be formulated to achieve different forms. For example, fragmented biomass can be mixed with sodium alginate and then gelled onto bare roots (Deacon and Fox, 1988), or formed into beads (Le Tacon et al., 1985; Mauperin et al., 1987), or granulated (Kropacek et al., 1989). The most advanced form of inoculum is that where the biomass is cultured within hydrogel beads (Kuek et al., 1992; Kuek, et al., 1990; Jeffries and Dodd, 1991). Mycelium cultured in beads should be physiologically superior to fragmented mycelia and therefore act more rapidly as a propagule. Unlike entrapped fragmented mycelium, growth within beads results in mycelium being established immediately below the surface of the beads where it can readily grow outwards (Kuek et al., 1992). Mycelium grown within beads is protected at all stages of the production process and inoculation procedure. Bead size can be varied and be as small as 0.5 mm in diameter. Using this technology, several species and isolates of *Laccaria, Hebeloma, Descolea, Elaphomyces* and *Pisolithus* have already been successfully produced as inoculum and other genera are being investigated (Kuek et al., 1992).

Inocula with lower volumes are less costly to produce, store, transport and apply. Those having low volume to biomass ratios are beads and slurries, whereas inocula produced via solid-substrate culture are bulky. Hydrogel bead inocula (Kuek et al., 1992) for 10000 containerized seedlings (ten beads per seed) would occupy only 2 L. It has been estimated (Malajczuk et al., 1993) that mycelial-slurry or solid-state cultured inocula (Gagnon et al., 1988; Boyle et al., 1987; Cordell et al., 1987; Le Tacon et al., 1988) are between twenty to one hundred times the volume of hydrogel bead inocula at comparable dosage rates. Thus, in physical form, inocula can be prepared in a volume-efficient manner with submerged aerobic culture.

Submerged aerobic culture is a better approach to large-scale production than solid-substrate culture although the choice of process will often be made on the basis of the price that the inoculum can be sold for and the minimum acceptable level of inoculum quality. Inocula produced via submerged aerobic culture will be costlier than those via solid-substrate culture since the establishment of process equipment for the former is more capital intensive. However, if cost/benefit analysis such as those above are used, there is little doubt that in Western Australia at least if not elsewhere, submerged aerobic culture can be used profitably for the commercial production of inocula.

Storage life of inocula. Apart from maintaining quality, retention of viability during storage is advantageous because large requirements for inoculum can be met by storing the output from smaller batches. This means that production equipment can be smaller in scale, resulting in a lower capital cost for the production process.

The form of an inoculum influences its storage life. Bead inoculum in which mycelia of *Hebeloma westraliensis* and *Laccaria laccata* were grown, retained their ability to form colonies on agar plates after 7 and 5 months storage in water at 4°C respectively (Kuek et al., 1992). Manipulation of the composition of the medium used to suspend the beads is likely to further increase storage life. Fragmented mycelia of *Hebeloma crustuliniforme* entrapped in hydrogel had 90% of initial viability after five months in cold storage (Mauperin et al., 1987). Solid-substrate cultured inocula of *Pisolithus tinctorius* or *Hebeloma crustuliniforme* can be stored for a month at 2–4°C (Hung and Molina, 1986; Cordell et al., 1987; Le Tacon et al., 1988). Slurries of mycelium in water have been stored for as long as four months (Boyle et al., 1987). Thus, if stored under refrigeration, inocula should have storage lives in excess of operational requirement (time for warehousing, shipment, and application).

Application of inocula. Row-drill applicators have been used for a solid-substrate cultured inoculum (Cordell et al., 1987). For containerized seedlings, such inocula would be readily applied using the same machines which sow seed. The practicality of hydrogel bead inoculum (Kuek et al., 1992) was demonstrated when 80000 containerized seedlings were inoculated via fluid-drilling in a commercial nursery in Western Australia in 1990. The resultant seedlings were nearly 100% mycorrhizal (Hardy et al., 1992). Mycelial slurries can be delivered via fluid drilling or custom-built injectors (Fortin et al., 1988). Thus, inocula are readily handled mechanically and therefore, the application of inocula is not likely to be a problem for adoption of inoculation technology.

Conservation of genotype. It is clear that an ectomycorrhizal fungus verified to be efficacious by several years of field trial results would be a valuable commercial resource. Yet, one of the weakest points in the thrust towards the adoption of ectomycorrhizal technology is the lack of effort put into research on the long-term storage of master cultures. Due to the unsuitability of other preservation techniques, conservation of ectomycorrhizal fungi is largely dependent on continued sub-culture on agar media. The potential for contamination through repeated sub-culture, and for spontaneous mutation and selection applied by culture and storage conditions (Kidby, 1974) poses a real danger, the end result of which is loss of valuable genotypes. More stable methods of storage characterized by fewer sub-cultures and lower selection pressures are urgently required. Cryopreservation techniques are an obvious choice but these should be investigated to provide an understanding as to why not all ectomycorrhizal fungi can be preserved this way. A profitable avenue may be to adapt for ectomycorrhizal fungi, the cryopreservation techniques successfully developed for embryology.

In conclusion

It is reasonable to conclude that there are few technical barriers to the large-scale production and use of inocula of specific ectomycorrhizal fungi. Indeed, the main remaining items on the technological timetable for practicalization of inoculation technology are concerned with scale-up of processes. Such studies are not likely to proceed seriously until there is commercial justification for the investment in funds which would be necessary. This leads to the concluding remark that the technological timetable has been uncoupled from the evidence timetable i.e. production technology is better prepared for the advent of inoculation technology on a broadscale than is the overall scientific evidence to justify the use of the technology in increasing the economic productivity of plantations. On the positive side, this means that when the field evidence is sufficiently developed and accepted widely enough, ectomycorrhizal technology can be rapidly implemented.

References

Beckjord P R, Adams R E and Smith D W 1980 Effects of nitrogen fertilization on growth and ectomycorrhizal formation of Red Oak. For. Sci. 26, 529–536.

Beckjord P R, Melhuish J H and McIntosh M S 1984 Effects of nitrogen and phosphorus fertilization on growth and formation of ectomycorrhizae of *Quercus alba* and *Q. rubra* seedlings by *Pisolithus tinctorius* and *Scleroderma auranteum*. Can. J. Bot. 63, 1677–1680.

Bougher N L, Grove T S and Malajczuk N 1990 Growth and phosphorus acquisition of karri (*Eucalyptus diversicolor* F. Muell.) seedlings inoculated with ectomycorrhizal fungi in relation to phosphorus supply. New Phytol. 114, 77–85.

Boyle C D, Robertson W J and Salonius P O 1987 Use of mycelial slurries of ectomycorrhizal fungi as inoculum for commercial tree seedling nurseries. Can. J. For. 17, 1480–1486.

Cordell C E, Marx D H, Maul S B and Owen J H 1987 Production and utilization of ectomycorrhizal fungal inoculum in the eastern United States. *In* Mycorrhizae in the Next Decade: Practical Applications and Research Priorities, Proc. 7th NACOM. Eds. D M Sylvia, L L Hung and G H Graham. pp 287–289. Institute of Food and Agricultural Science, University of Florida, Gainesville, Florida.

Deacon J W and Fox F M 1988 Delivery of microbial inoculants into the root zone of transplant crops. *In* Proc. Brighton Crop Protection Conference, November 1988, Vol 2. pp 645–653. British Crop Protection Council, United Kingdom.

Dixon R K, Garrett H E, Cox G S, Marx D H and Sander I L 1984 Inoculation of three *Quercus* species with eleven isolates of ectomycorrhizal fungi. I. Inoculation success and seedling growth relationships. For. Sci. 2, 364–371.

Fleming L V 1985 Experimental study of sequences of ectomycorrhizal fungi on birch (*Betula* spp.) seedling root systems. Soil Biol. Biochem. 17, 591–600.

Fortin C, Fortin J-A, Gaulin G, Jomhe N and Lemay S 1988 Large-scale ectomycorrhizal inoculation of containerized-grown seedlings. *In* Canadian Workshop on Mycorrhizae in Forestry. Eds. M Lalonde and Y Piché. pp 115–118. Universite Laval, Ste-Foy.

Gagnon J, Langlois C G and Fortin J A 1987 Growth of containerized jack pine seedlings inoculated with different ectomycorrhizal fungi under a controlled fertilization regime. Can. J. For. Res. 17, 840–845.

Gagnon J, Langlois C G and Fortin J A 1988 Growth and ectomycorrhiza formation of containerized black spruce seedlings as affected by nitrogen fertilization, inoculum type, and symbiont. Can. J. For. Res. 18, 922–929.

Garbaye J, Delwaulle J C and Diangana D 1988 Growth response of eucalypts in the Congo to ectomycorrhizal inoculation. For. Ecol. Management 24, 151–157.

Gardner J H and Malajczuk, N 1988 Recolonization of rehabilitated bauxite mine sites in Western Australia by mycorrhizal fungi. Forest Ecol. Manag. 24, 27–42.

Grove T S, Malajczuk N, Burgess T, Thomson B D and Hardy G 1991a Growth responses of plantation eucalypts to inoculation with selected ectomycorrhizal fungi. *In* Proc. IUFRO Symp. Intensive Forestry: the Role of Eucalypts, Durban, South Africa, Sept. 1991. pp 86–93.

Grove T S, Malajczuk N and Dell B 1991b Effects of applied nitrogen and phosphorus on ectomycorrhizal formation, plant growth and nutrient uptake in *Eucalyptus diversicolor* seedlings. *In* Productivity in Perspective. Ed. P J Ryan. p 209. Forest Commission, Sydney.

Hardy G E ST J, Thomson B D, Malajczuk N and Grove T S 1992 Commercial inoculation of eucalypts with ectomycorrhizal fungi. *In* International Symposium on Management of Mycorrhizas in Agriculture, Horticulture and Forestry, Abstracts . p 110. Aust. Inst. of Agric. Sci., Perth.

Harley J L and Smith S E 1983 Mycorrhizal Symbiosis. Academic Press, London. 483 p.

Harvey L M, Smith J E, Kristiansen B, Neill J and Senior E 1988 The cultivation of ectomycorrhizal fungi. *In* Biotechnology of Fungi for Improving Plant Growth. Eds. J M Whipps and R D Lumsden. pp 27–39. Cambridge University Press, Cambridge.

Hung L-L and Molina R 1986 Temperature and time in storage influence the efficacy of selected isolates of fungi in commercially produced ectomycorrhizal inoculum. For. Sci. 32, 534–545.

Jeffries P 1987 Use of mycorrhizae in agriculture. Crit. Rev. Biotech. 5, 319–357.

Jeffries P and Dodd J C 1991 The use of mycorrhizal inoculants in forestry and agriculture. *In* Handbook of Applied Mycology, Volume 1: Soil and Plants. Eds. D K Arora, B Rai , K G Mukerji G and G R Knudsen. pp 155–185. Marcel Dekker, New York.

Kendrick B and Berch S 1985 Mycorrhizae: Applications in agriculture and forestry. *In* Comprehensive Biotechnology, Volume 4. Eds. C W Robinson. and J A Howell. Chapter 8. Pergamon Press, Oxford.

Kidby D K 1977 Culture maintenance and productivity. *In* Extracellular Microbial Polysaccharides. Eds. P A Sandford and A Laskin. pp 1–13. American Chemical Society, Washington, D C.

Kropacek K, Cudlin P and Mejstrik V 1989 The use of granulated ectomycorrhizal inoculum for reforestation of deteriorated regions. Agric. Ecos. Environ. 28, 263–269.

Kropp B R and Langlois C-G 1990 Ectomycorrhizae in reforestation. Can. J. For. Res. 20, 438–451.

Kuek C, Tommerup I C and Malajczuk N 1990 Hydrogel bead inocula for the production of ectomycorrhizal eucalypts for plantations. *In* Abstracts 8th NACOM, Jackson, Wyoming, Sept. 1990. p 175.

Kuek C, Tommerup I C and Malajczuk N 1992 Hydrogel bead inocula for the production of ectomycorrhizal eucalypts for plantations. Mycol. Res. 96, 273–277.

Larroche C and Gros J-B 1986 Spore production of *Penicillium roqueforti* in fermentors filled with buckwheat seeds: Batch and semi-continuous cultivation. Appl. Microbiol. Biotech. 24, 134–139.

Laukevics J J, Apsite A F, Viesturs U E and Tengerdy R P 1984 Solid substrate fermentation of wheat straw to fungal protein. Biotechnol. Bioeng. 26, 1465–1474.

Le Tacon F, Jung G, Mugnier J, Michelot P and Mauperin C 1985 Efficiency in a forest nursery of an ectomycorrhizal fungus inoculum produced in a fermentor and entrapped in polymeric gels. Can. J. Bot. 63, 1664–1668.

Le Tacon F, Garbaye J, Bouchard D, Chevalier G, Olivier J M, Giumberteau J, Poitou N and Frochot H 1988 Field results from ectomycorrhizal inoculation in France. *In* Canadian Workshop on Mycorrhizae in Forestry. Eds. M Lalonde and Y Piché. pp 51–74. Universite Laval, Ste-Foy.

Malajczuk N, Grove T, Tommerup I C, Bougher N L, Thomson B D, Dell B and Kuek C 1993 Ectomycorrhizas. *In* Microorganisms that Promote Plant Productivity. Eds. P Daft and J Dawson. Martinus Nihoff, Amsterdam. (*in press*).

Marx D H 1980 Ectomycorrhizal fungus inoculations: a tool for improving forestation practices. *In* Tropical Mycorrhiza Research. Ed. P Mikola. pp 13–71. Oxford University Press, London.

Marx D H and Cordell C E 1988 Specific ectomycorrhizae improve reforestation and reclamation in the eastern United States. *In* Canadian Workshop on Mycorrhizae in Forestry. Eds. M Lalonde and Y Piché. pp 75–86. Universite Laval, Ste-Foy.

Marx D H and Kenney D S 1982 Production of ectomycorrhizal fungus inoculum. *In* Methods and Principles of Mycorrhizal Research. Ed. N C Schenck. pp 131–146. American Society of Pathology, St. Paul.

Mauperin C, Mortier F, Garbaye J, Le Tacon F and Carr G 1987 Viability of an ectomycorrhizal inoculum produced in a liquid medium and entrapped in a calcium alginate gel. Can. J. Bot. 65, 2326–2329.

Moo-Young M, Moreira A R and Tengerdy R P 1983 Principles of solid-substrate fermentation. *In* The Filamentous Fungi Volume Four. Eds. J E Smith, D R Berry, and B Kristiansen. pp 117–144. Edward Arnold, London.

Ruehle J L 1982 Field performance of container-grown loblolly pine seedlings with specific ectomycorrhizae on a reforestation site in South Carolina. Southern J. Appl. For. 6, 30–33.

Silman R W 1980 Enzyme formation during solid-substrate fermentation in rotating vessels. Biotechnol. Bioeng. 22, 411–420.

Tommerup I C, Kuek C and Malajczuk N 1987 Ectomycorrhizal inoculum production and utilization in Australia. *In* Mycorrhizae in the Next Decade: Practical Applications and Research Priorities, Proc. 7th NACOM. Eds. D M Sylvia, L L Hung and J H Graham. pp 293–295. Institute of Food and Agricultural Science, University of Florida, Gainesville, Florida.

Valdes M 1986 Survival and growth of pines with specific ectomycorrhizae after 3 years on a highly eroded site. Can. J. Bot. 64, 885–888

Production of inoculum and inoculation with arbuscular mycorrhizal fungi

DAVID M. SYLVIA and AMIEL G. JARSTFER
Soil and Water Science Department, University of Florida, Gainesville, FL 32611-0290, USA

Key words: culture, inoculation, inoculum, mycorrhiza

Abstract

As more efficient technologies become available, VAM inoculum production should progress rapidly toward commercial application. The aim of this paper is to provide a status report on current VAM inoculum production and utilization technologies.

Culture of VAM fungi on plants growing in disinfested soil has been the most frequently used technique for increasing propagule numbers. Culture of VAM fungi in soilless media avoids the need to treat soil to remove detrimental organisms and allows better control over the physical and chemical characteristics of the growth media. However, the availability of P in most soilless media requires prudent management of nutrients.

Colonized roots, spores and hyphae of VAM fungi may be produced in hydroponic or aeroponic cultures. The key to successful colonization in these systems is maintenance of low P concentrations. The colonized-root inocula produced in these systems is free of any substrate so they can be sheared, resulting in very high propagule numbers.

Ri-plasmid transformed-root cultures offer a clean method for growing colonized roots. However, successful cultural conditions for root-organ cultures are a compromise between adequate nutrients for rapid root growth and nutrient concentrations that do not inhibit fungal spread in the roots. Therefore, it appears unlikely that this approach will lead to an economical technology for mass production of VAM inoculum.

An inoculum must be infective and effective in the agricultural system for which it is intended. The ideal inoculum should be small in dose and large in response. Any of several methods can be used to deliver propagules of VAM fungi to the rhizosphere. The most common application method is to place the inoculum below the seed or seedling prior to planting. Seed coating and somatic embryo encasement with inoculum are methods that offer promise with some crops. Alternatively, sheared-roots can be suspended in a hydrogel for application as a flowable inoculum.

Several areas of VAM-inoculum production and application technology merit further investigation. Scaling up nutrient-film and aeroponic-culture technologies, along with sheared-root processing of colonized roots produced in these systems, should provide readily available low-cost, high-quality inocula for nursery inoculation and field testing. Little is known about the storage of VAM inocula. Further knowledge of the storage properties of VAM inocula would encourage commercial production and the development of new formulations. New and innovative inoculation technologies are needed to provide the most efficient application of available inocula in a wide array of crop production systems.

Introduction

As more efficient technologies become available for the production and use of arbuscular mycorrhizal (AM) inoculum, additional applications for these fungi should become commercially feasible. Soilless culture and adaptations of hydroponics can be used to produce high quality inoculum with propagule numbers many times greater than the soil-based

Published as Florida Agricultural Experiment Station Journal Series No. R-02301

A.D. Robson, L.K. Abbott & N. Malajczuk (eds.), Management of mycorrhizas in agriculture, horticulture and forestry, 231–238.

inoculum used in the past. The resulting lower cost of inoculum and relative ease of production should make these technologies applicable in less-developed agricultural areas as well as in highly-industrialized agricultural systems which currently use large amounts of phosphate fertilizers. The purpose of this chapter is to provide a brief synopsis of current AM inoculum production and utilization technologies, and to provide recommendations for further development of these technologies. For further details on these methods we refer the reader to Jarstfer and Sylvia (1992).

Production of inoculum

General recommendations

Culture of AM fungi on plants growing in disinfested soil has been the most frequently used technique for increasing propagule numbers (Menge, 1984). The recommendations presented here are derived mainly from experiments in soil-based pot cultures. However, they also apply more generally to the newer technologies for inoculum production discussed below.

A highly susceptible host plant should be used; it should produce root mass quickly and tolerate the high-light conditions required for the fungus to reproduce rapidly. Host plants should be screened to insure that maximum inoculum levels are achieved (Bagyaraj and Manjunath, 1980). Hosts plants which can be propagated from seed are preferable to cuttings since seeds can be more readily disinfested than cuttings.

All components of the culture system should be disinfested prior to initiation of a culture of AM fungi. Host disinfestation generally is accomplished more easily than disinfestation of soil or propagules from soil. The objective of soil disinfestation is to kill existing AM fungi, pathogenic organisms and weed seeds while preserving a portion of the non-pathogenic microbial community. We commonly pasteurize soil by heating it to 85°C for two 8-h periods, 48 h apart.

Containers used for pot cultures should be shielded from contaminated soil, splashing water and crawling insects. In addition, specific isolates of AM fungi should be kept well separated to reduce cross contamination. Container size should match the potential volume of the root system within practical space constraints. Larger containers may result in higher spore concentrations (Ferguson and Menge, 1982).

Light quality and irradiance, soil water content, and temperature markedly influence root colonization and spore production (Menge, 1984). Non-shaded greenhouses or high-intensity metal-halide and sodium vapour lamps used for supplemental light give good results, provided that adequate cooling can be achieved. A moderate watering regime often supports optimal spore production (Nelsen and Safir, 1982). A good strategy is to apply water regularly to well drained soils or growth media. Since sporulation is positively correlated with temperatures from 15 to near 30°C for many AM fungi (Schenck and Smith, 1982), a warm growth environment should be maintained.

Fungal responses to P and N fertilization are strain dependent. Frequent applications of a high-N fertilizer has increased spore production by several isolates of AM fungi, yet sporulation of an isolate of *Glomus intraradices* was severely reduced by this treatment (Douds and Schenck, 1990a). Colonization is often depressed by high P concentration (Same et al., 1983). Furthermore, the relative amounts of N and P directly affect root colonization by AM fungi. A strategy that we have used to increase colonization is to provide plants with a nutrient regime low in P but high in N (Sylvia and Neal, 1990).

Pesticides have been tested for their effects on AM colonization and sporulation. Some soil-applied biocides greatly reduce or eliminate AM fungi (Trappe et al., 1984), while others increase colonization and sporulation (Sreenivasa and Bagyaraj, 1989). Though the use of selected pesticides in inoculum production has been advocated, they have rarely been used to protect against or reduce organisms contaminating pot cultures (Menge, 1983).

Soil inoculum from pot cultures of many *Glomus* spp. may be stored for several months or even years. However, for more reliable storage of a wider range of AM fungi, the fungi can be dried in situ with the host and then frozen at -70°C (Douds and Schenck, 1990b).

Soilless media pot culture

Culturing AM fungi in soilless media avoids the need for sterilization of soil to remove detrimental organisms and allows better control over the physical and chemical characteristics of the growth media. Soilless media are more uniform in composition, weigh less and facilitate better aeration than soil.

At least 15 species of AM fungi from four genera have been cultured in soilless media (Table 1). The ideal soilless mixture should hold sufficient water for plant growth but also allow good aeration. Bark, calcined clay, expanded clay and perlite provide good aeration. Peat and vermiculite hold more water than the former materials but allow air to penetrate better than sand. When dry, perlite and vermiculite are very light while sand has a high bulk density.

The availability of P in most soilless media requires prudent management of nutrients. Frequent addition of dilute, soluble nutrient solutions (Douds and Schenck, 1990a; Waterer and Coltman, 1988), incorporation of time-release fertilizer (Coltman et al., 1988) or the use of less-available forms of P (Thompson, 1986) are strategies for nutrient management in soilless media.

Hydroponics and aeroponics

Colonized roots, spores and hyphae of AM fungi may be produced in hydroponic or aeroponic culture. Plants are usually inoculated in sand or vermiculite before they are transferred to these systems. Plants have also been inoculated directly in the aeroponic system (Hung et al.,

Table 1. Soilless media used in the culture of arbuscular mycorrhizal fungi. References to the literature are presented in Jarstfer and Sylvia (1992)

Material	AM Fungi
Bark—shredded douglas fir	*Glomus fasciculatum*
Calcined Montmorillonite Clay	*G. macrocarpum, G. monosporum, G. versiforme, Scutellispora calospora*
Expanded Clay Aggregates	*Acaulospora laevis, G. constrictum, G. etunicatum, G. fasciculatum, G. intraradix, G. macrocarpum, G. mosseae, Scutellispora pellucida*
Peat	
hypnum	*G. fasciculatum*
sphagnum	*A. spinosa, G. fasciculatum, G. mosseae*
+ perlite and vermiculite	*G. intraradix, G. fasciculatum*
+ pumice (2:1)	*Gigaspora margarita, G. fasciculatum, G. tenue*
Perlite	*G. fasciculatum*
+ soilrite[a] (1:1)	*G. fasciculatum*
Sand	
basaltic	*G. aggregatum*
river	*A. spinosa, G. fasciculatum, G. mosseae*
silica	*G. fasciculatum, G. mosseae, G. fasciculatum*
+ vermiculite (3:1)	*G. mosseae, G. etunicatum, G. fasciculatum*
+ grit (2:1)	*G. mosseae, G. fasciculatum*
Vermiculite	*A. spinosa, G. fasciculatum, G. mosseae*

[a] (perlite, vermiculite and peat; 1:1:1)

1991). The key to successful colonization in these systems is maintenance of low P concentrations (3 to 24 μmol).

Inoculum can be produced using the nutrient-film technique (NFT) by growing precolonized plants in a defined nutrient solution which flows over the host roots (Elmes and Mosse, 1984; Mosse and Thompson, 1984). The ebb-and-flow system used by Dehne and Backhaus (1986) is similar to the NFT but incorporates inorganic expanded clay pellets as an easily aerated substrate.

A culture system which applies a fine mist of defined nutrient solution to the roots of the host plant is termed aeroponic culture (Zobel et al., 1976). This system was adapted for growing AM fungi (Hung and Sylvia, 1988). Greater concentrations of spores have been produced in aeroponic cultures than in comparable soil-based pot cultures. Because the colonized-root inoculum produced in this system is free of any substrate it can be sheared, resulting in very high propagule numbers (Sylvia and Jarstfer, 1992).

Tissue culture

No one has successfully grown AM fungi in single-organism culture. Spores are easily germinated but hyphae cannot be subcultured. However, several tissue-culture techniques for producing contaminant-free AM inoculum have succeeded. Ri-plasmid transformed-root cultures offer an efficient method for growing colonized roots as no plant growth regulators are required for sustained growth (Mugnier and Mosse, 1987). However, N and sucrose supply (Mugnier and Mosse, 1987), P concentration (Bécard and Fortin, 1988) and pH (Mosse and Hepper, 1975) have all been shown to influence the spread of AM fungi in cultured roots. We believe that successful cultural conditions for root-organ cultures are a compromise between adequate nutrients for rapid root growth and nutrient concentrations that do not inhibit fungal spread to the new roots. For example, Bressan and Sylvia (unpublished results) found that conditions which stimulated root growth depressed germination of spores in agar cultures (Fig. 1). Therefore, it appears that there are some fundamental obstacles to be overcome before approach will lead to an economical technology for mass production of AM inoculum.

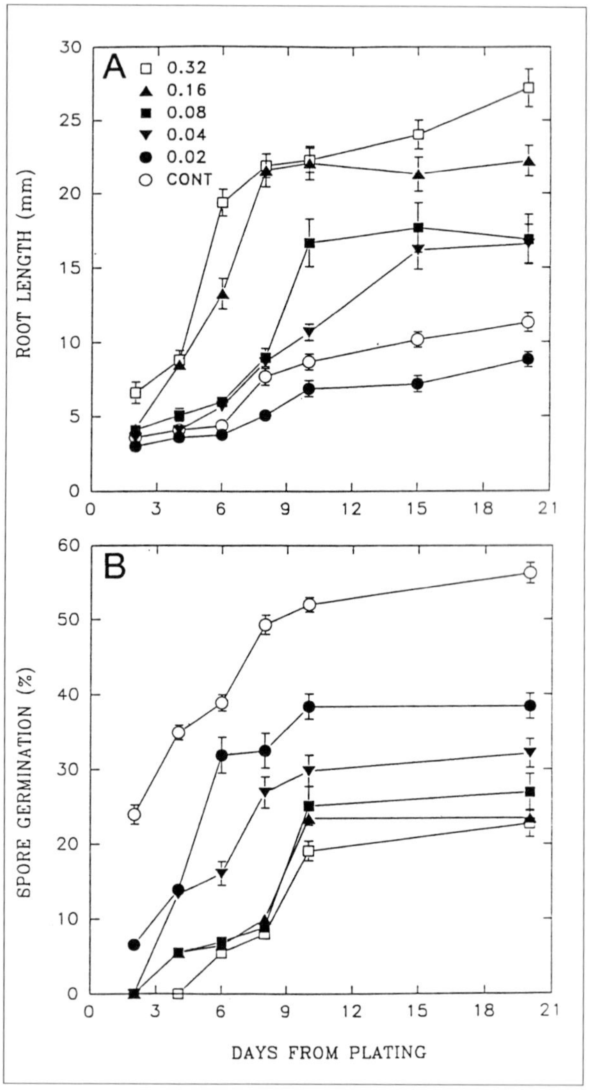

Fig. 1. Effect of P/N ratio on root growth and spore germination. (A) Growth of *Anthylis vulneraria* L. roots, transformed with the Ri plasmid of *Agrobacterium rhizogenes*, on M media (Bécard and Fortin, 1988) (control, P/N = 0.022) and with P/N ratios adjusted up to 0.32. (B) Spore germination of *Glomus etunicatum* on water agar (control) and M media with adjusted P/N ratios. Increasing P/N ratios enhanced root elongation but decrease spore germination. Symbols represent the mean of six replicates + S.E.

Commercial production of inoculum

There have been several attempts to produce AM inoculum on a commercial scale. Jeffries and Dodd (1991) cite some examples in their review paper, though the inoculum produced by Native Plants, Inc. in Salt Lake City, Utah, USA is no longer available. Two additional products which have recently become available are Sorevex MYCORI-MIX (Premier Enterprises, 326 Main Street, Red Hill, PA 18076 USA), a growing medium that contains *Glomus intraradices,* and VAMINOC (Agricultural Genetics Company, NPPL, Rothamsted Experimental Station, Herts. Al5 2JQ, UK), which incorporates several AM fungi in a granular carrier.

Wood (1991) has reviewed the potential markets for AM inoculum as well as obstacles to successful commercialization. The failures of large-scale commercial production can be attributed to: (i) poor understanding of the biology of the fungi resulting in unreliable inoculum, (ii) public ignorance of mycorrhiza resulting in little demand for the product, (iii) fragmented markets due to the fact that many potential consumers of inoculum are currently using culture practices (e.g., high fertilizer rates and broad spectrum fungicides) which are incompatible with AM fungi and (iv) unreasonable expectations (of inoculum performance and market size) resulting in large financial commitments that demand too high a return on investments.

A company that anticipates marketing AM inoculum should consider the following strategy. First and foremost, reliable field data must be obtained for the inoculum (see below). This can be readily accomplished by working closely with the research community. An additional benefit of working with researchers is that they will become good 'salesmen' for a proven product.

Inoculation with AM fungi

Required characteristics of high quality inoculum

An inoculum must be infective and effective in the agricultural system for which it is intended. In addition, inocula should be concentrated, pathogen-free, and have a shelf life which allows processing, distribution and use without a significant loss of viability.

Infectiveness is the ability of a fungus to penetrate and spread in roots (Abbott and Robson, 1981). Many of the benefits of AM fungi begin early, often within the first 3 weeks in the plant growing cycle. Effectiveness is the ability of AM fungi to enhance growth or stress tolerance of the host. Effectiveness can vary dramatically among AM isolates so screening studies should be undertaken before large quantities of inoculum are produced for field use (Hung et al., 1990). Effectiveness should be tested under crop production conditions since indigenous AM fungi, pathogens, fertility levels and edaphic factors will influence effectiveness.

Researchers should strive to produce an inoculum that is small in dose but large in response. The application of 20 to 30 metric tons of AM inoculum per hectare has produced growth responses but these rates are unacceptably high for practical agriculture (Powell, 1984). The production of concentrated AM inoculum should yield a product with low bulk density where each propagule represents only a small fraction of the manufacturing cost. Hayman et al. (1981) concentrated pot culture soil by wet sieving to remove clay and sand particles from the inoculum. With aeroponic culture of roots it is possible to produce inoculum with hundreds of thousands of propagules per dry gram of root since no substrate is present with the inoculum (Sylvia and Jarstfer, 1992).

Inoculum composition

The most widely reported inoculum compositions are mixtures of spores, colonized roots, hyphae and the supporting substrate from pot cultures grown in pasteurized soil or soilless media. These media have been shredded (Warner, 1985) and cubed (Owusu-Bennoah and Mosse, 1979) in attempts to facilitate efficient distribution of the inoculum. Propagule numbers from pot cultures range from two (Wilson, 1984) to several hundred (Liyanage, 1989) per gram of inoculum. Natural or indigenous inoculum consisting of raw soil collected at a site known to produce consistent AM colonization can also be used to inoculate plants. However, propagule numbers in natural inoculum are generally less than ten propagules per gram of soil and may contain pathogens.

Pot-culture inoculum has been separated into its various components for application. A commercial inoculum (Nutri-Link®) was available until recently which consisted of spores separated from pot cultures and mixed with atapulgite clay. More commonly, pot cultures are sieved, concentrating the inoculum by removing much of the mineral component. Roots collected on coarser sieves have been reduced in size with a blender (Azcon et al., 1978). Roots alone, as well as vesicles and spores separated from roots, may be used for inoculum (Biermann and Linderman, 1983). Freeze-dried roots have also been used as inoculum (Ganry et al., 1985). In addition, spores, hyphae, vesicles and colonized roots sieved from pot cultures have been pelletized with alginate (Ganry et al., 1985; Strullu and Plenchette, 1991).

Colonized roots from NFT culture have been used with success for field inoculation (Elmes et al., 1984). Colonized roots from aeroponic cultures could also be used in a similar manner or sheared for more efficient application (Sylvia and Jarstfer, 1992). Expanded-clay pellets colonized by AM fungi have been used directly or air-dried before use (Dehne and Backhaus, 1986). Mycorrhizal seed pellets are another inoculum form which has been proposed (Warner, 1985).

Application methods

Any method by which viable propagules of AM fungi are delivered efficiently to the rhizosphere can produce colonized plants. The most common application method is to place the inoculum below the seed or seedling prior to planting (Hayman, 1987). For example, expanded clay pellets were applied successfully through the same tube used to drop maize seeds from an air-powered seed drill (Baltruschat, 1987). As another example, Native Plants, Inc. recommended that Nutri-Link® be incorporated in seedling beds or placed in furrows for field applications. As these methods suggest, applications will be most efficient when existing machines and practices are used or only slightly modified for incorporating AM fungi. Seed coating (Hattingh and Gerdemann, 1975) and somatic embryo encasement (Strullu et al., 1989) with inoculum are additional methods worthy of further investigation. Seed coating may be improved if an alginate or carrageenan matrix is used to thicken the coating and protect the inoculum. Peat-based seed pellets have a greater bulk than coated seeds but may provide more inoculum with the seed. Somatic embryos require much more water than is provided by the coatings and would probably work best if fluid-drilled with inoculum suspended in a hydrogel. Sheared-root inocula from aeroponic cultures have been suspended successfully in a hydrogel for application (Sylvia and Jarstfer, 1992).

Conclusion

In response to increased concern for environmental quality, sustainable technologies need to be incorporated into agricultural systems. Management of AM fungi is an important aspect of such an approach (Linderman and Bethlenfalvay, 1992). However, the obligate, symbiotic nature of AM fungi and limited inoculum supplies continue to impede research aimed at managing these beneficial fungi. High production costs, as well as problems associated with soil-based pot cultures, have reduced enthusiasm for commercial development. Nonetheless, the inability to use fermentation processes to produce AM inocula should not discourage further attempts to develop innovative technologies to mass produce AM fungi.

Several areas of AM-inoculum production and application technology merit further investigation. Scaling up nutrient-film and aeroponic-culture technologies, along with sheared-root processing of colonized roots produced in these systems, could provide low-cost, high-quality inocula for nursery inoculation and field testing. Little is known about the storage of AM inocula. Further knowledge of the storage properties of AM inocula would encourage commercial production and the development of new formulations. New and innovative inoculation technologies are needed to provide the most efficient application of available inocula in a wide array of crop production systems.

References

Abbott L K and Robson A D 1981 Infectivity and effectiveness of five endomycorrhizal fungi: competition with indigenous fungi in field soils. Aust. J. Agric. Res. 32, 621–630.

Azcon R, Azcon-Aguilar C and Barea J M 1978 Effects of plant hormones present in bacterial cultures on the formation and responses to VA endomycorrhiza. New Phytol. 80, 359–364.

Bagyaraj D J and Manjunath A 1980 Selection of a suitable host for mass production of vesicular-arbuscular mycorrhizal inoculum. Plant Soil 55, 495–498.

Baltruschat H 1987 Field inoculation of maize with vesicular-arbuscular mycorrhizal fungi by using expanded clay as carrier material for mycorrhiza. J. Plant Dis. Protect. 94, 419–430.

Bécard G and Fortin J A 1988 Early events of vesicular-arbuscular mycorrhiza formation on Ri T-DNA transformed roots. New Phytol. 108, 211–218.

Biermann B J and Linderman R G 1983 Use of vesicular-arbuscular mycorrhizal roots, intraradical vesicles and extraradical vesicles as inoculum. New Phytol. 95, 97–106.

Coltman R R, Waterer D R and Huang R S 1988 A simple method for production of *Glomus aggregatum* inoculum using controlled-release fertilizer. Hortic. Sci. 23, 213–215.

Dehne H W and Backhaus G F 1986 The use of vesicular-arbuscular mycorrhizal fungi in plant production. I. Inoculum production. J. Plant Dis. Protect. 93, 415–424.

Douds D D and Schenck N C 1990a Increased sporulation of vesicular-arbuscular mycorrhizal fungi by manipulation of nutrient regimes. Appl. Environ. Microbiol. 56, 413–418.

Douds D D and Schenck N C 1990b Cryopreservation of spores of vesicular-arbuscular mycorrhizal fungi. New Phytol. 115, 667–674.

Elmes R P and Mosse B 1984 Vesicular-arbuscular endomycorrhizal inoculum production. II. Experiments with maize (*Zea mays*) and other hosts in nutrient flow culture. Can. J. Bot. 62, 1531–1536.

Elmes R P, Hepper C M, Hayman D S and O'Shea J 1984 The use of vesicular-arbuscular mycorrhizal roots by the nutrient film technique as inoculum for field sites. Ann. Appl. Biol. 104, 437–441.

Ferguson J J and Menge J A 1982 Factors that affect production of endomycorrhizal inoculum. Proc. Fla. State Hortic. Soc. 95, 37–39.

Ganry F, Diem H G, Wey J and Dommergues Y R 1985 Inoculation with *Glomus mosseae* improves N2 fixation by field-grown soybeans. Biol. Fert. Soils 1, 15–23.

Hattingh M J and Gerdemann J W 1975 Inoculation of Brazilian sour orange seed with an endomycorrhizal fungus. Phytopathology 65, 1013–1016.

Hayman D S 1987 VA mycorrhizas in field crop systems. *In* Ecophysiology of VA Mycorrhizal Plants. Ed. G Safir. pp 171–192. CRC Press, Inc., Boca Raton, Florida.

Hayman D S, Morris E J and Page R J 1981 Methods for inoculating field crops with mycorrhizal fungi. Ann. Appl. Biol. 99, 247–253.

Hung L L and Sylvia D M 1988 Production of vesicular-arbuscular mycorrhizal fungus inoculum in aeroponic culture. Appl. Environ. Microbiol. 54, 353–357.

Hung L L, Sylvia D M and O'Keefe D M 1990 Isolate selection and phosphorus interaction of vesicular-arbuscular mycorrhizal fungi in biomass crops. Soil Sci. Soc. Am. J. 54, 762–768.

Hung L L, O'Keefe D M and Sylvia D M 1991 Use of a hydrogel as a sticking agent and carrier of vesicular-arbuscular mycorrhizal fungi. Mycol. Res. 95, 427–429.

Jarstfer A G and Sylvia D M 1992 Inoculum production and inoculation strategies for vesicular-arbuscular mycorrhizal fungi. *In* Soil Microbial Technologies: Applications in Agriculture, Forestry and Environmental Management. Ed. B Metting. pp 349–377. Marcel Dekker, Inc., New York.

Jeffries P and Dodd J C 1991 The use of mycorrhizal inoculants in forestry and agriculture. *In* Handbook of Applied Mycology. Vol I: Soil and Plants. Eds. D K Arora, B Rai, K G Mukerji and G R Knudsen. pp 155–185. Marcel Dekker, Inc., New York.

Linderman R G and Bethlenfalvay G J 1992 VA mycorrhiza and sustainable agriculture. Soil Science Society of America, Madison, WI.

Liyanage H D 1989 Effects of Phosphorus Nutrition and Host Species on Root Colonization and Sporulation by Vesicular-Arbuscular (VA) Mycorrhizal Fungi in Sand-Vermiculite Medium. M.S. Thesis, University if Florida, Gainesville.

Menge J A 1983 Utilization of vesicular-arbuscular mycorrhizal fungi in agriculture. Can. J. Bot. 61, 1015–1024.

Menge J A 1984 Inoculum production. *In* VA Mycorrhiza. Eds. C Ll Powell and D J Bagyaraj. pp 187–203. CRC Press, Inc., Boca Raton, Florida.

Mosse B and Hepper C M 1975 Vesicular-arbuscular mycorrhizal infections in root organ cultures. Physiol. Plant Pathol. 5, 215–223.

Mosse B and Thompson J P 1984 Vesicular-arbuscular endomycorrhizal inoculum production. I. Exploratory experiments with beans (*Phaseolus vulgaris*) in nutrient flow culture. Can. J. Bot. 62, 1523–1530.

Mugnier J and Mosse B 1987 Vesicular-arbuscular mycorrhizal infection in transformed root–inducing T-DNA roots grown axenically. Phytopathology 77, 1045–1050.

Nelsen C E and Safir G R 1982 Increased drought tolerance of mycorrhizal onion plants caused by improved phosphorus nutrition. Planta 154, 407–413.

Owusu-Bennoah E and Mosse B 1979 Plant growth response to vesicular-arbuscular mycorrhizae. XI. Field inoculation responses in barley, lucerne, and onion. New Phytol. 83, 671–679.

Powell C Ll 1984 Field inoculation with VA mycorrhizal fungi. *In* VA Mycorrhiza. Eds. C Ll Powell and D J Bagyaraj. pp 205–222. CRC Press, Inc., Boca Raton, Forida.

Same B I, Robson A D and Abbott L K 1983 Phosphorus, soluble carbohydrates and endomycorrhizal infection. Soil Biol. Biochem. 15, 593–597.

Schenck N C and Smith G S 1982 Responses of six species of vesicular-arbuscular mycorrhizal fungi and their effects on soybean at four soil temperatures. New Phytol. 92, 193–201.

Sreenivasa M N and Bagyaraj D J 1989 Use of pesticides for mass production of vesicular-arbuscular mycorrhizal inoculum. Plant Soil 119, 127–132.

Strullu D G and Plenchette C 1991 The entrapment of *Glomus* sp. in alginate beads and their use as root inoculum. Mycol. Res. 95, 1194–1196.

Strullu D G, Romand C, Callac P, Teoule E and Demarly Y 1989 Mycorrhizal synthesis in vitro between *Glomus* spp. and artificial seeds of alfalfa. New Phytol. 113, 545–548.

Sylvia D M and Jarstfer A G 1992 Sheared-root inocula of vesicular-arbuscular mycorrizal fungi. Appl. Environ. Microbiol. 58, 229–232.

Sylvia D M and Neal L H 1990 Nitrogen affects the phosphorus response of VA mycorrhiza. New Phytol. 115, 303–310.

Thompson J P 1986 Soilless cultures of vesicular-arbuscular mycorrhizae of cereals: Effects of nutrient concentration and nitrogen source. Can. J. Bot. 64, 2282–2294.

Trappe J M, Molina R and Castellano M A 1984 Reactions of mycorrhizal fungi and mycorrhizal formation to pesticides. Annu. Rev. Phytopathol. 22, 331–359.

Warner A 1985 Mycorrhizal Seed Pellets. United States Patent 4,551,165.

Waterer D R and Coltman R R 1988 Phosphorus concentration and application interval influence growth and mycorrhizal infection of tomato and onion transplants. J. Am. Soc. Hortic. Sci. 5, 704–708.

Wilson J M 1984 Competition for infection between vesicular-arbuscular mycorrhizal fungi. New Phytol. 97, 427–435.

Wood T 1991 VA Mycorrhizal Fungi: Challenges for Commercialization. *In* Fungal Biotechnology, Handbook of Applied Mycology, vol. 4. Eds. D K Arora, R P Elander and K G Mukerji. pp 823–847. Marcel Dekker Inc., New York.

Zobel R W, Del Tredici P and Torrey J G 1976 Method for growing plants aeroponically. Plant Physiol. 57, 344–346.